T0382681

A CONCISE GUIDE TO GEOPRESSURE

Geopressure drives fluid flow and is important for hydrocarbon exploration, carbon sequestration, and designing safe and economical wells. This concise guide explores the origins of geopressure and presents a step-by-step approach to characterizing and predicting pressure and least principal stress in the subsurface. The book emphasizes how geology, and particularly the role of flow along permeable layers, drives the development and distribution of subsurface pressure and stress. Case studies, such as the Deepwater Horizon blowout, and laboratory experiments are used throughout to demonstrate methods and applications. It succinctly discusses the role of elastoplastic behavior, the full stress tensor, and diagenesis in pore pressure generation, and it presents workflows to predict pressure, stress, and hydrocarbon entrapment. It is an essential guide for academics and professional geoscientists and petroleum engineers interested in predicting pressure and stress, and understanding the role of geopressure in geological processes, well design, hydrocarbon entrapment, and carbon sequestration.

PROFESSOR PETER B. FLEMINGS holds the Leonidas T. Barrow Centennial Chair in Mineral Resources with the Jackson School of Geosciences at The University of Texas at Austin. He leads "UT GeoFluids," a long-running industry-sponsored effort dedicated to developing new concepts and approaches to predict pressure and stress in the subsurface. He served on the U.S. Department of Energy's well integrity team during the Deepwater Horizon blowout, he is a Geological Society of America fellow, and he was a distinguished lecturer for the American Association of Petroleum Geologists and the International Ocean Drilling Program. He has served as chief scientist on numerous scientific drilling expeditions.

A CONCISE GUIDE TO GEOPRESSURE

Origin, Prediction, and Applications

PETER B. FLEMINGS
The University of Texas at Austin

CAMBRIDGE
UNIVERSITY PRESS

University Printing House, Cambridge CB2 8BS, United Kingdom

One Liberty Plaza, 20th Floor, New York, NY 10006, USA

477 Williamstown Road, Port Melbourne, VIC 3207, Australia

314–321, 3rd Floor, Plot 3, Splendor Forum, Jasola District Centre, New Delhi – 110025, India

79 Anson Road, #06–04/06, Singapore 079906

Cambridge University Press is part of the University of Cambridge.

It furthers the University's mission by disseminating knowledge in the pursuit of education, learning, and research at the highest international levels of excellence.

www.cambridge.org
Information on this title: www.cambridge.org/9781107042346
DOI: 10.1017/9781107326309

First published 2021

Printed in the United Kingdom by TJ Books Limited, Padstow Cornwall

A catalogue record for this publication is available from the British Library.

Library of Congress Cataloging-in-Publication Data
Names: Flemings, Peter Barry, 1960– author.
Title: A concise guide to geopressure : origin, prediction, and applications / Peter B. Flemings, University of Texas, Austin.
Description: Cambridge, United Kingdom ; New York, NY : Cambridge University Press, 2021. | Includes bibliographical references and index.
Identifiers: LCCN 2020041265 (print) | LCCN 2020041266 (ebook) | ISBN 9781107042346 (hardback) | ISBN 9781107326309 (ebook)
Subjects: LCSH: Petroleum – Geology. | Sedimentary basins.
Classification: LCC TN870.5 .F54 2021 (print) | LCC TN870.5 (ebook) | DDC 553.2/8–dc23
LC record available at https://lccn.loc.gov/2020041265
LC ebook record available at https://lccn.loc.gov/2020041266

ISBN 978-1-107-04234-6 Hardback

Additional resources for this publication at www.cambridge.org/flemings.

Contents

Acknowledgments

I am fascinated with how depositional processes, stratigraphy, and flow couple to drive geological processes in sedimentary basins. I dedicate this book to the graduate students who joined me in this journey. It is an honor and a joy to try to keep up with your sharp and creative minds. In 2000, I began the GeoFluids Consortium, an industry-sponsored effort to understand pressure and stress in sedimentary basins. Almost every example from this book stems from this collaborative effort to understand the subsurface. My industry colleagues have patiently taught our group, provided data to further our efforts, and been outspoken in their ideas. In 2002, I joined the MIT geotechnics group for a sabbatical. Jack Germaine (Dr. G.) took me under his wing, guided my experimental efforts, and patiently answered every question. We are lifelong friends. At MIT, I took a theoretical soil mechanics course and a young doctoral student, Maria Nikolinakou, was my teaching assistant. Maria and I are now colleagues at U.T., and I learn from her every day. Also at U.T., Mahdi Heidari and I have found exciting problems at the interface of geology and engineering. Mahdi patiently reviewed many chapters of this book. Paul Hicks began my education in multiphase flow behavior. Today, David DiCarlo and I collaborate in multiphase flow. Niven Shumaker generously shared examples and methods for pore pressure prediction from seismic data. This journey began at Penn State and continues at the University of Texas. I am grateful to the Nittany Lions and the Longhorns for the freedom to pursue this journey. Aaron Price and Constantino Panagopulos were invaluable editors, compilers, and figure generators. I could not have gotten this done without their perseverance. My father Merton Flemings, inspired me to develop an understanding of the individual processes that drive pore pressure and to understand how these processes link. My running friends provided helpful reviews and emotional support. Lastly, I am grateful for the support and patience of my wife, Ann, and my son, Nicholas.

Nomenclature

This list contains definitions of symbols, dimensions, and the section of the book where they are first used. All symbols are defined in the text. There is, inevitably, some duplication.

English Units

Symbol	Name	Dimensions	SI Unit	Reference
A	Parameter in velocity-effective stress equation	$\frac{L^2T}{M}$	$\frac{m}{Pa \cdot s}$	Chapter 5 (Eq. 5.7)
A	Parameter in velocity-density equation	$\frac{L^4}{MT}$	$\frac{m^4}{kg \cdot s}$	Chapter 6 (Eq. 6.7)
A	Surface area	L^2	m^2	Chapter 2 (Eq. 2.12)
A	Reservoir area	L^2	m^2	Chapter 10 (Eq. 10.2)
A_e	Parameter in velocity full effective stress equation	$\frac{L^2T}{M}$	$\frac{m}{Pa \cdot s}$	Chapter 7 (Eq. 7.14)
B	Parameter in velocity-effective stress equation	-	-	Chapter 5 (Eq. 5.7)
B	Parameter in velocity-density equation	-	-	Chapter 6 (Eq. 6.7)
B	Skempton coefficient	-	-	Chapter 4 (Eq. 4.7)
B_e	Parameter in velocity vs. equivalent effective stress equation	-	-	Chapter 7 (Eq. 7.14)
BSE	Backscattered electron			Chapter 3 (Fig. 3.4)
c_α	Coefficient of secondary consolidation	-	-	Chapter 3 (Eq. 3.5)
c_b	Bulk compressibility	$\frac{L^2T}{M}$	$\frac{1}{Pa}$	Chapter 2 (Eq. 2.8)
c_f	Fluid compressibility	$\frac{L^2T}{M}$	$\frac{1}{Pa}$	Chapter 4 (Eq. 4.8)
c_s	Solid compressibility	$\frac{L^2T}{M}$	$\frac{1}{Pa}$	Chapter 2 (Eq. 2.8)

(cont.)

Symbol	Name	Dimensions	SI Unit	Reference
c_v	Coefficient of consolidation	$\frac{L^2}{T}$	$\frac{m^2}{s}$	Chapter 4 (Eq. 4.24)
C	Parameter in Butterfield equation	-	-	Chapter 5 (Table 5.1)
C	Loading efficiency	-	-	Chapter 4 (Eq. 4.9)
C	Parameter in Eaton equations	$\frac{L^2T}{M}$	$\frac{m}{Pa \cdot s}$	Chapter 5 (Equations 5.6 and 5.12)
C_c	Compression index (conventional log space)	-	-	Chapter 3 (Eq. 3.1)
C_e	Expansion index (conventional log space)	-	-	Chapter 3 (Fig. 3.5) Chapter 6 (Eq. 6.2)
dt_{ma}	Matrix travel time	$\frac{T}{L}$	$\frac{s}{m}$	Chapter 5 (Eq. 5.3)
dt	Travel time	$\frac{T}{L}$	$\frac{s}{m}$	Chapter 5 (Eq. 5.3)
e	Void ratio	-	-	Chapter 3 (Eq. 3.1)
e_{min}	Minimum void ratio	-	-	Chapter 6 (Eq. 6.1)
e_u	Void ratio, unloaded	-	-	Chapter 6 (Eq. 6.2)
e_λ	Void ratio at $s' =$ unity along any stress ratio η	-	-	Chapter 3 (Eq. 3.21)
$e_{\lambda_{K_0}}$	Void ratio at $s' =$ unity under uniaxial strain conditions	-	-	Chapter 3 (Eq. 3.25)
$e_{\lambda_{iso}}$	Void ratio at $s' =$ unity under isostatic (t=0) conditions	-	-	Chapter 3 (Eq. 3.22)
$e_{\lambda\tau}$	Void ratio at $s' =$ unity under Coulomb failure conditions	-	-	Chapter 3 (Fig. 3.17)
e_0	Void ratio at σ'_{v0}	-	-	Chapter 3 (Eq. 3.1)
e_1	Void ratio at the end of primary consolidation	-	-	Chapter 3 (Eq. 3.5)
e_{100}	Void ratio at stress 100 kPa	-	-	Section 3.2.1
EI-330	Eugene Island 330 oil field, Gulf of Mexico			Chapter 5
EMW	Equivalent mud weight	$\frac{M}{L^2T^2}$	$\frac{kg}{m^2s^2}$	Chapter 2 (Eq. 2.10)
ESP	Effective stress path			Chapter 3 (Fig. 3.9)
f	Parameter in Issler equation	-	-	Chapter 5 (Eq. 5.3)
F	Applied load	$\frac{M}{LT^2}$	Pa	Chapter 4 (Eq. 4.2)
FBP	Fracture breakdown pressure	$\frac{M}{LT^2}$	Pa	Chapter 8 (Fig. 8.8)
FCP	Fracture closure pressure	$\frac{M}{LT^2}$	Pa	Chapter 8 (Fig. 8.8)
FIT	Formation integrity test	$\frac{M}{LT^2}$	Pa	Chapter 8 (Fig. 8.8)
FPP	Fracture propagation pressure	$\frac{M}{LT^2}$	Pa	Chapter 8 (Fig. 8.8)
FWL	Free water level			Chapter 9 (Fig. 9.2)
FOL	Free oil level			Chapter 9 (Fig. 9.6)
g	Acceleration of gravity	$\frac{L}{T^2}$	$\frac{m}{s^2}$	Chapter 2 (Eq. 2.1)
G	Flow focusing ratio	-	-	Section 10.2.2
GR	Gamma ray			Section 5.3.1

(*cont.*)

Symbol	Name	Dimensions	SI Unit	Reference
GWC	Gas-water contact	-	-	Chapter 2 (Fig. 2.1)
h_{FWL}	Height above free water level	L	m	Chapter 9 (Fig. 9.5)
h_{FOL}	Height above free oil level	L	m	Chapter 9 (Fig. 9.6)
h_{GWC}	Height of gas-water contact above free water level	L	m	Chapter 2 (Eq. 2.23)
h_{OWC}	Height of oil-water contact above free water level	L	m	Chapter 2 (Eq. 2.24)
$ISIP$	Instantaneous shut-in pressure	$\frac{M}{LT^2}$	Pa	Chapter 8 (Fig. 8.8)
k_{mr}	Mudrock permeability	L^2	m^2	Chapter 10 (Eq. 10.2)
k	Permeability	L^2	m^2	Chapter 4 (Eq. 4.18)
K	Principal stress ratio	-	-	Chapter 3 (Eq. 3.15)
K_0	Lateral stress ratio for one dimensional strain	-	-	Chapter 3 (Eq. 3.3)
K_{0NC}	Normally consolidated lateral stress ratio for one dimensional strain	-	-	Chapter 3 (Fig. 3.8)
K_f	Stress ratio at Coulomb failure	-	-	Chapter 3 (Eq. 3.8)
LL	Liquid limit	-	-	Section 3.2.1
LOP	Leak-off pressure	$\frac{M}{LT^2}$	Pa	Chapter 8 (Fig. 8.8)
M	Slope of the Coulomb failure line in s'-t space	-	-	Chapter 3 (Equations 3.4 and 3.18)
m	Sedimentation rate	$\frac{L}{T}$	$\frac{m}{s}$	Chapter 4, (Eq. 4.30)
m_v	Coefficient of volume compressibility	$\frac{LT^2}{M}$	$\frac{1}{Pa}$	Chapter 4 (Eq. 4.11)
$mbsf$	Meters below seafloor	L	M	Chapter 3
n	Porosity	-	-	Chapter 3 (Eq. 3.2)
n_0	Reference porosity	-	-	Chapter 5 (Eq. 5.1)
n_m	Fraction of porosity that is bound water	-	-	Chapter 6 (Eq. 6.10)
N'	Normal interparticle force	$\frac{M}{LT^2}$	Pa	Chapter 4 (Eq. 4.1)
NCT	Normal compaction trend			Chapter 5
OWC	Oil-water contact			Chapter 2 (Fig. 2.6)
PPG	Pounds per gallon	$\frac{M}{L^2T^2}$	$\frac{kg}{m^2s^2}$	Chapter 2 (Eq. 2.10)
q	Darcy flow velocity	$\frac{L}{T}$	$\frac{m}{s}$	Chapter 4 (Eq. 4.18)
q	Deviatoric stress	$\frac{M}{LT^2}$	Pa	Chapter 3 (Fig. 3.9)
Q	Volumetric flux	$\frac{M}{L^3}$	$\frac{m^3}{s}$	Chapter 4 (Eq. 4.19)
r	Capillary tube radius	L	M	Chapter 2 (Eq. 2.16)
r_t	Threshold pore throat radius	L	M	Chapter 2 (Fig. 2.12)
R	Radius of curvature	L	M	Chapter 2 (Eq. 2.16)

(cont.)

Symbol	Name	Dimensions	SI Unit	Reference
R	Resistivity	$\frac{L^3M}{T^3I^2}$	$\Omega . m$	Chapter 5 (Eq. 5.10)
R	Bubble radius	L	M	Chapter 2 (Eq. 2.13)
R_h	Resistivity at equivalent hydrostatic effective stress	$\frac{L^3M}{T^3I^2}$	$\Omega . m$	Chapter 5 (Eq. 5.10)
RBBC	Resedimented Boston blue clay			Chapter 3 (Fig. 3.2)
RGoM-EI	Resedimented Gulf of Mexico – Eugene Island mudrock			Chapter 3 (Fig. 3.2)
RPC	Resedimented Presumpscot clay			Chapter 3 (Fig. 3.2)
s'	Average or principal effective stresses in plane of shearing	$\frac{M}{LT^2}$	Pa	Chapter 3 (Eq. 3.10)
s_e'	Average stress under isotropic conditions	$\frac{M}{LT^2}$	Pa	Chapter 3 (Eq. 3.20)
S_t	Storage coefficient	$\frac{L^2T}{M}$	$\frac{1}{Pa}$	Chapter 4 (Eq. 4.32)
S	Storage coefficient	$\frac{L^2T}{M}$	$\frac{1}{Pa}$	Chapter 4 (Eq. 4.35)
S	Smectite			Chapter 6 (Fig. 6.10)
$S+I$	Smectite plus illite			Chapter 6 (Fig. 6.10)
S_w	Wetting phase saturation	-	-	Chapter 2 (Fig. 2.12)
Sigma	Present effective stress	$\frac{M}{LT^2}$	Pa	Chapter 6 (Fig. 6.8)
Sigmax	Maximum past effective stress	$\frac{M}{LT^2}$	Pa	Chapter 6 (Fig. 6.8)
t	Maximum shear stress in plane of shearing	$\frac{M}{LT^2}$	Pa	Chapter 3 (Eq. 3.11)
t	Time	T	S	Chapter 4 (Eq. 4.19)
T	Tensile strength	$\frac{M}{LT^2}$	Pa	Chapter 8 (Eq. 8.1)
T	Dimensionless time factor for sedimentation	-	-	Chapter 4 (Eq. 4.30)
T_v	Dimensionless time factor for pore pressure dissipation	-	-	Chapter 4 (Eq. 4.27)
TSP	Total stress path			Chapter 3 (Fig. 3.9)
TVDrkb	True vertical depth below kelly bushing	L	M	Chapter 6 (Fig. 6.8)
TVD_{SS}	True vertical depth below sea surface	L	M	Chapter 2 (Eq. 2.1)
u	Pressure	$\frac{M}{LT^2}$	Pa	Chapter 2 (Eq. 2.3)
u_b	Borehole pressure	$\frac{M}{LT^2}$	Pa	Chapter 8 (Eq. 8.1)
u_c	Capillary pressure	$\frac{M}{LT^2}$	Pa	Chapter 2 (Eq. 2.20)
u_{cgo}	Gas-oil capillary pressure	$\frac{M}{LT^2}$	Pa	Chapter 2 (Fig. 2.6)
$u_{cHg-air}$	Mercury-air capillary pressure	$\frac{M}{LT^2}$	Pa	Chapter 2 (Eq. 2.22)
u_{cmig}	Minimum capillary pressure for migration of non-wetting phase	$\frac{M}{LT^2}$	Pa	Chapter 2 (Eq. 2.24)

(*cont.*)

Symbol	Name	Dimensions	SI Unit	Reference
u_{cow}	Oil-water capillary pressure	$\frac{M}{LT^2}$	Pa	Chapter 2 (Eq. 2.22, Fig. 2.6)
u_{crit}^{res}	Critical reservoir pressure at which seal leakage occurs	$\frac{M}{LT^2}$	Pa	Chapter 9 (Eq. 9.3)
u_d	Displacement pressure	$\frac{M}{LT^2}$	Pa	Chapter 2 (Fig. 2.12)
u_{de}	Extrapolated displacement pressure	$\frac{M}{LT^2}$	Pa	Chapter 2 (Fig. 2.12)
u_g	Gas phase pressure	$\frac{M}{LT^2}$	Pa	Chapter 2 (Fig. 2.6)
u_h	Hydrostatic pressure	$\frac{M}{LT^2}$	Pa	Chapter 2 (Eq. 2.1)
u^m	Pore pressure induced by mean stress	$\frac{M}{LT^2}$	Pa	Chapter 4 (Eq. 4.34)
u_o	Oil phase pressure	$\frac{M}{LT^2}$	Pa	Chapter 2 (Eq. 2.13)
u^q	Pore pressure induced by shear stress	$\frac{M}{LT^2}$	Pa	Chapter 4 (Eq. 4.34)
$u_{s'}$	Pore pressure induced by average stress	$\frac{M}{LT^2}$	Pa	Chapter 6 (Fig. 6.14)
u_t	Threshold pressure	$\frac{M}{LT^2}$	Pa	Chapter 2 (Fig. 2.12)
u_w	Water phase pressure	$\frac{M}{LT^2}$	Pa	Chapter 2 (Eq. 2.13)
u_w^{res}	Water pressure in reservoir	$\frac{M}{LT^2}$	Pa	Chapter 9 (Eq. 9.4)
u^*	Overpressure	$\frac{M}{LT^2}$	Pa	Chapter 2 (Eq. 2.3)
u_{res}^*	Reservoir overpressure	$\frac{M}{LT^2}$	Pa	Chapter 10 (Eq. 10.2)
u_{mr}^*	Mudrock overpressure	$\frac{M}{LT^2}$	Pa	Chapter 10 (Eq. 10.2)
U	Average degree of consolidation	-	-	Chapter 4 (Eq. 4.28)
U	Cohesion energy per molecule	$\frac{ML^2}{T^2}$	J	Section 2.6.1
U	Slope of the velocity-effective stress unloading curve	$\frac{L^2T}{M}$	$\frac{m^2s}{kg}$	Chapter 6 (Eq. 6.4)
UCS	Unconfined compressive strength	$\frac{M}{LT^2}$	Pa	Chapter 8 (Eq. 8.3)
V	Velocity	$\frac{L}{T}$	$\frac{m}{s}$	Chapter 5 (Eq. 5.4)
V	Volume	L^3	m^3	Chapter 2 (Eq. 2.12)
V_h	Velocity at equivalent hydrostatic effective stress	$\frac{L}{T}$	$\frac{m}{s}$	Chapter 5 (Eq. 5.4)
V_{max}	Velocity at preconsolidation stress	$\frac{L}{T}$	$\frac{m}{s}$	Chapter 6 (Eq. 6.5)
V_n	Interval velocity	$\frac{L}{T}$	$\frac{m}{s}$	Chapter 7 (Eq. 7.8)
V_{NMO}	Normal moveout velocity	$\frac{L}{T}$	$\frac{m}{s}$	Chapter 7 (Eq. 7.7)
V_o	Volume of oil	L^3	m^3	Chapter 2 (Eq. 2.12)
V_0	Reference velocity in velocity density cross plot	$\frac{L}{T}$	$\frac{m}{s}$	Chapter 6 (Eq. 6.7)
V_p	Vertical velocity	$\frac{L}{T}$	$\frac{m}{s}$	Chapter 7 (Eq. 7.9)
V_{rms}	Root mean square velocity	$\frac{L}{T}$	$\frac{m}{s}$	Chapter 7 (Eq. 7.3)
V_u	Unloaded velocity	$\frac{L}{T}$	$\frac{m}{s}$	Chapter 6 (Eq. 6.4)
V_w	Velocity of sound in water	$\frac{L}{T}$	$\frac{m}{s}$	Chapter 5 (Eq. 5.7)

(*cont.*)

Symbol	Name	Dimensions	SI Unit	Reference
V_w	Volume of water	L^3	m^3	Chapter 2 (Eq. 2.12)
v	Specific volume	-	-	Section 5.4.1 (Table 5.1)
v_0	Reference specific volume	-	-	Chapter 5 (Table 5.1)
w_L	Liquid limit	-	-	Section 3.2.1
W	Work done	$\frac{ML^2}{T^2}$	$\frac{kg.m^2}{s^2}$	Chapter 2 (Eq. 2.12)
z	Depth	L^1	m	Chapter 2 (Eq. 2.4)
$\bar{z}$	Centroid depth	L	m	Chapter 10 (Eq. 10.1)
z_{base}	Depth of reservoir base	L	m	Chapter 10 (Eq. 10.3)
z_{crest}	Depth of reservoir crest	L	M	Chapter 10 (Eq. 10.1)
Z	Relative depth of centroid	-	-	Chapter 10 (Eq. 10.1)
Z_{wd}	Water depth	L	m	Chapter 2 (Eq. 2.2)

Greek Units

Symbol	Name	Dimensions	SI Unit	Reference
α	Pore pressure coefficient	-	-	Chapter 2 (Eq. 2.7)
α	Fitting parameter in Eaton velocity equation	$\frac{1}{T}$	s	Chapter 5 (Eq. 5.5)
α	Fitting parameter in Eaton resistivity equation	$\frac{L^2M}{T^3I^2}$	Ω	Chapter 5 (Eq. 5.11)
α	Fitting parameter for Gardner velocity-density transform	$\frac{T}{L}$	$\frac{s}{m}$	Chapter 7 (Eq. 7.10)
α_b	Bulk sediment thermal expansion coefficient	$\frac{1}{K}$	$\frac{1}{°C}$	Chapter 4 (Eq. 4.33)
α_f	Fluid thermal expansion coefficient	$\frac{1}{K}$	$\frac{1}{°C}$	Chapter 4 (Eq. 4.33)
α_m	Matrix thermal expansion coefficient	$\frac{1}{K}$	$\frac{1}{°C}$	Chapter 4 (Eq. 4.33)
α_s	Solid thermal expansion coefficient	$\frac{1}{K}$	$\frac{1}{°C}$	Chapter 4 (Eq. 4.33)
β	Compaction coefficient	$\frac{LT^2}{M}$	$\frac{1}{Pa}$	Chapter 5 (Eq. 5.1)
β	Fitting parameter for Gardner velocity-density transform	-	-	Chapter 7 (Eq. 7.10)
γ	Interfacial tension	$\frac{M}{T^2}$	$\frac{mN}{m}$	Chapter 2 (Eq. 2.14)
γ	Fitting parameter in Eaton velocity equation	$\frac{L}{T}$	$\frac{m}{s}$	Chapter 5 (Eq. 5.5)
γ	Fitting parameter in Eaton resistivity equation	$\frac{L^3M}{T^3I^2}$	$\Omega \cdot m$	Chapter 5 (Eq. 5.11)
γ_{Hg-air}	Mercury air interfacial tension	$\frac{M}{T^2}$	$\frac{mN}{m}$	Chapter 2 (Eq. 2.22)

(*cont.*)

Symbol	Name	Dimensions	SI Unit	Reference
γ_{nws}	Non-wetting fluid-solid interfacial tension	$\frac{M}{T^2}$	$\frac{mN}{m}$	Chapter 2 (Eq. 2.14)
γ_{ow}	Oil-water interfacial tension	$\frac{M}{T^2}$	$\frac{mN}{m}$	Chapter 2 (Eq. 2.12)
γ_{ws}	Wetting fluid-solid interfacial tension	$\frac{M}{T^2}$	$\frac{mN}{m}$	Chapter 2 (Eq. 2.14)
δ	Thomsen delta	-	-	Chapter 7 (Eq. 7.9)
$\Delta e_{s'}$	Change in void ratio due to average stress change	-	-	Chapter 3 (Eq. 3.26)
Δe_q	Change in void ratio due to shear stress change	-	-	Chapter 3 (Eq. 3.26)
η	Slope of line in an s'-t plot	-	-	Chapter 3 (Eq. 3.14)
η	Slope of line in a σ'_m-q plot	-	-	Chapter 7 (Eq. 7.15)
η	Bulk viscosity	$\frac{M}{LT}$	$Pa \cdot s$	Chapter 4 (Eq. 4.38)
η_{cs}	Slope of line in an s'-t plot at critical state	-	-	Chapter 3 (Eq. 3.18)
η_{K_0}	Slope of line in an s'-t plot under uniaxial compression	-	-	Chapter 3 (Eq. 3.19)
η_τ	Slope of line in an s'-t plot at Coulomb failure	-	-	Chapter 3 (Eq. 3.17)
θ	Contact angle	-	*radians*	Chapter 2 (Eq. 2.14)
θ	Inclination of a surface relative to the plane upon which the principal stress is acting	-	*radians*	Chapter 3 (Fig. 3.12)
θ_{cr}	Inclination of Coulomb failure surface relative to the plane upon which the principal stress is acting	-	*radians*	Chapter 3 (Eq. 3.9)
θ_{Hg-air}	Mercury-air contact angle	-	*radians*	Chapter 2 (Eq. 2.22)
θ_{ow}	Oil-water contact angle	-	*radians*	Chapter 2 (Eq. 2.22)
λ	Slope on a plot of void ratio vs. natural log of average stress	-	-	Chapter 3 (Eq. 3.21)
λ^*	overpressure ratio	-	-	Chapter 2 (Eq. 2.11)
μ	Viscosity	$\frac{M}{LT}$	$Pa \cdot s$	Chapter 4 (Eq. 4.18)
ξ_m	Mean stress loading efficiency	$\frac{LT^2}{M}$	$\frac{1}{Pa}$	Chapter 4 (Eq. 4.34)
ξ_q	Shear stress loading efficiency	$\frac{LT^2}{M}$	$\frac{1}{Pa}$	Chapter 4 (Eq. 4.34)
μ	Unloading coefficient	-	-	Chapter 6 (Eq. 6.8)
v	Poisson's ratio	-	-	Chapter 8 (Eq. 8.5)
ρ	Density	$\frac{M}{L^3}$	$\frac{kg}{m^3}$	Chapter 2 (Eq. 2.4)
ρ_b	Bulk density	$\frac{M}{L^3}$	$\frac{kg}{m^3}$	Chapter 2 (Eq. 2.5)
ρ_g	Gas density	$\frac{M}{L^3}$	$\frac{kg}{m^3}$	Section 2.2
ρ_{max}	Density at preconsolidation stress	$\frac{M}{L^3}$	$\frac{kg}{m^3}$	Chapter 6 (Eq. 6.8)

(cont.)

Symbol	Name	Dimensions	SI Unit	Reference
ρ_o	Oil density	$\frac{M}{L^3}$	$\frac{kg}{m^3}$	Section 2.2
ρ_{pw}	Pore water density	$\frac{M}{L^3}$	$\frac{kg}{m^3}$	Chapter 2 (Eq. 2.2)
ρ_{sw}	Seawater density	$\frac{M}{L^3}$	$\frac{kg}{m^3}$	Chapter 2 (Eq. 2.2)
ρ_v	Density for an observed velocity under normal compaction	$\frac{M}{L^3}$	$\frac{kg}{m^3}$	Chapter 6 (Eq. 6.8)
ρ_0	Reference density in velocity-density cross plot for normal compaction	$\frac{M}{L^3}$	$\frac{kg}{m^3}$	Chapter 6 (Eq. 6.7)
σ'	Effective normal stress	$\frac{M}{LT^2}$	Pa	Chapter 4 (Eq. 4.1)
σ	Total normal stress	$\frac{M}{LT^2}$	Pa	Chapter 4 (Eq. 4.2)
σ'_{ff}	Normal effective stress on failure plane at Coulomb failure	$\frac{M}{LT^2}$	Pa	Chapter 3 (Eq. 3.7)
σ'_h	Minimum horizontal effective stress	$\frac{M}{LT^2}$	Pa	Chapter 3 (Eq. 3.3)
σ'_e	Equivalent stress	$\frac{M}{LT^2}$	Pa	Chapter 7 (Eq. 7.15)
σ_h	Minimum horizontal stress	$\frac{M}{LT^2}$	Pa	Chapter 3 (Eq. 3.13)
σ_H	Maximum horizontal stress	$\frac{M}{LT^2}$	Pa	Chapter 3 (Fig. 3.14)
σ_m	Mean total stress	$\frac{M}{LT^2}$	Pa	Chapter 4 (Eq. 4.7)
σ'_m	Mean effective stress	$\frac{M}{LT^2}$	Pa	Chapter 3 (Fig. 3.9)
σ'_p	Preconsolidation stress	$\frac{M}{LT^2}$	Pa	Chapter 6 (Eq. 6.2)
σ'_u	Unloaded vertical effective stress	$\frac{M}{LT^2}$	Pa	Chapter 3 (Fig. 3.5, Eq. 6.2)
σ_v	Vertical total stress	$\frac{M}{LT^2}$	Pa	Chapter 2 (Eq. 2.6)
σ'_v	Vertical effective stress	$\frac{M}{LT^2}$	Pa	Chapter 2 (Eq. 2.7)
σ'_{vh}	Vertical effective stress if pore pressure is hydrostatic	$\frac{M}{LT^2}$	Pa	Chapter 5 (Eq. 5.4)
σ_1	Maximum principal stress	$\frac{M}{LT^2}$	Pa	Section 3.3
σ'_1	Maximum principal effective stress	$\frac{M}{LT^2}$	Pa	Chapter 3 (Eq. 3.8)
σ_2	Intermediate principal stress	$\frac{M}{LT^2}$	Pa	Section 3.3
σ_3	Least principal stress	$\frac{M}{LT^2}$	Pa	Section 3.3
σ'_3	Least principal effective stress	$\frac{M}{LT^2}$	Pa	Chapter 3 (Eq. 3.8)
σ_3^{seal}	Least principal stress in seal above reservoir	$\frac{M}{LT^2}$	Pa	Chapter 9 (Eq. 9.3)
σ'_θ	Normal effective stress to a plane at angle theta to the principal stress	$\frac{M}{LT^2}$	Pa	Chapter 3 (Fig. 3.12)
τ	Shear stress	$\frac{M}{LT^2}$	Pa	Chapter 3 (Eq. 3.6)
τ_{ff}	Shear stress on failure surface at Coulomb failure	$\frac{M}{LT^2}$	Pa	Chapter 3 (Eq. 3.7)

(*cont.*)

Symbol	Name	Dimensions	SI Unit	Reference
τ_θ	Shear stress on a plane at angle θ to the principal stress	$\frac{M}{LT^2}$	*Pa*	Chapter 3 (Fig. 3.12)
$\tau_{\theta f}$	Shear stress at failure along critical failure plane	$\frac{M}{LT^2}$	*Pa*	Chapter 3 (Fig. 3.13a)
ϕ'	Friction angle based on effective stress	-	*radians*	Chapter 3 (Eq. 3.7)

1
Introduction

1.1 The Deepwater Horizon Blowout

The Deepwater Horizon blowout of the Macondo well in Mississippi Canyon Block 252 in the deepwater Gulf of Mexico began on April 20, 2010 (Figures 1.1 and 1.2). Eleven people died and approximately four million barrels of oil leaked into the Gulf of Mexico (Boebert & Blossom, 2016). Through this event, the general public became aware of the enormous pressures encountered in sedimentary basins and of the extraordinary complexity and risk associated with finding and producing hydrocarbons in the deep ocean.

In fact, the Macondo well was a dramatic but not unusual illustration of the conditions encountered when drilling in deepwater basins. It was not in particularly deep water, nor was its total depth particularly great (Fig. 1.3) (Deepwater Horizon Study Group, 2011). However, the pressures and stresses encountered in this well (Fig. 1.3) record many of the processes that are the focus of this book.

At Macondo, and in most sedimentary basins, pore pressure, u, is bounded below by the hydrostatic pressure, u_h, (Fig. 1.3, dashed purple line) and above by the overburden stress, σ_v (Fig. 1.3c, green line). u_h records the pressure due to a static column of water from the sea surface, while σ_v approximately records the stress due to the weight of the overlying sediment and water. The overpressure, u^*, is the pressure above the hydrostatic pressure $(u^* = u - u_h)$. The difference between the overburden stress and the pore pressure is the vertical effective stress $\left(\sigma_v' = \sigma_v - u\right)$.

At Macondo, pore pressures (black line, Fig. 1.3c) roughly parallel the overburden stress (green line, Fig. 1.3c); the vertical effective stress is the difference between these two and is almost constant from near the seafloor to 17 640 ft (5377 m). The reservoir pore pressures result from both elevated water pressure and the buoyancy of the hydrocarbons trapped within the reservoir. Petroleum reservoirs are multiphase systems and can be composed of water, oil, and gas, each of which can have discrete pressures. To unravel the pore pressure distribution, and understand the implications for how hydrocarbons are trapped, the

Figure 1.1 Fire boat response crews battle the remnants of the Deepwater Horizon drillship. Multiple Coast Guard helicopters, planes, and cutters responded to rescue the Deepwater Horizon's 126 person crew. Photo: U.S. Coast Guard. Source: USGS.

distribution of the different hydrocarbon phases and their pressures must be understood. I review how to describe pressures in multiphase systems in gravity and capillary equilibrium in Chapter 2. I review how water phase overpressures such as those at Macondo are generated in Chapter 4, and I illustrate how to estimate the trap integrity as a function of these water phase pressures for hydrocarbons or CO_2 in Chapter 9.

Petrophysical measurements (e.g., density, resistivity, and velocity) are commonly used to predict pore pressure in mudrocks, and this approach was used at Macondo (Pinkston & Flemings, 2019). This is possible because the mudrocks are very compressible and their compaction state records the effective stress as is discussed in Chapter 3. In Chapters 5, 6, and 7, I discuss methodologies to estimate pore pressure from the compaction state from log, core, and seismic data.

As the main reservoir target, the M56 sandstone, is approached, pore pressures drop abruptly by 1 200 psi (8.3 MPa) over 370 ft (113 m) (Fig. 1.3c). The decrease in pore pressure at the base of the Macondo well was one of the challenges

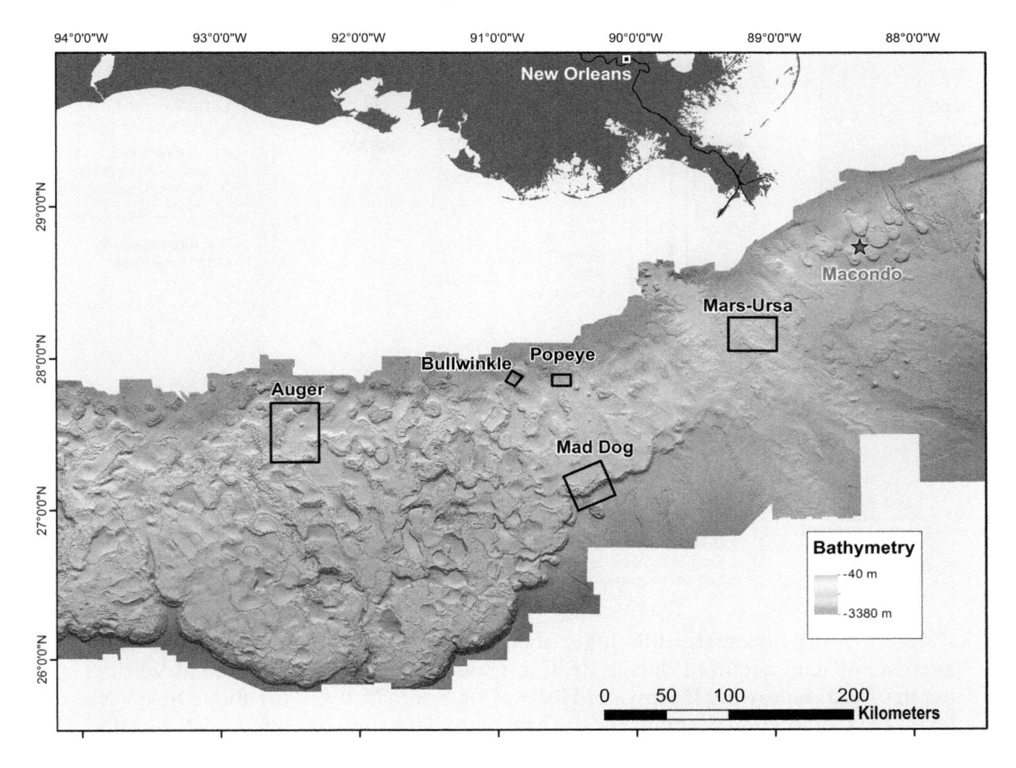

Figure 1.2 Bathymetry of the slope of the Gulf of Mexico. Multiple locations studied in this book are shown. Bathymetry is based on a 3D seismic deepwater bathymetry grid of the northern Gulf of Mexico made available for public use by the Bureau of Ocean Energy Management.

encountered when drilling and completing the well (Pinkston & Flemings, 2019). Pressure regressions like this are common and result from two- and three-dimensional flow of pore water in the subsurface. In Chapters 10 and 11, I present two- and three-dimensional conceptual and quantitative models to describe this flow. These models illustrate how permeable interconnected reservoirs trapped within overpressured mudrocks can result in dramatic pressure variation. I show how to predict this variation regionally.

The red line in Figure 1.3c is an estimate of the least principal stress. During drilling, the pressure in the open borehole is generally maintained to be above the pore pressure and below the least principal stress. When the pore pressure exceeds the wellbore pressure, flow from the formation into the borehole can occur (a kick) and when the borehole pressure approaches the least principal stress, borehole fluid can be lost through fractures into the formation (a loss to the formation). The small difference between pore pressure and least principal stress is the reason why so

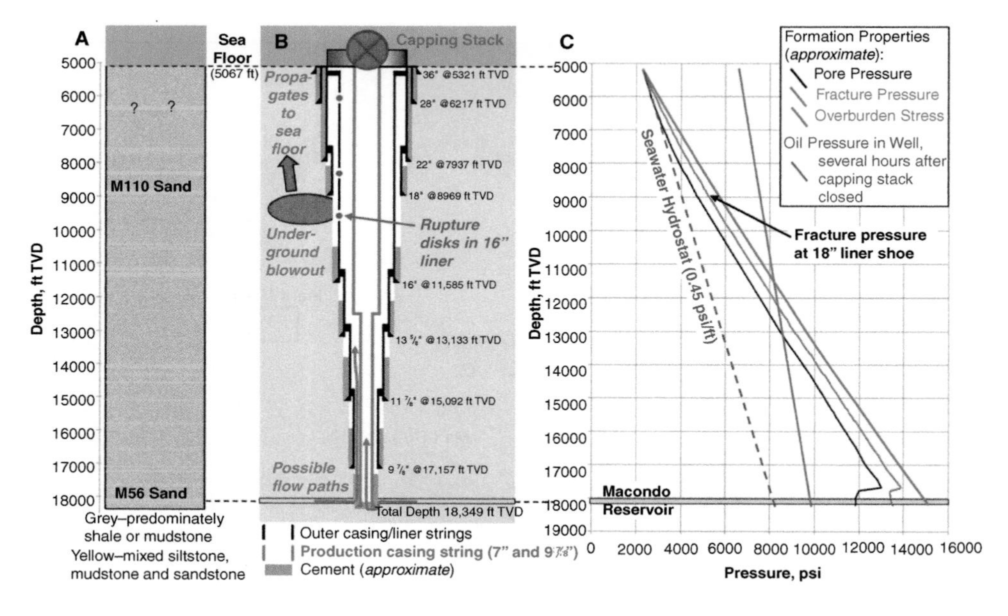

Figure 1.3 (a) Schematic lithologic section for the Macondo well based upon analysis of data acquired during drilling (Source: BP). Depths are total vertical depth (TVD) below the Deepwater Horizon rig floor, 75 ft (23 m) above mean sea level. (b) Completion diagram for the Macondo well showing outer nested casing and liner strings cemented in place during drilling, and production casing cemented across the Macondo Reservoir (M56 sand, yellow). Possible oil flow paths during blowout (shown in red) were either inside the production casing, between the production casing and the outer casing/liner strings, or both. (c) Approximate in situ pore pressure (black), fracture pressure (red), and overburden stress (green). The blue line shows the approximate oil pressure in the wellbore corresponding to a capping stack pressure of 6 600 psi, which was observed several hours after the well was shut in (calculated for an oil pressure gradient in the well of 0.25 psi/ft). From Hickman et al. (2012) with permission of the National Academy of Sciences.

many casing strings were required on the Macondo well (Fig. 1.3b). Least principal stress is also a primary control on hydrocarbon trapping in the subsurface. In Chapter 3, experiments and models are presented to describe the magnitude of least principal stress. In Chapter 8, methods to estimate least principal stress are described. In Chapter 9, I describe how hydrocarbons are trapped by least principal stress and estimate the column of hydrocarbons or CO_2 that can be trapped in the subsurface. Finally, in Chapter 11, models are presented to describe how reservoir pressures can exceed the least principal stress in bounding mudrocks, cause trap failure, and drive fluid venting at the seafloor. This insight can be used to predict optimal locations where hydrocarbons are trapped.

1.2 Audience and Application

I wrote this book for those in industry who study subsurface pressure and stress and for the graduate student with a passion for understanding how pore pressure drives geological processes. As a geoscientist, I hope this book will inspire the geoscientist to advance their understanding of Earth processes through insights provided by geotechnical and petroleum engineering science. In turn, I hope this book will inspire the engineer to extend engineering concepts to the larger scales, stresses, and depths of geological systems.

In the energy industry, there are a myriad of reasons to understand subsurface pressure and stress. In the exploration phase, an understanding of pressure and stress provides a tool to predict where hydrocarbons are trapped and how and to where hydrocarbons migrate in the subsurface. To design and drill a well safely and economically, it is necessary to constrain subsurface pressures and stresses. In the development phase, an understanding of pressure is an important tool to understand fluid distribution and reservoir connectivity, and it provides insight into the impact and risk of subsurface injection to maintain pore pressure (Naruk et al., 2019). The concepts presented also apply to understanding the state of pressure and stress in unconventional basins (Couzens-Schultz et al., 2013) and can be used to gain insight into the processes by which injection-induced earthquakes occur, whether by hydraulic fracturing or for waste water injection (Ellsworth, 2013).

As we look to the future, it is clear that subsurface CO_2 sequestration will become one vital tool to limit the impact of continued use of carbon-based fuel on global climate change. This book has significant application to this practice. Subsurface CO_2 sequestration involves injection of an immiscible buoyant fluid (Benson & Cole, 2008). Structural trapping is one important component of CO_2 sequestration, and it is directly analogous to how hydrocarbons are trapped. In addition, the design of injection programs will need to involve a complete understanding of the pressure and stress conditions in the subsurface.

Finally, this book provides a platform to understand how pore pressure and stress drive fascinating geological processes. Pore pressure and stress are driving forces in earthquakes and faulting (Cruz-Atienza et al., 2018; Rogers & Dragert, 2003; Saffer & Tobin, 2011). Submarine landslides can be driven by elevated pore pressure (Dugan & Flemings, 2000a). Sandstone injections (Boehm & Moore, 2002) and seafloor vents record the interaction of pore pressure and stress, and complex biological communities thrive at vent locations (Brooks et al., 1987).

1.3 The Discipline of Pore Pressure Analysis

The field of pore pressure analysis lies at the interfaces of geoscience, petroleum engineering, and geotechnical engineering. Advances will lie at these interfaces. I direct my graduate students in geoscience to have a strong basis in basic geoscience (particularly understanding of sedimentary rocks, stratigraphy, and structural geology), to have an exposure to how rocks deform and fluids flow through them (particularly through hydrogeology, geotechnical engineering, and geomechanics), and to understand multiphase behavior (through courses in petroleum engineering). Multiphase behavior has not been emphasized in the geosciences but is increasingly recognized as a vital component of geological systems.

Dickinson (1953) characterized basin overpressures and attributed their origin to the inability of low permeability shales to expel their fluids during compaction. Hubbert and Rubey (1959) and Rubey and Hubbert (1959) quantitatively explored the origin of overpressure in basins and how to predict these pressures, and presented models to describe least principal stress. However, as noted by Rubey and Hubbert (1959), the foundation for understanding and predicting pore pressure is rooted in the study of consolidation as developed in geotechnical engineering (Terzaghi, 1923, 1943). A series of now-classic papers followed that developed practical approaches to apply these seminal concepts (Eaton, 1969, 1975). As I discuss in Chapter 4, pore pressure research continues to develop practical approaches that rely on these concepts. For example, new compaction models have been developed to describe an observed porosity-effective stress relationship, or a different empirical relationship has been presented to describe the relationship between least principal stress and pore pressure.

Two books review overpressure in basins (Fertl, 1976; Mouchet et al., 1989). They focus on the origins of overpressure, review compaction behavior, and discuss methods to detect and predict abnormal formation pressures with wireline logs. These studies emphasize a one-dimensional analysis (e.g., a well profile). Edited volumes have also provided overviews of different aspects of pore pressure research. For example, Dutta (1987) reprinted some of the classic historical papers on geopressure and Huffman et al. (2001) captured many important contributions. The remaining literature is extensive but distributed in the geological and engineering literature. Many of these publications are cited in this volume. However, the references herein are not intended to be exhaustive.

The field of pressure and stress analysis began to evolve rapidly in the early 1990s. This evolution was driven by the dramatic expansion of the energy industry into the deepwater. In this environment, drilling is brutally expensive, and new problems were encountered. To operate effectively in this environment, pressure and stress had to be better understood. To do so, new tools were developed, more and better data were acquired, and resources were devoted to understanding the geopressured system. Operators began to deploy integrated teams of geophysicists, geologists, and petroleum and civil engineers to characterize and predict subsurface pressure and stress in order to develop exploration targets and then design and drill wells safely and economically. As a result, the study of pore pressure was no longer restricted to the silo of the drilling engineer or the explorationist. One example of this integrated approach was the coupling of seismic and well data to provide an integrated understanding of the three-dimensional distribution of subsurface pressure and stress (see Chapter 7). This view advanced us beyond the study of single vertical wells and into the domain of understanding the entire subsurface system.

As a result of these efforts, we now know that dipping permeable reservoir layers trapped within overpressured mudrock set up a complex hydrodynamic system (Chapters 10 and 11). This system results in locally depressed or elevated pore pressures relative to what is predicted from one-dimensional analysis. This is a vital insight for well design, as it controls the entrapment of hydrocarbons and drives a range of geological processes. We now understand how the evolution of structure and stratigraphy controls the entrapment of hydrocarbons, the expulsion of fluids through seafloor seeps and mud volcanoes, and the generation of submarine landslides. Chapters 10 and 11 focus on this behavior.

Historically, the disciplines of pore pressure analysis and geomechanics were separated; pore pressure was calculated independently and provided as input to geomechanical models. Today, we often link the analysis of pressure and stress. In Chapters 6 and 7, I present coupled approaches to predict pressure and stress. In addition, most geomechanical approaches have relied on poroelastic material models with Coulomb failure. Today, we are beginning to incorporate more realistic material behavior including elastoplasticity, stress-dependent strength, and creep into pressure and stress analysis. I begin this discussion in Chapter 3, and continue it in Chapters 4, 6, and 7.

These approaches are supported by a new generation of experimental studies on material behavior of mudrocks. Historically, the field relied on insights from low stress experiments conducted in geotechnical engineering laboratories. Now, experiments are routinely run at pressures and stresses characteristic of those encountered in the energy industry. Through these experiments we have a much better understanding of how rocks deform as a function of their composition at geological stresses. In Chapter 3 I review the experimental behavior of mudrocks, and in Chapter 8 I discuss the implications of these experiments for the prediction of least principal stress.

In the future, the fields of geomechanics and pore pressure analysis will continue to merge through two- and three-dimensional Earth models. It is within our reach to routinely predict both pore pressure and the full stress tensor in three dimensions within sedimentary basins (Chapter 7). Finally, there is strong potential to link the study of pressure and stress with the velocity anisotropy that results during compaction. Thus, there is the potential both to use seismic data to more fully predict subsurface pressure and stress, and, in turn, to use our understanding of stress and pressure to improve subsurface imaging.

1.4 Nomenclature

Geomechanics, petroleum engineering, and geotechnical engineering all have an internal nomenclature and this nomenclature can conflict. I provide a nomenclature table and a reference to where each term is first presented (see *List of Nomenclature*). In many areas of conflict, I have adopted the geotechnical expression. I apologize in advance to those who will be frustrated by unfamiliarity with these terms.

1.5 Summary

The Deepwater Horizon blowout at the Macondo well was a catastrophic human, environmental, and economic disaster. It exposed to the public the enormous pressures encountered in the subsurface. In fact, the pressures and stresses encountered were not unusual but characteristic of those encountered in deepwater drilling. This book illuminates the underlying processes that lead to pressure and stress profiles such as those encountered at Macondo and it provides strategies to predict those conditions ahead of the drill bit.

Significant advances in understanding pressure and stress in sedimentary basins began in the 1950s, with the foundation for these advances stemming from the study of soil behavior in the 1920s. With the advent of deepwater drilling, our understanding of overpressure and our approaches to pressure prediction have dramatically advanced over the last thirty years.

Future advances in the field will result from the continued integration of geomechanics with pore pressure analysis, and increased study of the material behavior of mudrocks at high stresses and temperatures. In the near future, it will be routine to predict the full stress field and pore pressure in sedimentary basins, and there is strong potential for coupling seismic imaging more closely to the state of stress in sedimentary basins.

2
Reservoir Pore Pressure

2.1 Introduction

In this chapter, I describe how to characterize reservoir pore pressures that are under capillary and gravity equilibrium (e.g., Fig. 2.1). This is approximately the state of a geological reservoir prior to production. Pore pressure is commonly described with a pressure versus depth plot (Fig. 2.1b). The reservoir pressure (u_g and u_w, Fig. 2.1) commonly, but not always, lies between the hydrostatic pressure (u_h) and the lithostatic stress (σ_v) (Fig. 2.1). Pore pressure is also characterized with overpressure (u^*), and average equivalent density (mud weight) plots (Fig. 2.1c, d).

I illustrate the approach with an example from the Bullwinkle field in the Gulf of Mexico. Initially, I neglect the impact of capillary migration pressure (the minimum pressure necessary for the non-wetting phase to migrate through the rock). Later, I describe how to incorporate this effect. I close with a conceptual discussion of the saturation and pressure distribution within reservoirs at gravitational and capillary equilibrium.

2.2 Pore Pressure

Hydrostatic pressure (u_h) (Fig. 2.1b) is the pressure caused by a static column of water. It is commonly expressed by assuming a constant pore water density (ρ_{pw}) referenced from the sea surface:

$$u_h = \rho_{pw} g TVD_{ss}. \quad \text{Eq. 2.1}$$

TVD_{ss} is the vertical depth from the sea surface. The hydrostatic pressure is often described with a seawater density (ρ_{sw}) that is distinct from the pore water density (ρ_{pw}):

$$u_h = \rho_{sw} g Z_{wd} + \int_{Z_{wd}}^{TVD_{ss}} \rho_{pw} g dz. \quad \text{Eq. 2.2}$$

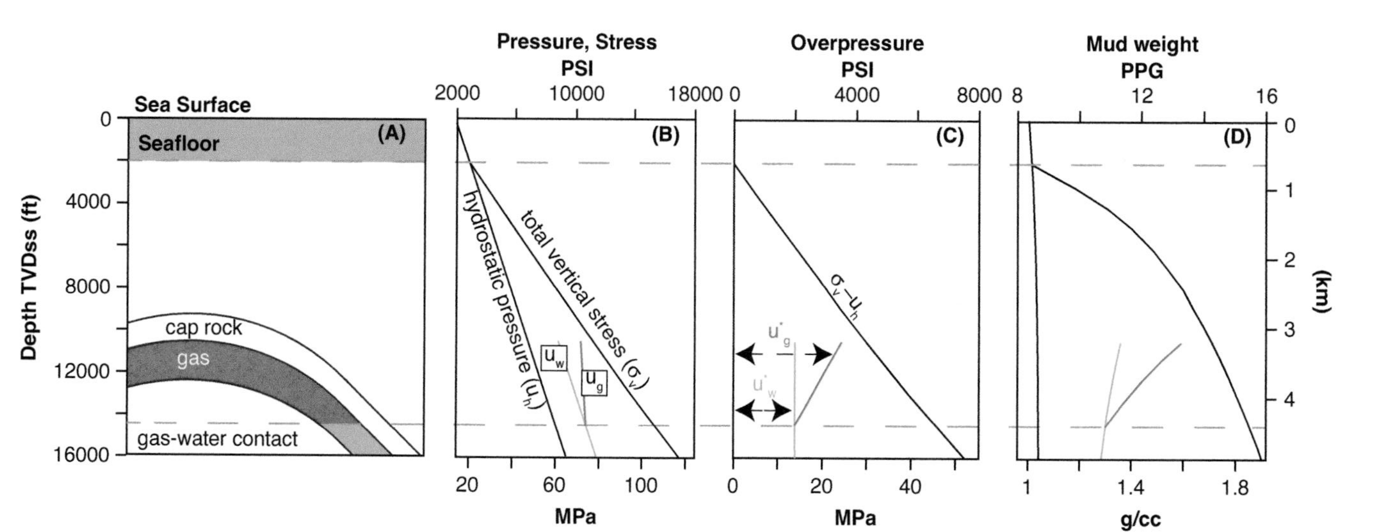

Figure 2.1 (a) Schematic diagram of gas trapped within a permeable and overpressured reservoir. Pore pressure is expressed through (b) pressure, (c) overpressure, and (d) average equivalent density (equivalent mud weight) plots. From Merrell et al. (2014). Reprinted by permission of the AAPG whose permission is required for further use. AAPG© 2014.

Table 2.1 *Typical density values for common fluids in the subsurface. Actual values can vary significantly as a function of composition, pressure, and temperature. In particular, gas density varies tremendously due to its compressibility and should be solved directly through either the ideal gas law or real gas law behavior as described in Dake (1978).*

Material	Density (kg/m^3)	Pressure Gradient (MPa/km)	Pressure Gradient (psi/ft)	Mud weight (PPG)
Seawater	1 023	10.0	0.444	8.5
Typical overburden	2 110	20.7	0.915	17.6
Typical brine	1 072	10.5	0.465	8.9
Typical gas	300	2.9	0.130	2.5
Typical oil	718	7.0	0.311	6.0
Water	1 000	9.8	0.434	8.3

Z_{wd} is the water depth, and g is acceleration of gravity. Seawater density is approximately equal to 1.023 g/cm^3 (Table 2.1) and is generally assumed constant. However, the pore water can vary regionally and with depth.

Overpressure (u^*) is the difference between the pore pressure and the hydrostatic pressure:

$$u^* = u - u_h. \qquad \text{Eq. 2.3}$$

In a static aquifer, the water pressure follows the hydrostatic gradient. For the case where the pore water and seawater density are equal, the water pressure parallels the hydrostatic gradient (Fig. 2.1b) and the overpressure (u^*) is constant (Fig. 2.1c).

Reservoirs, prior to production, are commonly described as in gravitational equilibrium. In this case, water pressures and hydrocarbon pressures follow their static pressure gradients (Fig. 2.1). Typical static gradients for fluids are shown in Table 2.1. The distribution of reservoir pore pressures is estimated by extrapolating along the static pressure gradient from measured values. For example, the pore pressure (u_{z_2}) at depth z_2, given a pressure measurement at depth z_1, is:

$$u_{z_2} = u_{z_1} + \rho g(z_2 - z_1). \qquad \text{Eq. 2.4}$$

Here, ρ is the fluid density, g is gravitational acceleration, and consequently ρg is the fluid pressure gradient.

In Figure 2.1b, both a water phase pressure (blue line) and a gas phase pressure (red line) are shown and they are assumed to be equal at the gas-water contact. We explore this assumption later in this chapter.

2.3 Vertical Stress and Vertical Effective Stress

The most common way to estimate the total vertical stress (σ_v) is to integrate the weight of the overlying material:

$$\sigma_v = \int_0^Z \rho_b(z) g dz, \qquad \text{Eq. 2.5}$$

where ρ_b is the bulk density of the sediment, which varies with depth. In the marine environment, Equation 2.5 is expanded to include the weight of the water column:

$$\sigma_v = \rho_{sw} g Z_{wd} + \int_{Z_{wd}}^{Z} \rho_b(z) g dz . \qquad \text{Eq. 2.6}$$

I discuss in Chapter 8 different approaches to characterize the overburden stress (σ_v). However, a common approach is to integrate log-derived bulk density measurements. The compressibility of water is so small that it is commonly assumed to be constant. However, the bulk density (ρ_b) varies greatly and cannot be assumed constant.

When a porous solid is externally loaded, the internal opposing forces are partitioned between water (or other fluids) and the porous solid (Fig. 2.2). Fluid pressure acts in opposition to external stresses over the fraction of the porous medium's boundary represented by the porosity (Fig. 2.2). Terzaghi (1923) first described how these forces are partitioned through the effective stress concept.

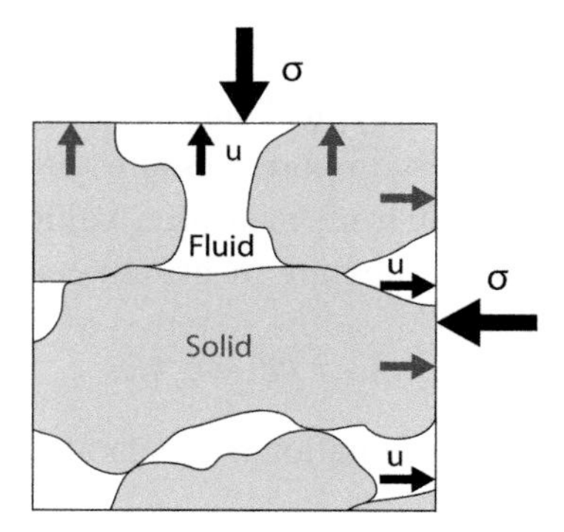

Figure 2.2 Magnified portion of a small block of porous medium showing external stresses and opposing stresses in the solid grains and opposing pressure in the pore fluid. From Ingebritsen et al. (2006). Reprinted with permission by Cambridge University Press.

Effective stress represents the average stress felt by the soil skeleton and was meant to describe the fraction of the stress that was 'effective' in deforming the rock. The effective vertical stress (σ_v') is the total vertical stress less some portion of the pore pressure:

$$\sigma_v' = \sigma_v - \alpha u. \qquad \text{Eq. 2.7}$$

α is a pore pressure coefficient, which varies between 0 and 1 as described by Ingebritsen et al. (2006). Equation 2.7 applies for both differential changes in stress and, as written, for absolute values of stress and pressure. Equation 2.7 is written for vertical stress, but can apply to any stress orientation. Terzaghi (1923) originally suggested α was equal to porosity. Subsequently, Nur and Byerlee (1971) showed that it was proportional to the ratio of the bulk compressibility (c_b) to solid compressibility (c_s):

$$\alpha = 1 - \left(\frac{c_s}{c_b}\right). \qquad \text{Eq. 2.8}$$

Under most conditions in sedimentary basins, the bulk compressibility is orders of magnitude greater than the solid compressibility and therefore α is equal to 1. This assumption breaks down in very stiff, and very low porosity material. Generally, I will assume $\alpha = 1$.

2.4 Overpressure, and Pressure Gradient Plots

Subsurface pressures and stresses are often presented as overpressure plots (Fig. 2.1c) or pressure gradient (mud weight) plots (Fig. 2.1d). In an overpressure plot, the hydrostatic pressure (u_h) is subtracted from the total pressures or stresses (Eq. 2.3). In this perspective, aquifers that are at a hydrostatic gradient have a constant overpressure (blue line, Fig. 2.1c). When the overpressure equals the reduced total vertical stress ($\sigma_v - u_h$), the pore pressure equals the total vertical stress ($u = \sigma_v$).

Finally, average density plots, or mud weight plots, express the pore pressure as the average density of a fluid measured from a datum that will balance the pressure at a particular depth (Fig. 2.1d). In this example, the sea surface is assumed to be the datum. However, generally the datum is the elevation of the rig floor. Average density as measured from the sea surface is calculated from pore pressure as follows:

$$\rho = \frac{u}{g * TVD_{ss}}. \qquad \text{Eq. 2.9}$$

This density can easily be converted to equivalent mud weight or a pressure gradient (e.g., Table 2.1). For example, equivalent mud weight in units of pounds per gallon (EMW) can be expressed as a function of pressure in units of pounds per square inch and depth in feet as:

$$EMW = \frac{u}{0.052 * TVD_{ss}}. \qquad \text{Eq. 2.10}$$

2.5 Example: The Bullwinkle J3 Sand

The Bullwinkle Oil Field is located on the western flank of a circular salt-withdrawal minibasin on the slope of the Gulf of Mexico, approximately 150 miles southwest of New Orleans, Louisiana in 1 350 ft water depth (412 m) (Figures 2.3 and 2.4). The field has been described extensively (Comisky, 2002; Flemings et al., 2001; Holman & Robertson, 1994; O'Connell et al., 1993; Swanston et al., 2003).

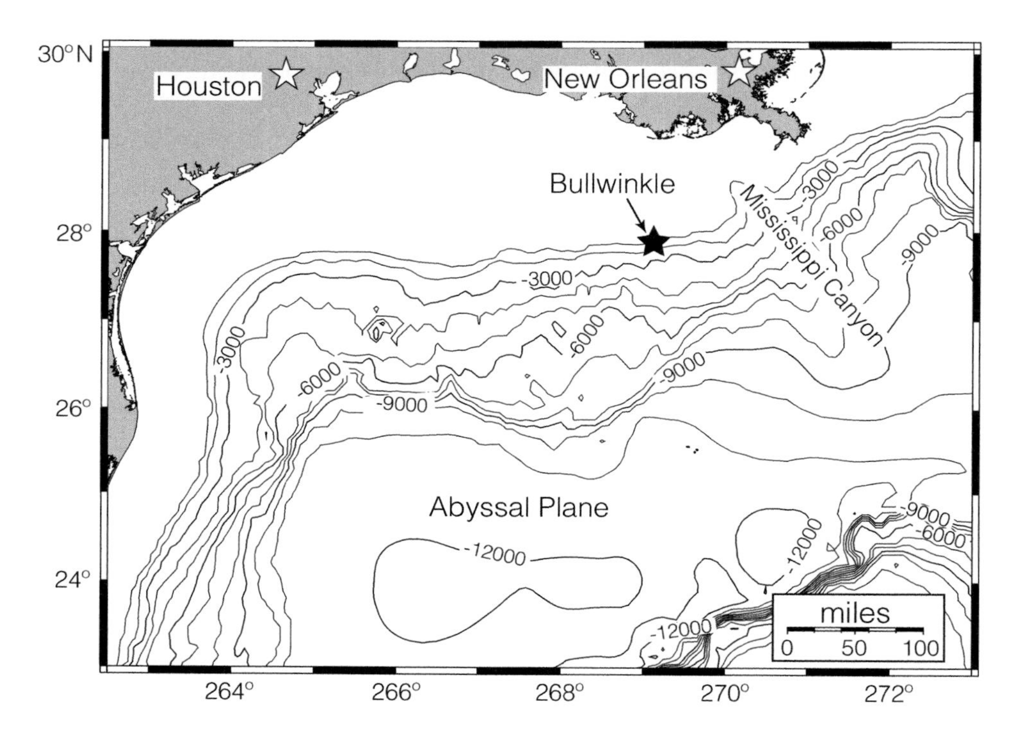

Figure 2.3 The Bullwinkle Basin (Green Canyon 65 and 109) is on the upper continental slope in approximately 1 300 ft water depth (400 m) approximately 150 miles southwest of New Orleans, LA. Reprinted from Flemings and Lupa (2004) with permission of Elsevier.

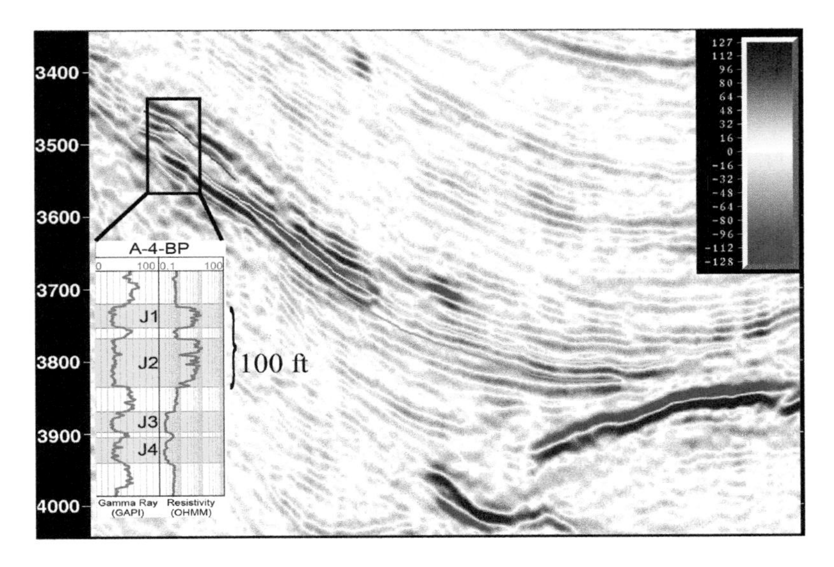

Figure 2.4 Seismic cross-section through the J sands at Bullwinkle. Cross-section is located in Figure 2.5. Inset on lower left illustrates the blocky character of the J sands.

The J sands are of early Nebraskan (3.35 Ma) age and host the majority of the reserves at Bullwinkle. The five J sands (J0–J4) are bowl-shaped interconnected channel and sheet turbidite sands that are interbedded with debris flow deposits and shales, and overlain by a thick section (500 ft) of bathyal shales. Pressure drawdown at each well followed the same depletion curve, indicating that the sands are in pressure communication. The J3 sand is very fine to fine-grained and has a blocky log character (Fig. 2.4, inset). The lithology and the grain size are relatively homogenous across the field. It is interpreted to be a ponded, internally amalgamated sheet sand. There is a small oil and gas pool at its crest, while the majority of it is brine saturated (Fig. 2.5).

The parameters necessary to characterize pressure in the J3 sand are shown in Table 2.2. Three critical measurements were made: (1) pressure was measured to be 56.7 MPa within the gas leg of the J3 sand at 3 493 m true vertical depth below sea surface (TVD_{SS}) with Schlumberger's Repeat Formation Tester (RFT); (2) the gas-oil contact was observed from logging data (neutron-density cross over) at 3 504 m TVD_{SS}; and (3) the oil-water contact was imaged from seismic data at 3 613 m TVD_{SS}. With these data and additional information about fluid densities (Table 2.2), bulk densities, and water depth, I characterize the initial reservoir pressure (Fig. 2.6).

The measured pore pressure is shown as the solid dot (Fig. 2.6). From this point, the gas pressure is extended along its static gradient (Eq. 2.4) upward to the crest

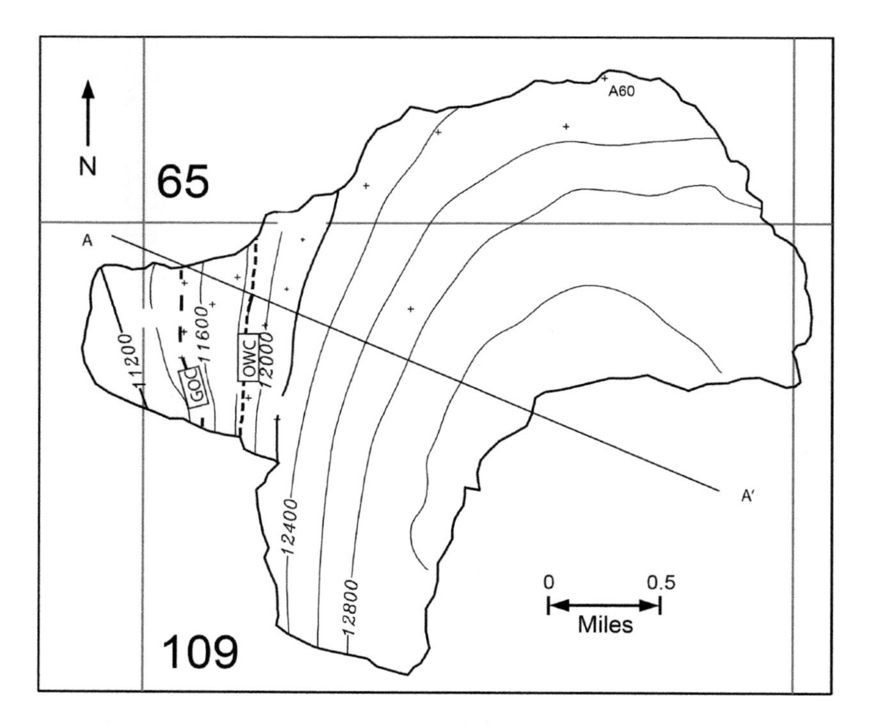

Figure 2.5 Map of the depth of the J3 sand in feet beneath the sea surface (TVD_{SS}). The polygon marks the boundary of the J3 sand. The structural crest of the sand is along the western margin. The original gas-oil contact (GOC) and the original oil-water contact (OWC) are shown with dashed lines. Line A-A' illustrates the location of the cross-section shown in Figure 2.4. Reprinted with permission of Offshore Technology Conference, from Flemings et al. (2001); permission conveyed through Copyright Clearance Center, Inc.

and downward to the gas-oil contact. Next, the oil pressure is assumed to equal the gas pressure at the gas-oil contact. We explore this assumption in the following section. The oil pressure is then extended along its static gradient upward from the gas-oil contact to the crest and downward to the oil-water contact. Finally, the water pressure is assumed to equal the gas pressure at the oil-water contact and from this, the water pressure is calculated by extending the water pressure along its hydrostatic gradient upward to the crest of the structure and down to the base of the structure.

At the crest of the structure, the water pressure is 55.1 MPa, the oil pressure is 55.9 MPa, and the gas pressure is 60.4 MPa (Table 2.3). The difference in pressure between adjacent, immiscible, phases is the capillary pressure (u_c). Thus, at the crest, the gas-oil capillary pressure (u_{cgo}) is 0.49 MPa and the oil-water capillary pressure is 0.8 MPa. In the overpressure plot (Fig. 2.6b), the water phase

Table 2.2 *Input parameters to describe pressure versus depth in the J3 sand at Bullwinkle.*

Parameter	Field Units	Metric Units
Seafloor Depth (TVD_{ss})	1 350 ft	411.6 m
Structural Crest (TVD_{ss})	11 100 ft	3 384.1 m
Pressure Measurement (TVD_{ss})	11 460 ft	3 493.9 m
Gas-oil Contact (TVD_{ss})	11 493 ft	3 504.0 m
OWC (TVD_{ss})	11 850 ft	3 612.8 m
Structural Base (TVD_{ss})	13 000 ft	3 963.4 m
Gas Density (ρ_g)	0.13 psi/ft	300 kg/m^3
Oil Density (ρ_o)	0.31 psi/ft	714 kg/m^3
Pore Water Density (ρ_{pw})	0.465 psi/ft	1 071 kg/m^3
Seawater Density (ρ_{sw})	0.44 psi/ft	1 013 kg/m^3
Local Overburden Density (ρ_b)	0.915 psi/ft	2 108 kg/m^3
Vertical Stress at Crest (σ_v)	9 683 psi	66.76 MPa
Pressure at Measurement (u_g)	8 224 psi	56.70 MPa

overpressure (blue line, Fig. 2.6b) is constant and equal to 19.5 MPa. Thus, the J3 sand has an overpressure (u^*) equal to 19.5 MPa.

Figures 2.6c and 2.6d illustrate the entire pressure depth profile at Bullwinkle. The water pressure in the reservoir (blue line, Fig. 2.6c) is parallel to the hydrostatic pressure, but elevated by the overpressure (u^*) of 19.5 MPa. The effective vertical stress (Eq. 2.7) is smallest at the crest of the structure (11.7 MPa) and greatest at the base of the structure (17.6 MPa) (Fig. 2.6, Table 2.3). This is because the overburden density is approximately twice the water density; thus overburden stress increases at approximately twice the rate of the static water pressure with depth. A common parameter used to characterize overpressure is the overpressure ratio (λ^*):

$$\lambda^* = \frac{u^*}{\sigma_v - u_h}. \qquad \text{Eq. 2.11}$$

λ^* describes the relative position of the pressure between the hydrostatic pressure and the lithostatic stress. At Bullwinkle, $\lambda^* = 0.63$ at the crest and 0.53 at the base of the J3 reservoir.

2.6 Capillary Pressure

2.6.1 Surface Tension

Capillary pressure was introduced by describing the gas-oil capillary pressure (u_{cgo}) as the difference between the gas pressure and the oil pressure, and the oil-water

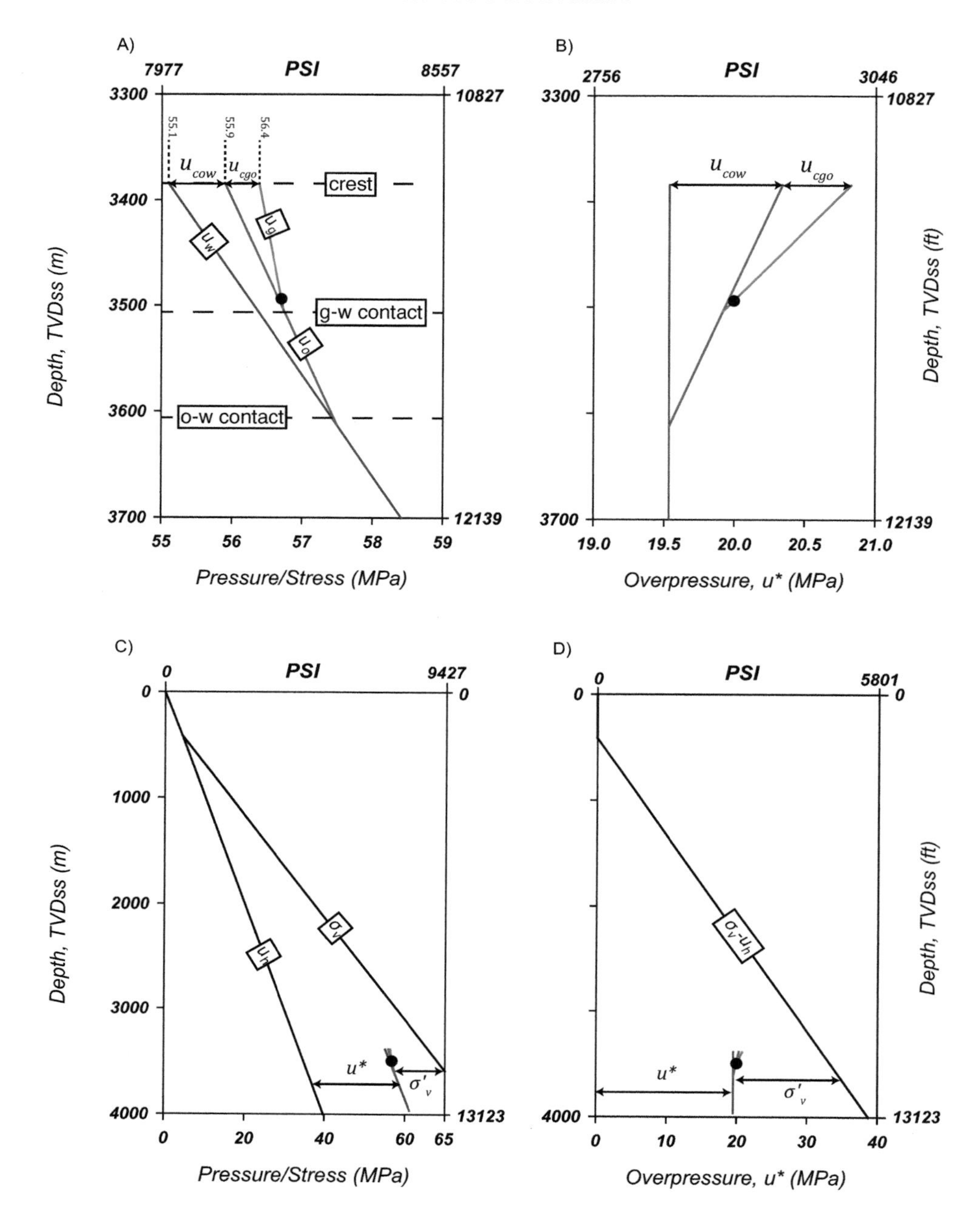

Figure 2.6 Pressure versus depth for the J3 reservoir in the Bullwinkle field. (a) and (b) pressure and overpressure plots over the reservoir interval. (c) and (d) pressure and overpressure from the sea surface to the reservoir.

Table 2.3 *Pressures in the Bullwinkle J3 reservoir assuming that the gas-oil capillary pressure (u_{cgo}) is zero at the gas-oil contact and that the oil-water capillary pressure is zero at the oil-water contact as illustrated in Figure 2.6.*

Parameter	Value	Unit
Overpressure (u^*)	19.5	MPa
Crestal Water Pressure (u_w)	55.1	MPa
Gas-oil Capillary Pressure @ Crest (u_{cgo})	0.5	MPa
Oil-water Capillary Pressure @ Crest (u_{cow})	0.8	MPa
Basal Water Pressure (u_w)	61.2	MPa
Effective Stress at Crest (σ_v')	11.7	MPa
Effective Stress at Base (σ_v')	17.6	MPa
Overpressure Ratio at Crest (λ^*)	0.63	-

capillary pressure (u_{cow}) as the difference between the oil and water pressures (Fig. 2.6, Table 2.3). I now explore capillary behavior more deeply because it controls reservoir pore pressures, trap integrity, and hydrocarbon migration.

Dake (1978b) gives an elegant overview of applications of capillary pressure in petroleum engineering. Smith (1966), Berg (1975), and Schowalter (1979) apply these concepts to understand reservoir pressures, trap integrity, and secondary migration. de Gennes et al. (2004) and Blunt (2017) provide wonderful descriptions that I rely heavily on in the ensuing section.

In a liquid, the molecules are attracted to each other. A molecule at the surface of a gas-liquid interface loses half of its cohesive interactions. If the cohesion energy per molecule is U inside the liquid, a molecule at the surface finds itself short by roughly $\frac{U}{2}$. The surface tension (γ) is a direct measure of this energy shortfall per unit surface area. If the molecule is of size a and a^2 is the exposed area on the surface, the surface tension is of order $\gamma \cong \frac{U}{2a^2}$. For most oils, for which the interactions are of the van der Waals type, at 25 degrees, $\gamma = 20 \frac{mJ}{m^2}$. Water involves hydrogen bonds and its cohesive energy is larger ($\gamma = 70 \frac{mJ}{m^2}$). Mercury, a strongly cohesive liquid metal, has $\gamma \cong 500 \frac{mJ}{m^2}$. The surface energy between two immiscible liquids A and B also has an interfacial tension, γ_{AB}, which is approximately the difference in the surface tensions of the individual components. For example, the oil-water interfacial tension is ~50 mJ/m.

2.6.2 Capillary Pressure

Surface tension causes the pressure inside a bubble to be greater than that outside of it. As described by de Gennes et al. (2004), consider a drop of oil (o) in water (w)

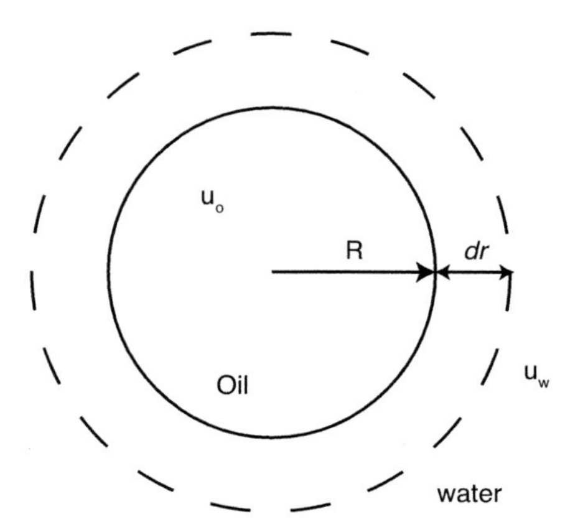

Figure 2.7 Pressure inside a drop of oil surrounded by water. The oil bubble that is at a pressure u_o expands by a distance dr into the surrounding water, which is at a pressure u_w. The pressure inside the bubble is shown to be greater than the pressure outside the bubble due to the effect of interfacial tension (Equations 2.12 and 2.13). From de Gennes et al. (2004). Reprinted by permission from Springer Nature© 2004.

(Fig. 2.7). The drop adopts a spherical shape of radius R to minimize its surface energy. If the o/w interface is displaced by an amount dR, the work done by the pressure and capillary force is:

$$\delta W = -u_o dV_o - u_w dV_w + \gamma_{ow} dA, \qquad \text{Eq. 2.12}$$

with $R \gg dR, dV_o = 4\pi R^2 dR = -dV_w$, and $dA = 8\pi R dR$. These are the increase in volume and surface area, respectively, of the drop. u_o and u_w are the pressures in the oil and the water, and γ_{ow} is the oil-water interfacial tension. At mechanical equilibrium, $\delta W = 0$ and Equation 2.12 reduces to

$$u_{cow} = \nabla u = u_o - u_w = \frac{2\gamma_{ow}}{R}. \qquad \text{Eq. 2.13}$$

Equation 2.13 is the Laplace equation for a fluid interface of spherical geometry. It shows that the smaller the drop, the larger the differential pressure, also known as the capillary pressure (u_c) or the Laplace pressure.

2.6.3 Contact Angle (θ)

The contact angle (θ) is the angle formed between two immiscible fluids that are in contact with a solid phase. It can be calculated from the interfacial tensions between the solids and the two fluids and by the interfacial tension of the fluid-fluid interface

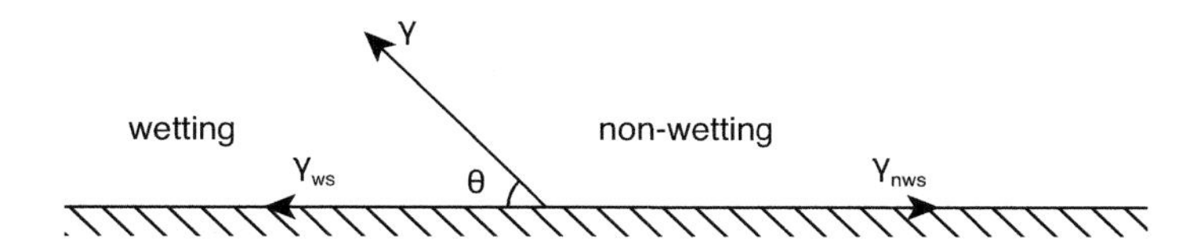

Figure 2.8 Two fluid phases in contact with a solid. The interfacial tension between the wetting phase and the solid is γ_{ws}. The interfacial tension between the non-wetting fluid and the solid is γ_{nws}. The interfacial tension between the two fluids is γ. From Blunt (2017). Reprinted with permission by Cambridge University Press.

(Fig. 2.8). Treating the interfacial tensions as forces and performing a horizontal force balance yields:

$$\gamma_{nws} = \gamma_{ws} + \gamma \cos\theta. \qquad \text{Eq. 2.14}$$

Equation 2.14 is rearranged to solve for the wetting angle:

$$cos\theta = \frac{\gamma_{nws} - \gamma_{ws}}{\gamma}. \qquad \text{Eq. 2.15}$$

By convention, the contact angle (θ) is measured through the denser fluid phase. Thus, in Figure 2.8, if the oil is the non-wetting phase and water is the wetting phase, the contact angle is measured through the water. The contact angle can vary from 0 to 180 degrees as a function of Equation 2.15 (Fig. 2.9). The wetting phase preferentially contacts the surface. If the contact angle is less than 90 degrees, the denser phase is wetting. If the contact is greater than 90 degrees, the denser phase is non-wetting. As described by Blunt (2017), while Equation 2.15 gives physical insight, it is seldom used. Instead, the contact angle is measured such as illustrated in Figure 2.9.

2.6.4 Capillary Rise

The pressure distribution in a reservoir can be viewed through the concept of capillary rise. Consider the case of a glass tube placed in water (Fig. 2.10). The water inside the tube rises to a height h above the water surface outside the tube (Fig. 2.11). The angle of the meniscus at the glass interface is the wetting angle (θ) and the meniscus within the tube is a portion of a sphere with a radius of curvature R. The relationship between the tube radius (r) and the radius of curvature (R) is:

$$r = Rcos\theta. \qquad \text{Eq. 2.16}$$

I consider the case where the water is overlain by oil. Equations 2.16 and 2.13 are combined to calculate the water pressure (u_w) at point A, just beneath the meniscus:

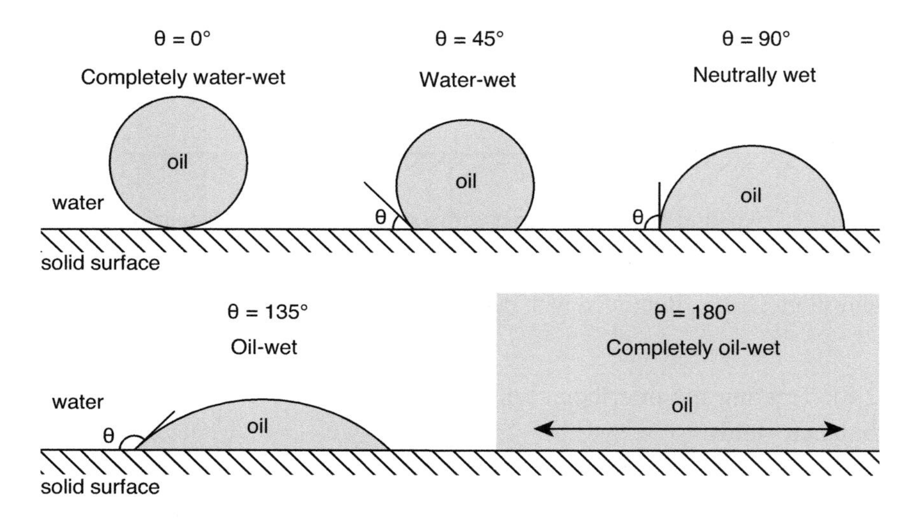

Figure 2.9 An illustration of different contact angles measured through water in an oil-water system. A drop of oil is placed on a solid mineral surface and surrounded by water. The contact angle is measured through the water (the denser phase). Buoyancy effects are ignored. A contact angle of 0 is complete wetting (top left): water spreads over the surface and oil makes no direct contact with the solid. A contact angle less than 90 degrees is water-wet; 90 degrees is neutrally wet (top right), while a contact angle greater than 90 degrees is oil-wet. Complete wetting of oil occurs for a contact angle of 180 degrees (lower right). Reprinted from Blunt (2017) with permission by Cambridge University Press. Source adapted from Morrow (1990) with special permission by the Society of Petroleum Engineers.

$$u_w = u_o - \frac{2\gamma_{ow} cos\theta}{r}. \qquad \text{Eq. 2.17}$$

u_o is the oil pressure immediately above the interface. At the level interface, far from the capillary tube, the tube radius can be considered infinite and the water and oil pressures are equal to each other and equal to u_1. Inside the tube, the pressure at point A within the water is due to the weight of the water column:

$$u_1 - \rho_w gh = u_w(A). \qquad \text{Eq. 2.18}$$

Outside the tube, the oil pressure at point A (just above the interface) is:

$$u_1 - \rho_o gh = u_o(A). \qquad \text{Eq. 2.19}$$

By combining Equations 2.17, 2.18, and 2.19, the relationship between capillary radius and capillary rise is:

Figure 2.10 Example of capillary rise. Colored water is placed in a tube. The smaller tubes have more capillary rise.

$$u_c = u_o - u_w = \frac{2\gamma_{ow} cos\theta}{r} = \Delta\rho g h, \quad \text{Eq. 2.20}$$

where,

$$\Delta\rho = \rho_w - \rho_o. \quad \text{Eq. 2.21}$$

In the example in Figure 2.11b, the water rises 6.4 cm above the interface for an oil-water system with a 1 mm capillary tube. An equivalent calculation for an air-water system results in a capillary rise of ~1.6 cm, which is more similar to what is shown in Figure 2.10. The difference is largely due to the fact that the relative density ($\Delta\rho$) is much smaller in the oil-water case.

2.6.5 Interpreting Mercury Injection Capillary Pressure (MICP) Curves

Imbibition occurs when the wetting phase saturation increases, and drainage occurs when it decreases. Drainage occurs as hydrocarbons fill a reservoir over geologic

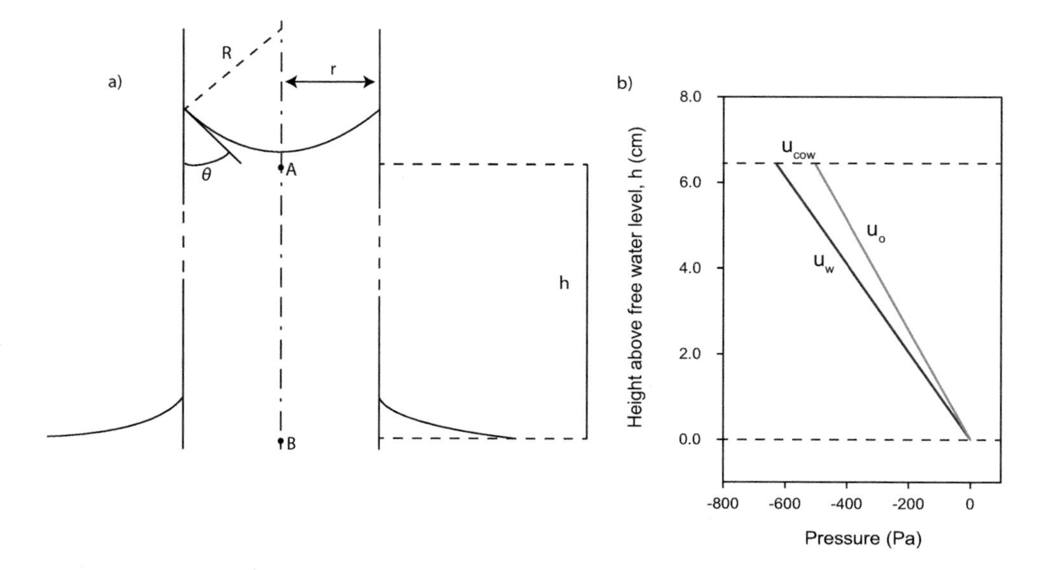

Figure 2.11 Capillary rise example. Oil overlies water. (a) Water inside the capillary tube rises to a point h from the free water surface away from the tube. (b) Pressure outside the capillary tube is assumed to equal zero at the free water surface. Oil pressure outside the tube rises at the oil hydrostatic gradient (grey line). Water pressure inside the tube rises at the water hydrostatic gradient (black line). The pressure difference between the oil and the water is balanced by the interfacial tension at point A. The following properties are assumed: r = 1mm, $\rho_w = 1000\frac{kg}{m^3}, \rho_o = 800\frac{kg}{m^3}, \theta = 24°$, and $\gamma = 73\frac{mN}{m}$. From de Gennes et al. (2004). Reprinted by permission from Springer Nature© 2004.

time through secondary migration, and it occurs during CO_2 injection. We estimate reservoir water phase pressure (see below) and trap integrity (Chapter 9) assuming reservoirs formed by drainage. The drainage curve is experimentally derived by forcing a non-wetting phase into a saturated sample (Fig. 2.12). This is commonly done by injecting mercury into an air-dried sample. Prior to interpretation, these measurements must be corrected for both apparatus compressibility and sample conformance. The conformance correction is the pressure required to allow the mercury to surround or conform to the sample exterior without intruding the pores (Comisky et al., 2011) .

Mercury injection tests for a mudrock and a siltstone that were shallowly buried are illustrated in Figure 2.12. The drainage experiment starts at the right edge ($S_w = 1$). As the pressure outside the sample is raised (the mercury pressure or capillary pressure), there is initially little change in the saturation until the displacement pressure (u_d) is reached. For these samples, as capillary pressure is increased above the displacement pressure, saturation increases rapidly. Finally, as capillary

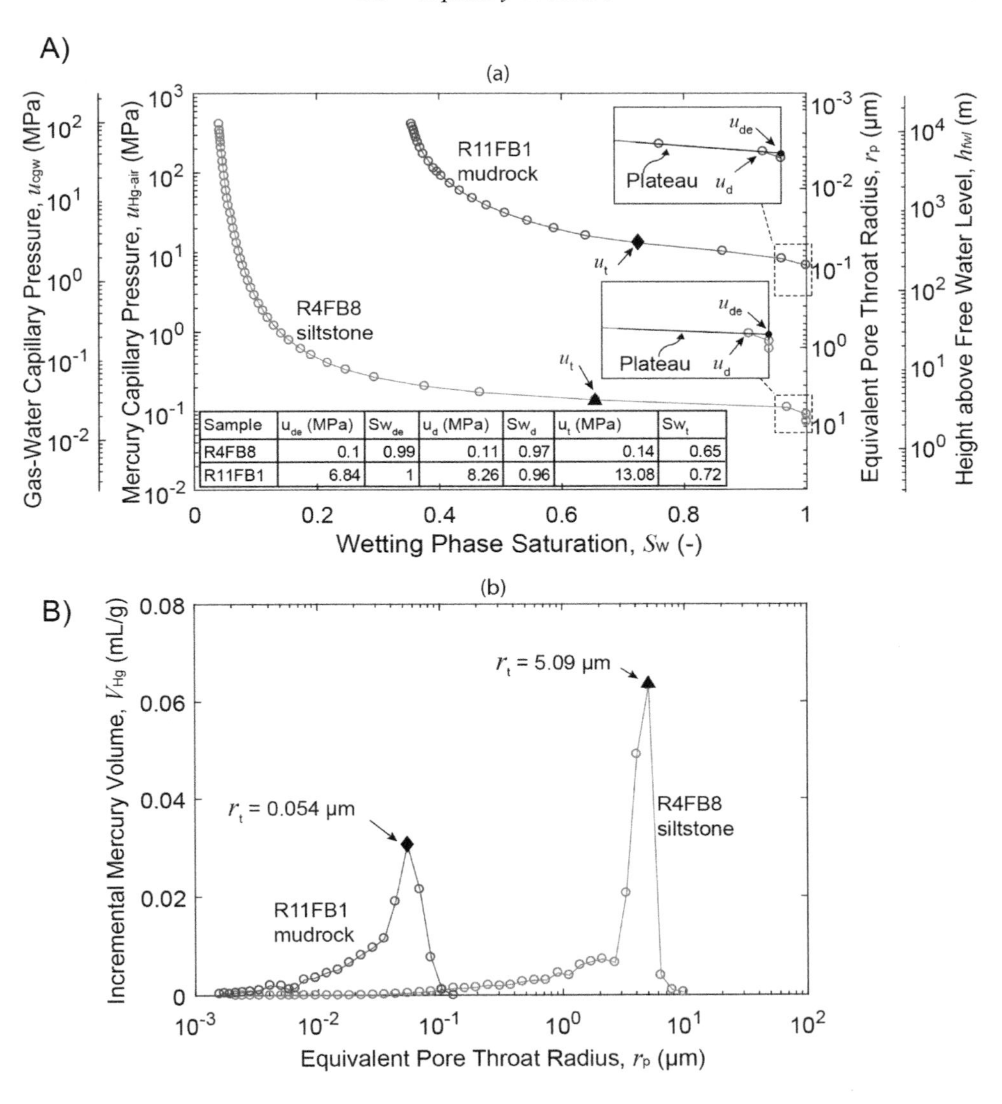

Sample	u_{de} (MPa)	Sw_{de}	u_d (MPa)	Sw_d	u_t (MPa)	Sw_t
R4FB8	0.1	0.99	0.11	0.97	0.14	0.65
R11FB1	6.84	1	8.26	0.96	13.08	0.72

Figure 2.12 (a) Mercury injection capillary pressure versus wetting phase saturation for a mudrock and siltstone compressed to ~4.0 MPa effective stress. Expanded views illustrate the interpreted displacement pressure (u_d) and extrapolated displacement pressure (u_{de}) for each sample. The table at the base of the figure records the capillary pressures and associated saturations. (b) Incremental mercury volume injected versus pore throat radius. To convert mercury-air capillary pressure to oil-water capillary pressure and height above free water level, the following parameters were assumed: $\gamma_{ow} = 30\frac{dyne}{cm}, \gamma_{hg_{air}} = 480\frac{dyne}{cm}, \theta_{ow} = 30°, \theta_{\mathrm{air-hg}} = 140°, \rho_o = 714\frac{\mathrm{kg}}{\mathrm{m}^3}, \rho_w = 1071\frac{\mathrm{kg}}{\mathrm{m}^3}, \rho_g = 300kg/m^3$. From Fang et al. (2020). Reprinted by permission of the AAPG whose permission is required for further use. AAPG© 2020.

pressure is increased to very high values, the saturation does not increase as rapidly. When the water saturation does not reduce further, the residual saturation is reached where the non-wetting phase saturation cannot be reduced further regardless of the capillary pressure.

Pittman (1992) and Comisky et al. (2011) discuss the interpretation of capillary curves. The pressure at which the non-wetting phase starts to enter the pores is commonly called the displacement pressure (u_d). I term the pressure at which the non-wetting fluid forms a connected filament across the sample the migration pressure (u_{cmig}). This is the pressure necessary for migration of the non-wetting fluid through the rock. I introduce the term migration pressure because there is a lot of ambiguity as to how to interpret the migration pressure from the capillary curve and because the migration pressure and the displacement pressure are often used interchangeably. For example, Smith (1966) describes the displacement pressure as the migration pressure. I define three terms that can be objectively determined from the capillary curve: the displacement pressure (u_d), the extrapolated displacement pressure (u_{de}), and the threshold pressure (u_t). The displacement pressure (u_d) is, approximately, the first data point where S_w drops below 1 (after correction for conformance). The extrapolated displacement pressure (u_{de}) is determined from a hyperbolic fit to the displacement curve and its projection to a horizontal asymptote (see Figure 1 of Thomeer (1960)). In practice, this value is approximately equal to the point where the plateau of the capillary curve is extended to 100% wetting saturation. Thomeer (1960) suggested this is the most accurate estimate of displacement pressure. Finally, the threshold pressure (u_t) was defined by Katz and Thompson (1986, 1987) to occur when the curvature of the capillary curve changes from concave downward to concave upward (u_t, Fig. 2.12). This is the point where the maximum injected pore volume occurs for a given increase in capillary pressure (Fig. 2.12b). u_t can be expressed as a threshold pore throat radius (Eq. 2.17), termed r_t.

I illustrate the interpretation of u_d, u_{de}, and u_t for a mudrock and a siltstone (Fig. 2.12). The extrapolated displacement pressure (u_{de}) for the mudrock is approximately half the threshold pressure (u_t) and the non-wetting phase saturation at the threshold pressure is significantly higher than at the displacement pressure (Fig. 2.12a). Given the ambiguity in the literature, I view u_t as an upper bound and u_{de} as a lower bound for the migration pressures (u_{cmig}). Schowalter (1979), based on experimental results, avoided this ambiguity and suggested that the migration pressure occurs at a non-wetting phase saturation of ~10%.

The mercury air capillary pressure (u_{Hg-air}) can be converted to describe capillary pressures in other fluid systems if the wetting angle (θ) and interfacial tension (γ) are known. For example, to convert mercury-air capillary pressures to equivalent oil-water capillary pressures:

Table 2.4 *Common values for contact angle and interfacial tension.*

System	Contact Angle (θ)	Interfacial Tension (γ) $\left(\frac{dyn}{cm}\right)$
Laboratory		
Air-water	0	72
Oil-water	30	48
Air-Mercury	140	480
Air-oil	0	24
Reservoir		
Water-oil	30	30
Water-gas	0	50*
Oil-gas	0	24

*pressure/temp dep.

$$u_{cow} = u_{cHg-air}\left(\frac{\gamma_{ow}\ cos\theta_{ow}}{\gamma_{Hg-air}\ cos\theta_{Hg-air}}\right). \quad \text{Eq. 2.22}$$

Typical values for interfacial tension and wetting angle are shown in Table 2.4. In Figure 2.12, the equivalent gas-water capillary pressure is illustrated on the right-hand side, and in Figure 2.13, the equivalent oil-water capillary pressure is shown on the right hand side.

A common and simple way to conceptualize a reservoir is through a 'bundle of tubes model.' The reservoir is assumed to comprise a range of different cylindrical tubes of different diameters. As the capillary pressure is raised, the non-wetting phase enters smaller and smaller tubes. In this conceptual view, each capillary pressure on the capillary curve can be related to an equivalent cylindrical radius through Equation 2.17 as is illustrated on the right side of Figures 2.12 and 2.13.

2.7 Capillary/Gravity Equilibrium

2.7.1 Two Phase Capillary/Gravity Equilibrium

I use an aquarium to illustrate the distribution of fluid phases and pore pressures present in a two-phase system at capillary and gravity equilibrium (Fig. 2.14). A fine-grained 'seal' composed of 0.5 mm sand overlies a coarse-grained 'reservoir' composed of ~5 mm gravel in an anticlinal geometry. As air is added from below, it fills the reservoir and displaces the water that was there originally (shown with black). The gas-water contact separates the zone above where gas is present (white) from the water leg below (black) (Fig. 2.14a). Pressure measurements within the gas leg are approximately constant (red line), whereas the water pressures follow

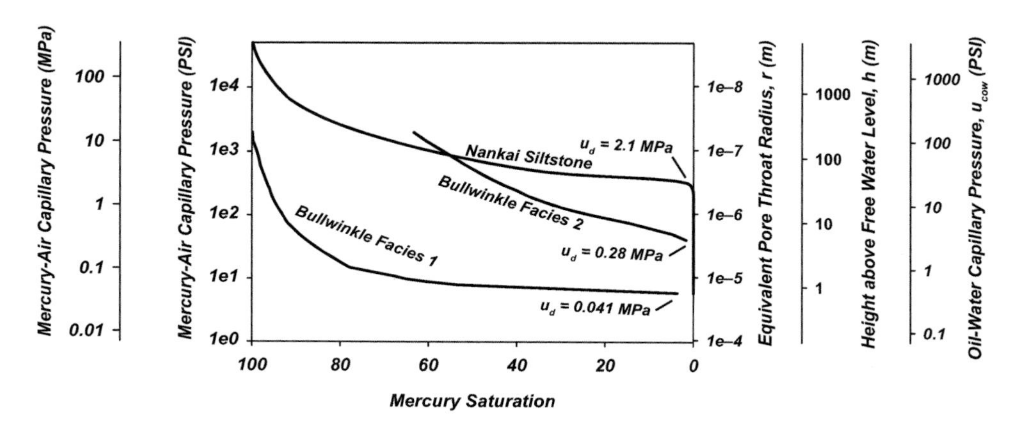

Figure 2.13 Mercury injection capillary curve for three reservoirs. The Bullwinkle Facies 1 and Facies 2 curves are from well-sorted, fine-grained sands in the Bullwinkle Oil Field (Best, 2002). In contrast, the Nankai Siltstone is from a clayey-silt mudstone from the Nankai Trough (CRS 88) (Schneider, 2011). To convert mercury-air capillary data to oil-water capillary pressure and height above free water level, the following parameters were assumed: $\gamma_{ow} = 30\frac{dyne}{cm}, \gamma_{hg_{air}} = 480\frac{dyne}{cm}, \theta_{ow} = 30°, \theta_{\text{air-hg}} = 140°, \rho_o = 714\frac{\text{kg}}{\text{m}^3}, \rho_w = 1071\frac{kg}{m^3}, \rho_g = 300kg/m^3$.

the hydrostatic gradient (blue line) (Fig. 2.14b). At the crest, the gas pressure is ~930 Pa more than the water pressure. At the gas-water contact, the gas and water pressures differ by 120 Pa.

The capillary pressure at the gas-water contact is the migration pressure ($u_{cmig} = 120$ Pa), the differential pressure necessary for the gas to migrate downward into the water-saturated reservoir. The pore throat radius given this capillary pressure (Eq. 2.17) is 1.2 mm. This is approximately 20% of the grain diameter (5 mm), which is in the range expected for the ratio of pore throat radius to grain diameter for a well-sorted sand.

The elevation of the gas-water contact above the free water level can be calculated directly from the capillary pressure observed at the gas-water contact:

$$h_{GWC} = \frac{u_{cgw}}{\Delta\rho g}. \qquad \text{Eq. 2.23}$$

It is found to be 1.2 cm as shown in Figure 2.14b.

I next describe the fluid distribution in the Bullwinkle J2 reservoir. In this oil reservoir, the depth of the crest, the depth of the oil-water contact, the fluid properties, and the pressure within the oil column were known. I combine these data with the reservoir capillary curves shown in Figure 2.13 to describe the distribution of fluid phases with depth (Fig. 2.15). The free water level is the

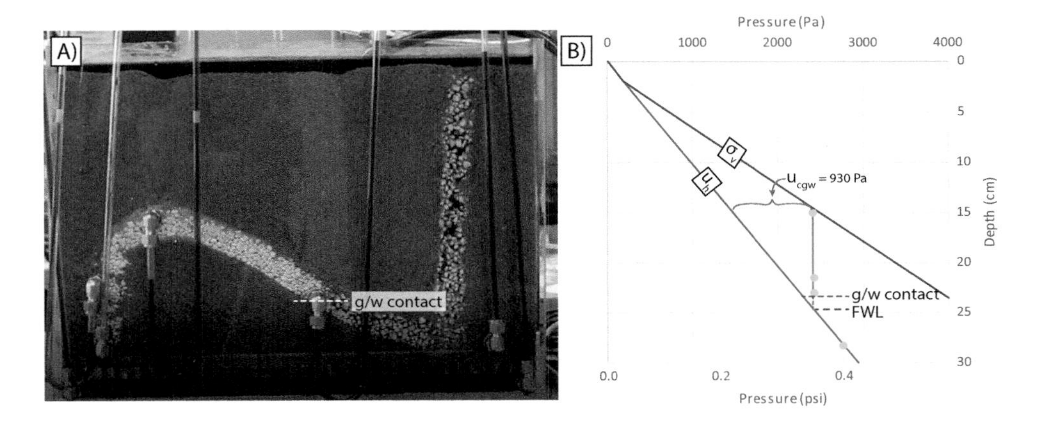

Figure 2.14 Aquarium example of fluid distribution and pressure in a two-phase system at capillary and gravitational equilibrium. (a) A fine-grained seal (composed of 0.5 mm sand) overlies a coarse-grained reservoir (composed of 3–8 mm gravel). Air is trapped within the structure. (b) Pressure versus depth. Pressure is measured at each circle. The capillary pressure at the crest of the structure is 930 Pa. The gas-water contact lies 1.2 cm above the free water level (FWL).

depth where the oil and water pressures are equal and thus the capillary pressure is zero ($u_o = u_w$) (Fig. 2.15). Above this, there is a zone where the non-wetting phase is absent, because the capillary pressure is not large enough for the non-wetting phase to enter the rock. The oil-water contact occurs where the capillary pressure equals the migration pressure for the reservoir (u_{cmig}):

$$u_{cmig} = \Delta\rho g h_{OWC}. \quad \text{Eq. 2.24}$$

h_{owc} is the height of the oil-water contact above the free water level. In this example, I assume the migration pressure is the displacement pressure (u_d), or the minimum pressure where significant saturation of the non-wetting phase occurs. This is easy to interpret in the examples given, where there is an abrupt break in slope of the capillary curve at low saturations and where the capillary curve is fairly flat above the displacement pressure (Fig. 2.12).

Like the aquarium example, this example shows that if the reservoir rock has a significant migration pressure, the depth where the oil and water pressures are equal (the free water level) is significantly below the fluid contact. If the reservoir has the capillary properties of the Nankai Siltstone and the migration pressure is 300 psi (0.15 MPa) (Fig. 2.12), then the free water level is ~42 meters below the oil-water contact (OWC) (horizontal dashed lines, Fig. 2.15). In this case, the water pressure is 300 psi less than the oil pressure at the OWC (note separation between the green and the blue solid lines at the OWC in Figure 2.15b). In contrast, if the reservoir has

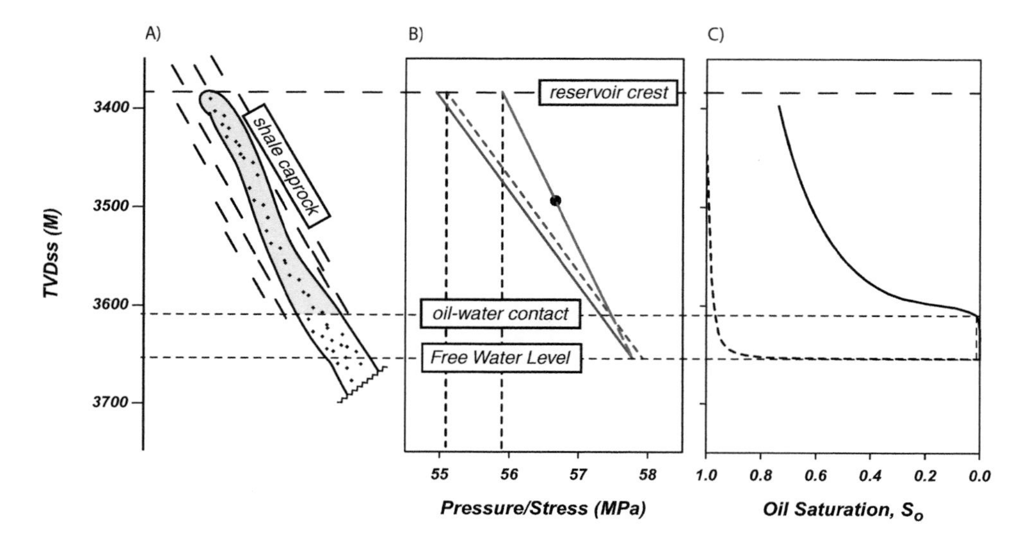

Figure 2.15 (a) Sketch of the Bullwinkle J2 oil reservoir extending from its structural crest at 3 384 m to the oil-water contact at 3 613 m. (b) The black dot illustrates the measured oil pressure. The oil pressure is extrapolated along its static gradient upward to the reservoir crest and downward to the oil-water contact. The solid blue line illustrates the water pressure assuming the (u_{cmig}) of the Nankai Siltstone whereas the dashed line is the water pressure assuming the migration pressure of the Bullwinkle Facies 1 (Fig. 2.13). (c) Oil saturation as a function of height above free water level for the Nankai Siltstone (solid) and Bullwinkle Facies 1 (dashed line). The illustrated free water level (lower horizontal dashed line) assumes the migration pressure (u_{cmig}) of the Nankai Siltstone.

the capillary properties of Facies 1 of the Bullwinkle sandstone with a migration pressure of only 6 psi (0.0029 MPa) (Fig. 2.13), then the free water level is only 0.8 meters below the OWC. If we compare the two examples (dashed versus solid blue line, Fig. 2.15), we see that the reservoir with a higher migration pressure results in a lower predicted water pressure (solid line) and a higher predicted capillary pressure at any depth.

As the energy industry explores and produces from increasingly fine-grained reservoirs, it is becoming important to account for reservoir capillary behavior to determine reservoir pressure. Without incorporating capillary behavior, the water pressure will not be calculated correctly.

2.7.2 Three Phase Capillary/Gravity Equilibrium

I close by returning to the original Bullwinkle J3 example to consider the presence of all three phases (Fig. 2.6). I use the capillary model of the Nankai Siltstone to

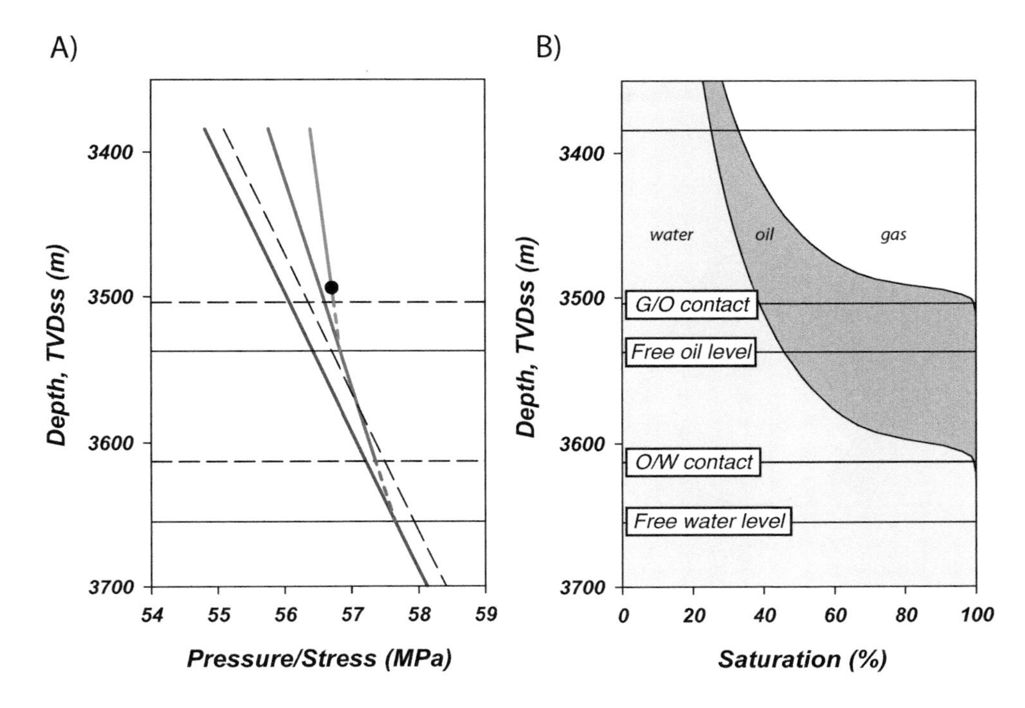

Figure 2.16 (a) Plot of oil, gas, and water pressures for the J3 Bullwinkle reservoir. In this example, the Nankai siltstone capillary curve was used (Fig. 2.13). (b) Fraction of pore space filled with water, oil and gas in the J3 Bullwinkle reservoir as calculated from the Nankai siltstone capillary curve (Fig. 2.13).

illustrate the behavior (Fig. 2.16). When three phases are present, the water saturation is calculated exactly as done previously from the oil-water capillary pressure (u_{cow}). However, in the case of the gas phase, the sum of the water saturation and the oil saturation is equal to the wetting phase on the gas-oil capillary curve.

The results are illustrated in Figure 2.16. The water saturation is everywhere the same as in the two-phase example (compare Fig. 2.15c with Fig. 2.16b). However, above the gas-water contact, the gas saturation increases dramatically at the expense of the oil saturation. It can be envisioned that with sufficient structural height, the oil saturation would ultimately go to zero. Blunt (2017) describes this process analytically. At the gas-oil contact, the oil pressure is offset from the gas pressure by the gas-oil migration pressure (0.135 MPa) and at the oil-water contact, the water pressure is offset from the oil pressure by the oil-water migration pressure (0.146 MPa). The difference between the blue line and the dashed black line is the difference between water pressure predicted with and without considering the effects of the migration pressure (0.281 MPa). The example given with the

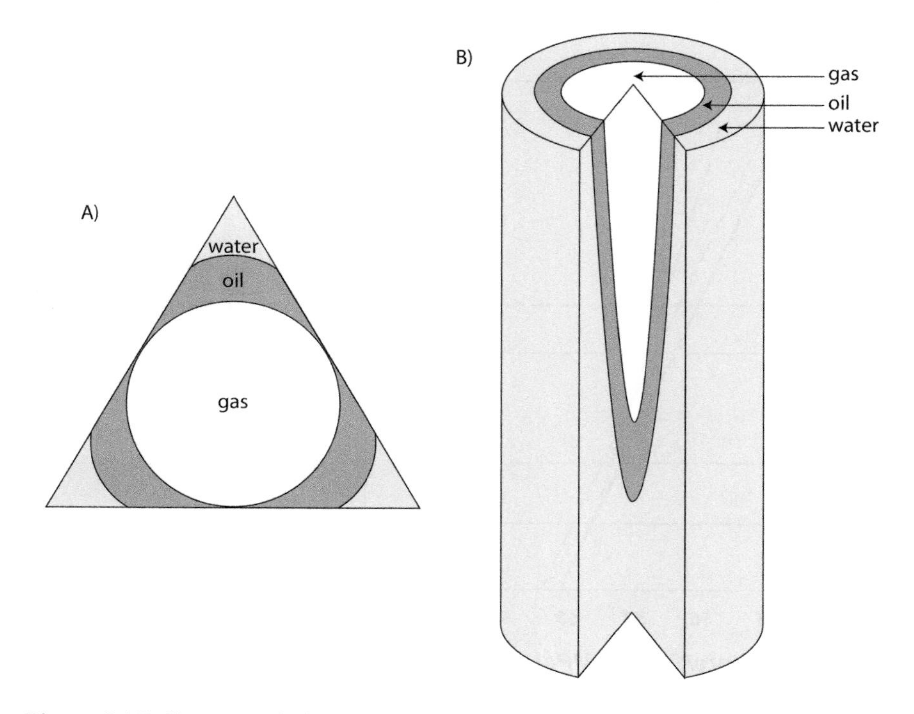

Figure 2.17 Conceptual view of distribution of fluid phases in a reservoir. (a) At the grain scale, water is interpreted to coat the grains and oil is interpreted to lie between the gas and water phases. (b) At the scale of the hydrocarbon column, this can be conceptually viewed as an interconnected tubular filament: the gas phase lies at its center, and it is bounded by the oil, and ultimately the water phase. The gas, oil, and water are separate immiscible phases with distinct pressures due to the interfacial tension at the boundary between the fluid phases.

Nankai Siltstone curve demonstrates the effect of a relatively high migration pressure on reservoir pressure estimation and on saturation. However, the basic physical processes will be present in reservoirs with much lower migration pressures.

Figure 2.17 emphasizes an important conceptual view of the reservoir which, although highly simplified, provides important insight. A pore scale conceptual view (Fig. 2.17a) is that the grains are surrounded by water and the gas phase is surrounded by oil. Each phase is connected along some pathway through the reservoir. At the reservoir scale, this system can be viewed as a vertical column. Water is present everywhere (Fig. 2.17b). The gas phase is the non-wetting fluid with respect to the oil phase. Pressures within each phase follow their own static pressure gradients. The pressures between the phases are different due to the interfacial tension present at the boundary between the fluid phases.

2.8 Summary

I have described how to characterize reservoir pore pressures that are under capillary and gravity equilibrium through reservoir examples and through a simple fish tank demonstration. Different immiscible fluid phases have distinct pressures. These pressures result from capillary behavior and through an understanding of this behavior, we can systematically describe these pressures and the associated pore fluid saturations. It is critical to account for capillary behavior to successfully recover the water phase pressure, which is one of the critical parameters we need to know in pore pressure analysis.

3

Mudrock Material Behavior

3.1 Introduction

Mudrock is the most abundant material in the uppermost 5 km of the Earth's crust (Petley, 1999). It has low permeability and undergoes enormous compaction during burial. When mudrock is loaded sufficiently rapidly (e.g., by burial or tectonic loading), the load is borne by both the solid grains (as effective stress) and the pore fluid (as overpressure) because the fluid cannot escape at the rate the loading occurs. The compressibility of the rock defines how much fluid will be expelled, and its permeability defines how rapidly that fluid can be expelled. Because of mudrock's high compressibility and low permeability, these overpressures can be maintained for geological timescales. Thus, the compaction behavior of mudrocks is one of the most important controls on whether overpressures will form. The degree of compaction is a sensitive indicator of the effective stress. I show in Chapters 5 and 6 how the compaction state is used to interpret the effective stress and ultimately the pore pressure.

Mudrocks are fine-grained siliciclastic sediments. They include fissile shales, non-fissile mudrocks, siltstones, and claystones. Milliken and Hayman (2020) describe mudrocks as composed by weight or volume of >50% particles that are <62.5 μm (the silt-sand boundary). In sediments, the most similar classification for clay-rich sediments is that of Shepard (1954). Sediments with more than 80% silt-sized or smaller grains are divided into four types based on the ratio of clay-sized particles to the sum of the clay plus silt-sized particles: (1) silt (<20% clay-sized), (2) clayey silt (20% to 50% clay-sized), (3) silty clay (50% to 80% clay-sized), and (4) clay (>80% clay-sized). It should be noted that the clay-silt size boundary is 4 μm in the geological community (Shepard, 1954), whereas in the engineering community this cut-off is commonly 2 μm (Lambe & Whitman, 1979). In this book, I follow the convention that the clay-silt boundary lies at 2 μm.

This chapter focuses on 'soft' mudrocks (Li & Wong, 2016). These types of mudrocks have not undergone significant cementation, although they can be buried to great depths and have undergone very large effective stresses during burial. They comprise a large proportion of rocks in sedimentary basins, and they house a large fraction of the world's overpressure.

We describe mudrocks through field observations, experimental analyses of intact core, and the analysis of resedimented material. Resedimentation is a particularly useful tool that simulates natural sedimentation (Casey et al., 2015; Santagata & Kang, 2007; Schneider et al., 2011; Sheahan, 1991). In this approach, material from disaggregated mudrock is mixed with water. The sediment slurry is fully disaggregated and uniformly mixed and then uniaxially incrementally loaded in a core liner. The approach eliminates sample variability, produces uniform specimens, and allows us to study the effects of composition, effective stress, and stress history on petrophysical properties through systematic experiments (Casey et al., 2016). We use resedimentation to gain insight into the petrophysical properties of mudrocks.

I review five characteristics of mudrock material behavior that impact pressure and stress in the subsurface. First, I describe how composition and texture impact compaction. Second, I review the elastoplastic nature of mudrocks. Third, I describe the relationship between vertical and horizontal stress under vertical uniaxial strain. Fourth, I discuss how we determine the strength of mudrocks and how pore pressure is generated during undrained shearing. Fifth, I review creep under uniaxial strain. I then summarize simple conceptual models for the Earth's stress state. Finally, I present a generalized compaction model for mudrocks and I apply this model to characteristic stress states.

3.2 Mudrock Material Behavior

3.2.1 Compaction

There is extraordinary variation in how mudrocks compact with depth (Fig. 3.1). Grain size distribution, shape distribution, effective stress, previous stress history, cementation, temperature, sedimentation rate, pore fluid chemistry, and clay mineralogy all play a role in mudrock compaction (Casey et al., 2019; Velde, 1996). However, before chemical compaction (diagenesis) begins, porosity loss is driven primarily by mechanical compaction from increasing effective stress (Casey et al., 2019; Velde, 1996).

The compression curve describes the decline in porosity (or void ratio) that occurs with increasing effective stress (Skempton, 1969) (Fig. 3.2). Myriad empirical models are used to describe one-dimensional compression as discussed in Chapter 5 (e.g., Table 5.1). The expression most commonly used in the geotechnical community describes the decline in void ratio as a logarithmic function of vertical effective stress,

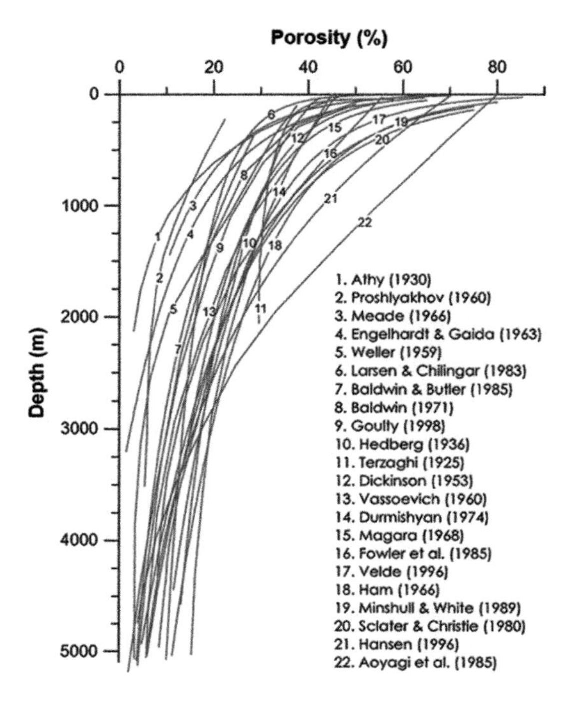

Figure 3.1 Mudrock porosity versus depth from published examples. Reprinted from Mondol et al. (2007) with permission of Elsevier.

$$e = e_0 - C_c \log(\frac{\sigma'_v}{\sigma'_{v0}}), \qquad \text{Eq. 3.1}$$

where e_0 is the initial value of the void ratio (e), σ'_{v0} is the initial stress, and C_c is the compression index, or slope of the compression curve in a log-stress plot. Commonly, a stress of unity is assumed for the initial stress state (σ'_{v0}). The void ratio is the ratio of the porosity (n) to the solid volume or,

$$e = \frac{n}{1-n}. \qquad \text{Eq. 3.2}$$

The compression behavior is strongly influenced by the fraction of clay-sized particles present and by the clay composition. Mudrocks with a large clay fraction have a higher initial porosity and a larger compression index than those with a small clay fraction. For example, the resedimented Gulf of Mexico – Eugene Island mudrock (RGoM-EI) (63% clay-sized particles) has a high initial porosity and a high compression index (steep slope), whereas the Presumpscot Clay (37% clay-sized particles) has a lower initial porosity and flatter slope (Fig. 3.2). In addition, when the clay is dominated by smectite, the initial porosity and the compression

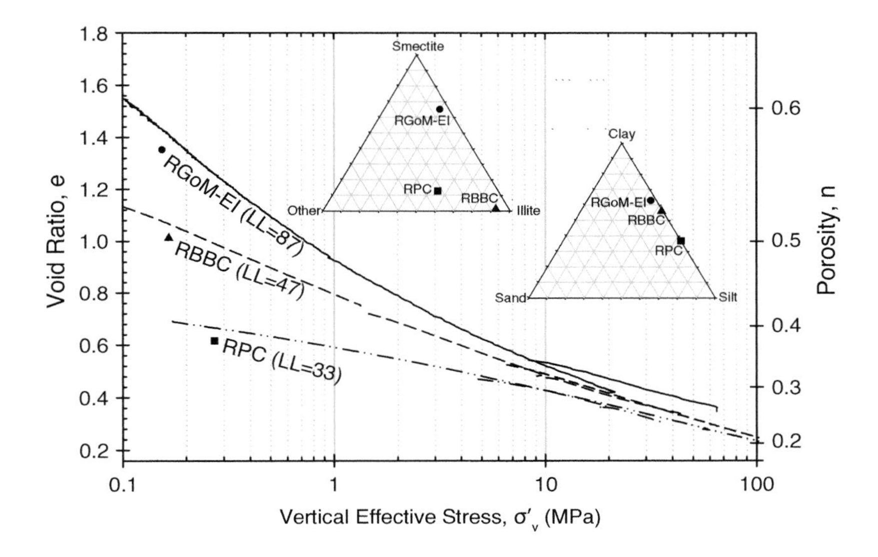

Figure 3.2 One-dimensional normal compression behavior for three mudrocks. The resedimented Gulf of Mexico mudrock (RGoM-EI) is smectite-rich, fine-grained, and has a very high liquid limit (LL). It has a high porosity at low effective stresses and compacts greatly with increasing stress. The resedimented Boston Blue Clay (RBBC) has about the same grain size as the RGoM-EI, but the clay fraction is dominated by illite; it has a lower initial porosity and compacts less with increasing stress. Finally, the resedimented Presumpscot Clay (RPC) is a siltstone with a small clay fraction. It has the lowest initial porosity and compacts the least with increasing stress.

index (C_c) are greater than with specimens that have more illite. For example, the resedimented Boston Blue Clay (RBBC) mudrock has about the same clay-sized fraction as the RGoM-EI mudrock, but is dominated by illite. It has a lower initial porosity and a smaller compression index than the RGoM-EI sample (Fig. 3.2).

The liquid limit (an easily measured engineering parameter) is the water content (the weight of the water divided by the weight of the solid) at which the sediment is at the boundary between a plastic and a liquid state; at higher water contents, the material behaves as a liquid and is unable to withstand shear stress (Lambe & Whitman, 1979). The liquid limit increases as the clay fraction increases and, for a given clay fraction, the liquid limit is greatest with clays composed of smectite, intermediate with illite, and lowest with kaolinite. This behavior can be understood from the perspective of the surface area of the material (Lambe & Whitman, 1979). Water attracted to the surface of particles will not behave as a liquid; thus, the greater the surface area, the greater the liquid limit. Because clay particles are so much smaller than silt particles, they have a much higher surface area, and a higher liquid limit. Thus, the RPC (Fig. 3.2), composed of only 37% clay-sized particles, has a liquid limit of 33, while the RBBC and RGoM-EI, with clay fractions of 56% and 63% clay-sized particles, have much higher liquid limits (Fig. 3.2). Furthermore, the surface area of smectites (circa 800 m^2/g) is much larger than illite (circa 80 m^2/g), which is in turn greater than that of kaolinite (circa 10 m^2/g) (Lambe & Whitman, 1979; Mitchell & Soga, 2005). As a result, for a given silt fraction, it is not surprising that a smectite-dominated rock (e.g., RGoM-EI) has a higher liquid limit than an illite dominated rock (RBBC) (Fig. 3.2).

Burland (1990) showed that the compression index C_c and the void ratio at an effective stress of 100 kPa (e_{100}) correlate with the liquid limit. This is qualitatively illustrated in Figure 3.2: RGoM with a liquid limit 87% has a very high value of both e_o and C_c, whereas the Presumpscot Clay (RPC) has a liquid limit of only 33% and much lower values of e_o and C_c. Aplin et al. (1995) and Yang and Aplin (1998, 2004) used this insight to quantify the relationship between clay fraction of mudrocks and the compression coefficients C_c and e_{100}. Casey et al. (2019) extended this approach and use the liquid limit to describe the compression behavior of mudrocks with a wide range of clay composition and silt fraction.

Figure 3.3 shows two images of intact mudrock core from the shallow Ursa Basin that are of similar composition and grain size. One sample is from 75 meters below seafloor (mbsf) and at an in situ vertical effective stress of 0.3 Mpa, whereas the other is from 570 mbsf at an in situ vertical effective stress of 4.0 MPa. As is typical of mudrocks, they are made of a mixture of quasi-spherical silt grains and a matrix

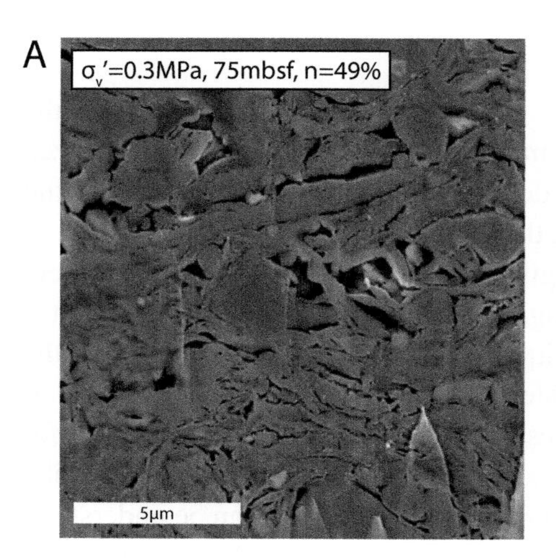

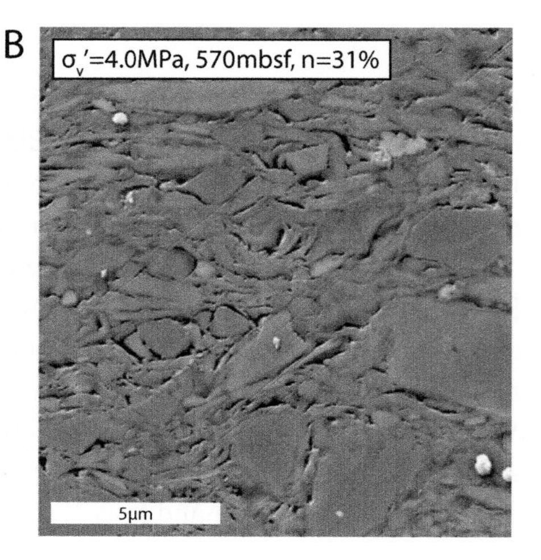

Figure 3.3 (a) Image of Ursa Basin mudrock buried 75 meters below seafloor (mbsf). (b) Image of Ursa Basin mudrock buried 570 mbsf. The 13-fold increase in effective stress results in a reduction of pore size, flattening of pores, and greater horizontal alignment of particles. These images are described in detail by Day-Stirrat et al. (2012), and are used here with permission of Elsevier.

of clay aggregates. The largest pores are interparticle pores adjacent to the largest (~5 μm diameter) silt particles; smaller, intraparticle pores are preserved within clay aggregates. Silt grains form bridges providing locations where large pores are preserved (Fig. 3.3a, b). The porosity declines from 0.49 to 0.31 as the vertical effective stress increases from 0.3 to 4.0 MPa. With increased vertical effective stress, the pores evolve from angular or rounded in shape to more elongate (Fig. 3.3a versus Fig. 3.3b). The flattened pores and the grains are more horizontally aligned in the deeper sample than the shallower sample.

The effect of silt fraction on compression is beautifully illustrated in a series of experiments performed by Schneider (2011). She combined mudrock from the Nankai Trough with varying fractions of silt particles, resedimented these materials to 0.1 MPa, and then uniaxially consolidated these sediments to 21 MPa (Fig. 3.4). As silt is admixed, the compression behavior systematically changes as described by Burland (1990) (Fig. 3.4a). As the silt fraction increases, the initial porosity decreases and the slope of the compression curve flattens. With increased stress, all of these compression curves converge. Thus, the effects of texture and composition may be most dramatic at lower effective stress. The images clearly show that larger pore throats are preserved with a higher silt fraction. It is interpreted that the silt grains more effectively form a bridging framework, shielding the large pores from the stress present (Schneider et al., 2011).

3.2.2 Elastoplastic Behavior of Mudrocks

The elastoplastic behavior of mudrock is illustrated with a uniaxial compression test on a specimen taken from 51 meters below seafloor in the Ursa Basin in the Gulf of Mexico (Fig. 3.5). Prior to testing, the effective stress was zero and the void ratio (e) was 1.33 (point A, Fig. 3.5). In the laboratory, the core was compressed uniaxially. From point A to point B, as the sample was reloaded, there was little loss in void ratio as stress was raised. As the stress approaches point B, the slope of the compression curve increases. Between point B and point C the sample follows an approximately linear path. At point C, the sample is unloaded to point D. The unloading slope (point C to point D) is less than the previous loading slope (point B to point C). The sample is then reloaded back to point C and then point E, before it undergoes its final unloading to point F.

This experiment demonstrates that mudrocks undergo enormous strain as effective stress increases. In this example, the effective stress is increased to 20 MPa, which represents a depth of approximately 2 km (6 560 ft) under hydrostatic conditions. The void ratio declines from ~2.0 ($n = 0.67$) near the seafloor to ~0.4 ($n = 0.29$), resulting in a bulk strain of 40% (Fig. 3.5).

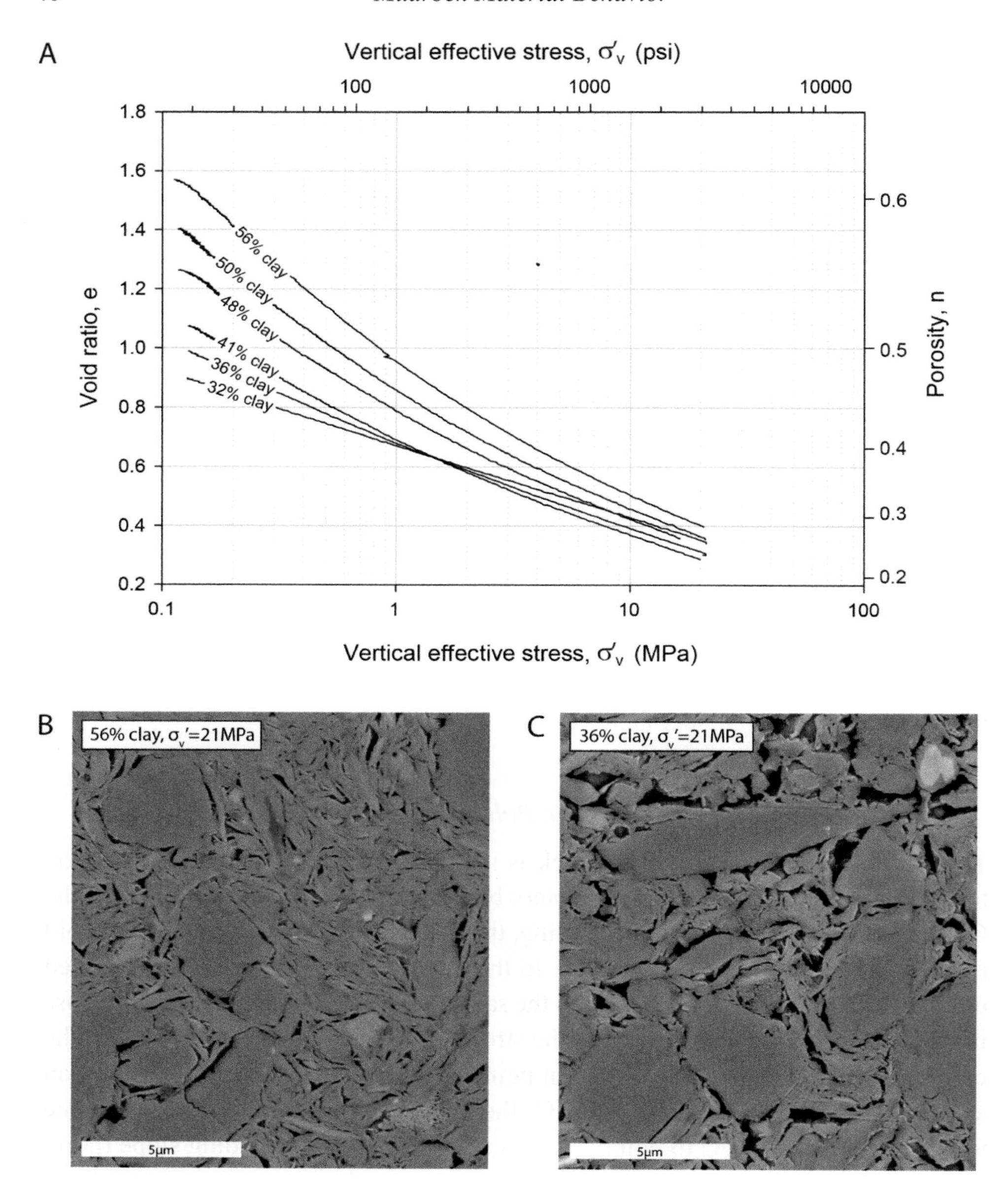

Figure 3.4 Resedimented Nankai Trough mudrock. (a) Compression curve for the parent material (56% clay) is compared to that of the parent material with progressively more silt added. (b) Backscattered electron (BSE) image of the parent material after compression to 21 MPa. (c) BSE image of material with admixed silt also compressed to 21 MPa. Images represent vertical cross-section of samples (i.e., load was applied from the top of the images). These experiments are discussed by Schneider (2011) and by Reece et al. (2013). Images reprinted with permission by Julia Reece.

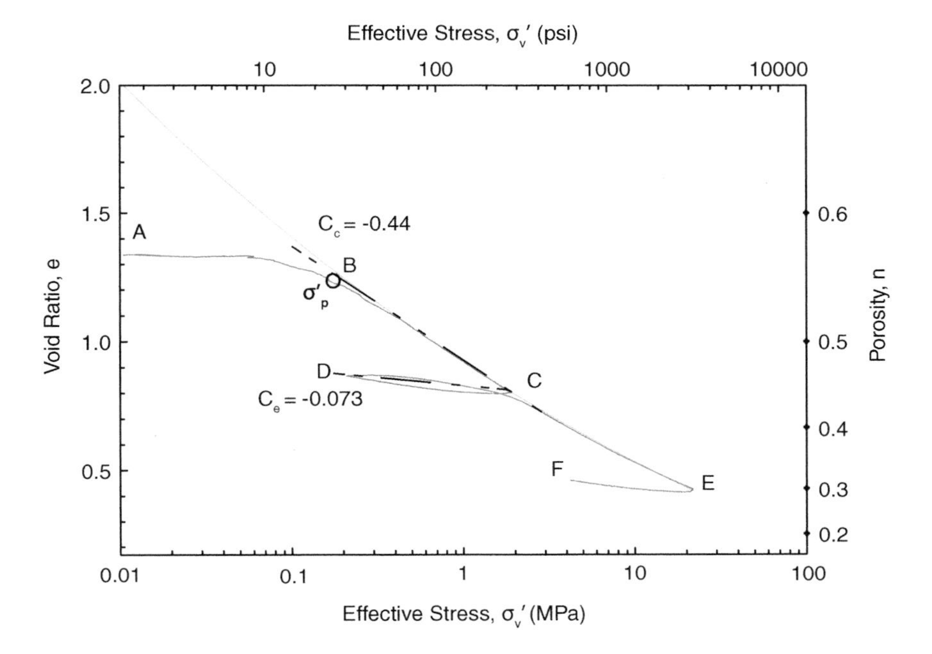

Figure 3.5 Compression test for specimen from U1324C-1H-1WR at 51.2 mbsf in the Ursa Basin, Gulf of Mexico. This specimen is nearby that presented in Figure 3.3a. The grey line is the virgin compression curve calculated with the Butterfield equation (Butterfield, 1979) (Table 5.1). The preconsolidation stress (σ_p') (open circle, 0.16 MPa) is the interpreted maximum past effective stress the sample has experienced. Figure modified from Long et al. (2011) with permission from Elsevier.

The preconsolidation stress (σ_p') is the maximum past stress the rock has experienced. The specimen is less stiff (the compression slope is greater) when it is loaded at its maximum stress (virgin compression) than it is when it is deformed at stresses less than its preconsolidation stress. For example, when loaded from point B to C, the compression slope is steeper than when it is either unloaded from point C to D or reloaded from point D to C. In addition, at stresses below the preconsolidation stress (σ_p'), the behavior is elastic: loading and unloading follow approximately the same path (e.g., the rock unloads from point C to D and then reloads from point D to C, Figure 3.5). In contrast, at stresses greater than the preconsolidation stress, the sample exhibits plastic behavior: the loading path is very different from the unloading path (e.g., the rock loads from point C to E but unloads from point E to F, Figure 3.5) with the result that a large fraction of the deformation is irrecoverable.

This elastoplastic behavior is used to interpret the preconsolidation stress. As discussed (Eq. 3.1), the slope on a semi-log plot of the virgin compression curve is described by C_c (dashed line, Fig. 3.5). The slope of the unload curve is defined by the expansion index, C_e. In this example, C_c is approximately six times greater than C_e. The difference in slope between the loading curve beneath the preconsolidation stress (e.g., point D to C) and the slope during virgin compression (point C to E) is commonly used to interpret the preconsolidation stress (open circle, point B) (Becker et al., 1987; Casagrande, 1936). Thus, insight into the geological loading history can be interpreted from a compression experiment on a core.

Elastoplastic behavior can be illustrated by compressing a piece of Styrofoam. To compress the Styrofoam, the stress needs to be raised to a critical level whereupon the pore structure will start to collapse. If the stress is then released, the Styrofoam will not expand on the path that it contracted. When the Styrofoam is reloaded, it first behaves stiffly until the stress reaches the previous maximum stress and then it deforms rapidly (plastically) once again. Elastoplastic behavior was studied most intensively in the metallurgical community (Taylor & Quinney, 1932) and it has been adopted by the geotechnical community to describe soil behavior (Wood, 1990). In Chapter 6, I discuss techniques developed to account for elastoplastic behavior in pressure prediction.

3.2.3 Horizontal Stresses During Uniaxial Normal Compression

I next consider the evolution of horizontal stresses during vertical uniaxial mudrock compression. In the laboratory, this is accomplished with a triaxial apparatus. As the vertical effective stress is increased, the confining stress is continually adjusted so that there is no lateral strain in the sample. The ratio of horizontal to vertical effective stress during uniaxial strain is described by K_0:

$$K_0 = \frac{\sigma'_h}{\sigma'_v}. \qquad \text{Eq. 3.3}$$

The variation in K_0 during uniaxial compression of resedimented Boston Blue Clay (RBBC) is presented in Figure 3.6. K_0 decreases during the initial recompression phase of the test prior to reaching the preconsolidation stress (σ'_p). At greater stresses, K_0 is fairly constant during normal consolidation. At the end of consolidation, K_0 may decrease by a small amount before equilibrating to a constant value (point A, Fig. 3.6). This decrease in K_0 is believed to be due to the dissipation of small excess pore pressures that developed within the specimen during the consolidation phase.

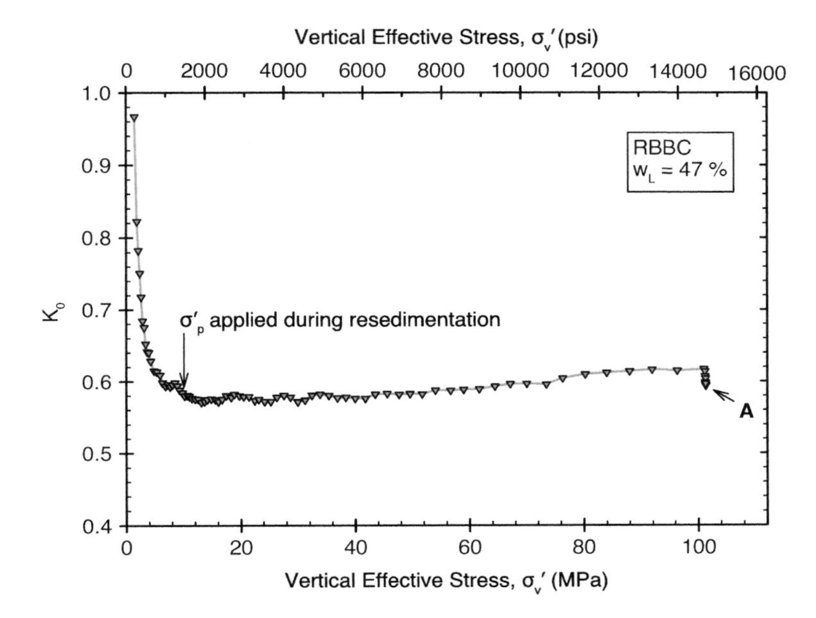

Figure 3.6 The variation in K_0 during uniaxial compression on resedimented Boston Blue Clay (RBBC). The preconsolidation stress (σ_p') imposed on the sample during resedimentation is indicated, as well as the point of equilibrium following pore pressure equilibration (point A). Reprinted from Casey et al. (2015) with permission from Elsevier.

The stress ratio (K_0) depends strongly on the texture of the material. Zablocki (2019) measured K_0 for mudrock material mixed with varying amounts of silt and compressed to 10 MPa and showed that K_0 increases dramatically as the clay fraction increases (Fig. 3.7). At values of 20% clay or less, the stress ratio is only 0.4; however at clay fractions of 70%, the K_0 value is approximately 0.7 (Fig. 3.7).

The stress ratio also depends on the clay composition; materials dominated by smectite have a higher value of K_0 than those dominated by illite. For example, Figure 3.8 illustrates the stress ratios from uniaxial compression tests for the same three lithologies presented in Figure 3.2. The RGoM-EI has a high initial stress ratio, which increases rapidly with stress (pink triangles, Fig. 3.8). In contrast, illite-rich mudrocks (RBBC) have slightly lower initial values of K_0 that increase modestly with effective stress (blue circles, Fig. 3.8). Finally, silt-rich rocks (more granular material) have lower K_0 values that stay constant or even decrease with increasing effective stress (orange squares, Fig. 3.8). The liquid limit has once again proven a good proxy to predict the K_0 behavior (Fig. 3.8). Materials with high liquid limits have high K_0 values. Casey et al. (2015) present a quantitative approach to predict the stress ratio as a function of the liquid limit.

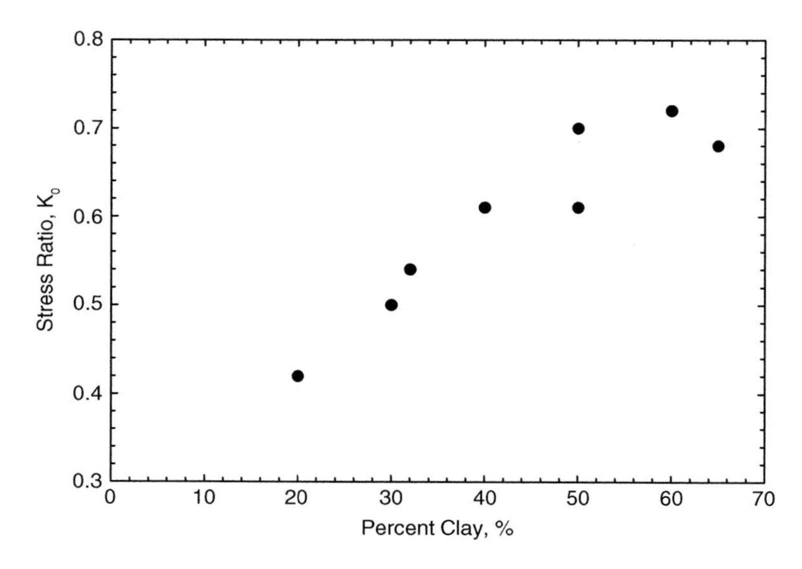

Figure 3.7 Values of the normally consolidated uniaxial stress ratio (K_0) from triaxial compression tests of resedimented Gulf of Mexico Eugene Island (RGoM-EI) sediment admixed with silt. The K_0 values shown were determined at a vertical effective stress of 10 MPa. From Zablocki (2019). Reprinted with permission by Mark Zablocki.

In summary, the stress ratio under normal uniaxial consolidation (K_0) varies both as a function of lithology (e.g., fraction of silt present) and also as a function of mineralogy (e.g., smectite-rich versus illite-rich). As a result, more granular rocks (e.g., siltstone) have a lower stress ratio than clay-rich rocks. Furthermore, more smectitic mudrocks are strongly stress-sensitive: the stress ratio increases with increasing stress level (Fig. 3.8, triangles). As I discuss in Chapter 8, this behavior suggests that the least principal stress will vary both by lithology and, for the case of clay-rich mudrocks, by stress state.

3.2.4 Strength Measurements and Stress-Induced Pressures

We explore mudrock strength and stress-induced pressures with a K_0-consolidated undrained shear test (Fig. 3.9). In this test, the specimen was uniaxially loaded (point A to point B, solid line) to an in situ mean effective stress (σ'_m) of 59.5 MPa. The specimen was then closed to drainage and subjected to an increasing vertical load while the confining (lateral) stress was held constant. The total stress path (TSP, Fig. 3.9) rises up and to the right: the deviatoric stress (q) increases at three times the rate of the mean effective stress. In contrast, the effective stress path (ESP, Fig. 3.9) first rises approximately vertically and then arches sharply to the left. The

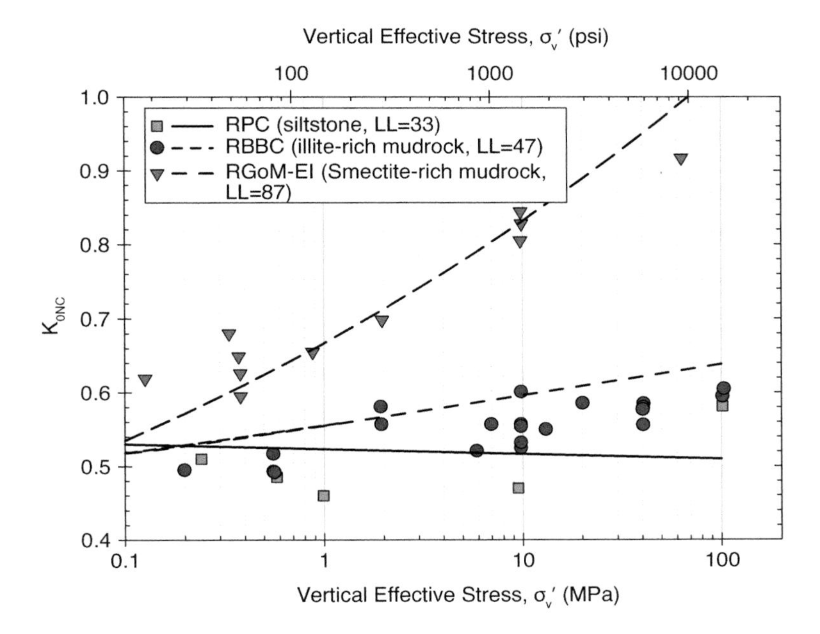

Figure 3.8 The variation in K_0 (the ratio of horizontal to vertical effective stress during uniaxial compression) as a function of stress state. Resedimented Presumpscot Clay (RPC) is a clayey siltstone with 37% clay-sized particles (<2 microns) composed of 76% illite and 10% illite-smectite (see Fig. 3.2). Resedimented Boston Blue Clay (RBBC) is a silty clay (56% clay-sized) composed of 65% illite and 28% illite-smectite (see Fig. 3.2). Resedimented Gulf of Mexico-Eugene Island (RGoM-EI) is a mudrock with 63% clay-sized particles composed of 87% smectite-illite and 8% illite (see Fig. 3.2). The symbols record experimental results. The three lines represent empirical fits to experimental data presented. Figure modified from Casey et al. (2015) with permission from Elsevier.

maximum value of the deviatoric stress (q) is the undrained strength. With further deformation, the stress path ultimately follows a constant slope toward the origin. A straight line is extended from this constant slope to the origin. The slope (M) of this line can be used to determine the critical state friction angle under triaxial compression, where

$$M = \frac{6sin\phi'}{3 - sin\phi'}. \qquad \text{Eq. 3.4}$$

Equation 3.4 assumes no cohesion. The difference between the change in total stress and the change in effective stress is the pore pressure that is generated. It is composed of (1) the pressure induced by the change in total mean stress (u_{mean}) and (2) the pressure induced by the change in deviatoric stress (u_{shear}) (Fig. 3.9). In this

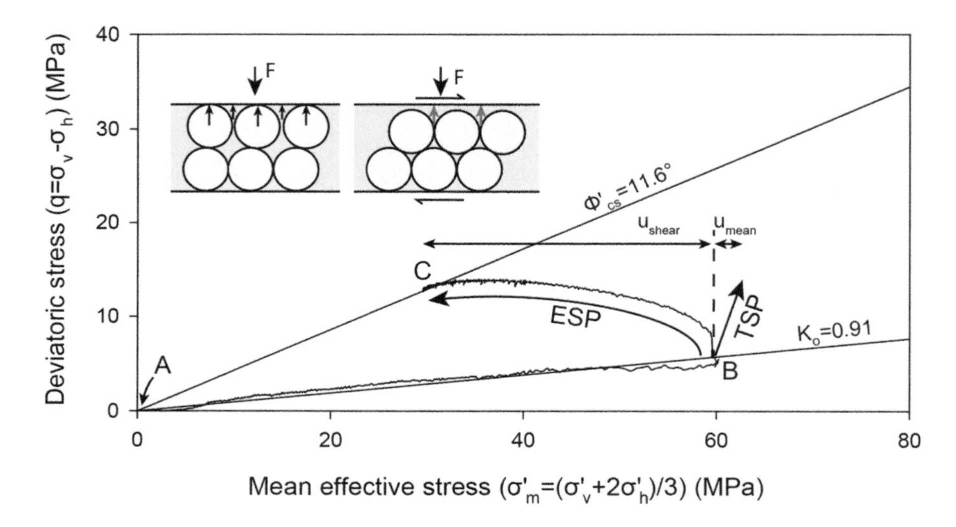

Figure 3.9 Stress path during uniaxial compression (point A to point B) and undrained shearing (point B to point C) of resedimented Eugene Island 330 sediment (RGoM-EI). The effective stress path (ESP) and the total stress path (TSP) are shown. The total stress path is illustrated with the hydrostatic pore pressure (the back pressure) subtracted to clearly illustrate the shear-induced pressure (u_{shear}) and the mean stress-induced pressure (u_{mean}). The inset illustrates the origin of shear-induced pressure: shearing induces closer grain packing, which results in the pore pressure bearing an increased fraction of the applied load.

example, the pressure induced by the change in shear stress is much greater than that induced by the change in mean stress. This illustrates an important property of most normally consolidated mudrocks: changes in either mean or shear stress can generate pore pressure. In response to an increase in mean stress, the rock would compact but the trapped pore fluid is incompressible and it bears the increase in load (Fig. 3.9, inset). In response to an increase in deviatoric stress (q), the mudrock weakens and cannot withstand the applied mean total stress (σ_m); as a result, the trapped fluid must bear the load. Below, I discuss how under drained conditions, there is a greater degree of compaction at higher shear stresses for a given mean effective stress. This behavior is not accounted for in poroelastic models (Wang, 2000).

As in the case of K_0, the friction angle depends on both lithology and clay composition; materials dominated by smectite have a lower friction than those dominated by illite. For example, Figure 3.10 illustrates the friction angle for the same three lithologies presented in Figures 3.2 and 3.8. The RGoM-EI has low initial friction angle, which declines rapidly with stress (pink triangles, Fig. 3.10). In contrast, illite-rich mudrocks (RBBC) have higher initial values of ϕ' that decrease modestly with effective stress (blue circles, Fig. 3.10). Finally, silt-rich rocks (RPC,

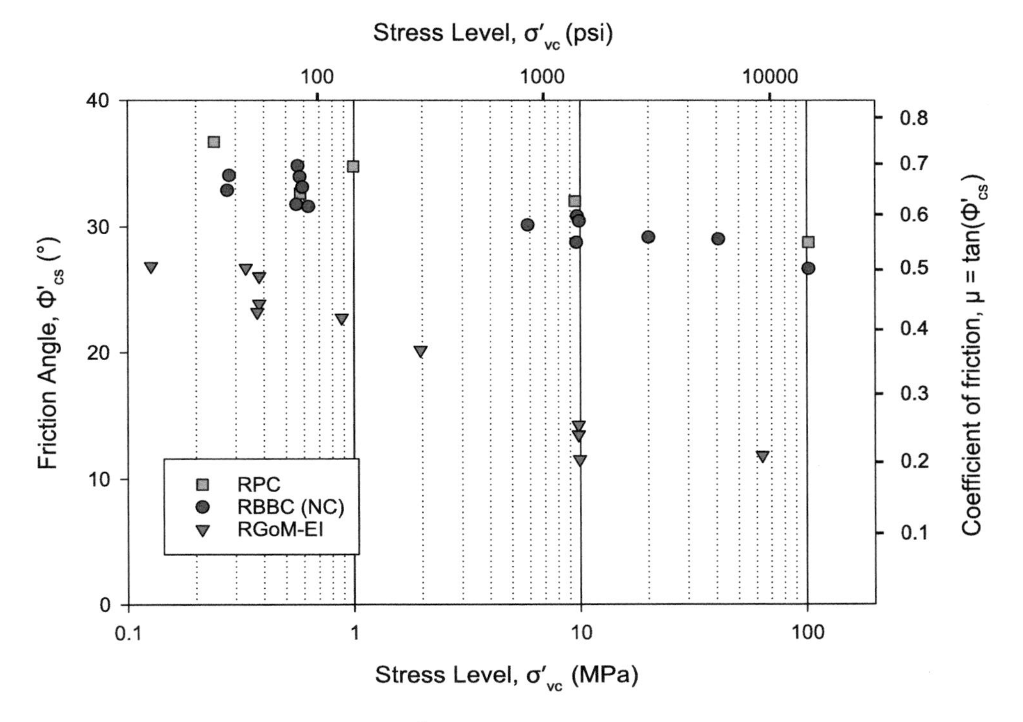

Figure 3.10. The variation in ϕ' with pre-shearing stress level for Resedimented Presumpscot Clay (RPC), Resedimented Boston Blue Clay (RBBC), Resedimented Gulf of Mexico-Eugene Island (RGoM-EI). The symbols record experimental results. Figure modified from Casey et al. (2016) with permission by John Wiley and Sons.

more granular material) have higher ϕ' that decline slightly with increasing effective stress (orange squares, Fig. 3.10). The liquid limit has once again proven a good proxy to predict the ϕ' behavior. Materials with high liquid limits have low ϕ' values that decline rapidly with increasing stress (Casey et al., 2016).

3.2.5 Viscoplastic Behavior (Creep)

All particulate materials experience time-dependent deformations (volumetric and shear) under constant effective stress conditions. This is termed creep and is also referred to as secondary compression, drained creep, and undrained creep. It is interpreted to record deformation resulting from readjustment of particle contacts at essentially constant effective stresses (Wood, 1990). Creep is commonly described as a logarithmic function of time:

$$e = e_1 - c_\alpha log\frac{t}{t_1}, \quad \text{Eq. 3.5}$$

where e_1 is the void ratio at the end of primary consolidation (i.e., after pore fluid has drained due to a loading increment), and t_1 is a reference time corresponding to the end of primary consolidation. c_α is the coefficient of secondary consolidation. It is found by plotting void ratio (or strain) versus log time and determining the slope. It is proportional to the compression index (C_c) (e.g., Fig. 3.5). Mesri and Godlewski (1977) suggest $\frac{c_\alpha}{C_c}$ is constant for a particular lithology and can be approximated as 0.04 (Wood, 1990).

To illustrate the impact of creep, Equation 3.5 is used to simulate a rock loaded to 10 MPa over one year and then allowed to creep for one million years (point B to point C, Fig. 3.11). I use the compression coefficient ($C_c = 0.49$), which is approximately that of the Ursa specimen (Fig. 3.5), and assume $\frac{c_\alpha}{C_c} = 0.04$. For each order of magnitude in time, the strain is 2% (the void ratio declines by ~0.02 ($C_c^*0.04 = 0.02$). This is termed the creep coefficient. After one million years, the void ratio declines from 0.43 to 0.31 (equivalent to a porosity change from 0.30 to 0.24) (point B to C, Fig. 3.11). Karig and Ask (2003) independently performed creep tests on very similar Gulf of Mexico material and also found a creep coefficient of 2%.

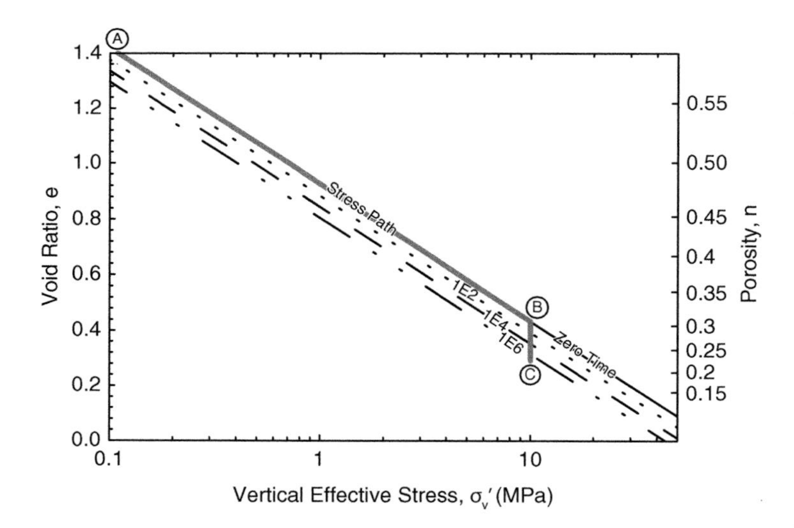

Figure 3.11 An example of creep behavior modeled from Equation 3.5. Mudrock (Fig. 3.5) is initially compressed over one year from point A to point B. Thereafter, stress is held constant for one million years (point B to C). Over this time period, the void ratio declines from 0.43 to 0.31 (equivalent to a porosity change from 0.30 to 0.24) as a result of creep.

Equation 3.5 is one of many forms of the creep law. Karig and Ask (2003) suggest that creep cannot be described with the simple parallel lines illustrated in Figure 3.11: the strain over a given decade in time increases with time, and more strain occurs at higher stresses than lower stresses. Sone and Zoback (2014) also suggest that strain at longer timescales is greater than predicted by Equation 3.4 and suggest a power-law approach is more appropriate. These studies predict even more creep than modeled here.

Creep impacts our ability to predict pressure in sedimentary basins. I discuss in Chapter 6 that deeper, hotter, and older mudrocks are often more compressed than younger mudrocks at the same effective stress in the shallow section. Creep can explain this behavior. Furthermore, creep may impact the state of stress in basins. For example, Warpinski et al. (1985) argue that mudrocks undergo creep and thus the stress ratio becomes isotropic with time. Similarly, Zoback and Kohli (2019) argue that stress relaxation is the cause of higher stress ratio (more isotropic) conditions in shales. In contrast, others argue that sediments can support large differential stresses indefinitely (Evans, 1989). In fact, there is relatively little experimental data that directly measures creep of in situ differential stresses. For example, Zoback and Kohli (2019) inferred stress relaxation from vertical strain due to increments in differential stress. In one of the few measurements of lateral stress, Karig and Ask (2003) observed that over almost three months, the stress ratios remained nearly constant. Mesri and Castro (1987) suggest that the rate of increase in horizontal stress will not approach 1.0 even over geological timescales but that it would have a measurable effect.

In the lab, creep follows primary consolidation. However, since rates of consolidation scale with the square of the drainage distance, field consolidation times are millions of times slower than the laboratory scale. Thus, it is unclear if creep occurs after primary consolidation or if it coexists with primary consolidation (Watabe & Leroueil, 2015). This can lead to very different deformation.

3.3 Stress State

3.3.1 Representation of Stress State

On any stressed point, there are three orthogonal planes upon which there are zero shear stresses. These are the principal stress planes, and the normal stresses that act upon them are the principal stresses. They are referred to as the maximum (σ_1), intermediate (σ_2), and minimum (σ_3) principal stresses.

The Mohr circle plots shear versus normal effective stress in the plane that contains the maximum and minimum principal stresses (Fig. 3.12). The normal effective (σ'_θ) and shear (τ_θ) stress along any plane oriented at an angle (θ) with

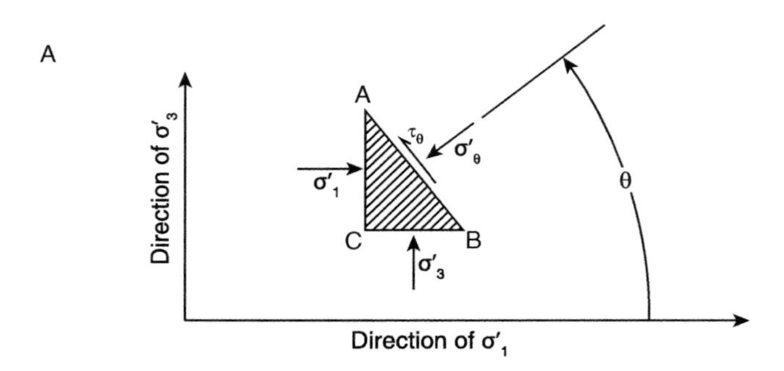

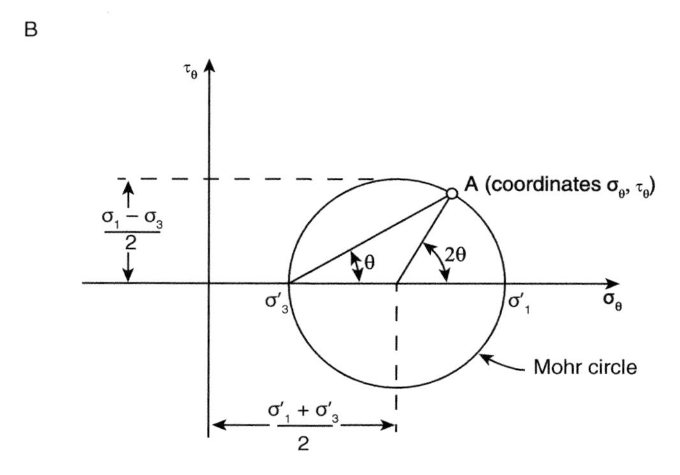

Figure 3.12 Representation of stress by the Mohr circle. (a) The shear and normal stresses along any plane (AB) are a function of the angle (θ) of the plane to the maximum principal stress (σ'_1). (b) Mohr circle describes the stress state for any plane perpendicular to the $\sigma'_1 - \sigma'_3$ plane. Modified from Lambe and Whitman (1979) with permission by John Wiley and Sons.

respect to the maximum principal stress direction can be graphically calculated (Fig. 3.12b). The x-axis intercepts of the circle (where shear stress is zero) record the maximum (σ'_1) and minimum (σ'_3) principal effective stresses. The maximum shear stress is thus:

$$\tau = \frac{\sigma_1 - \sigma_3}{2}. \qquad \text{Eq. 3.6}$$

Consider an initial stress state controlled by uniaxial compression (Eq. 3.3) (point A, Fig. 3.13). The vertical effective stress (maximum principal effective stress, σ'_1) is then increased while holding the lateral stress constant (Fig. 3.13a). As the vertical effective stress increases, the Mohr circle expands until it becomes tangent to the Mohr envelope (point B, Fig. 3.13a). At this point, the full strength of the rock has been reached and it is in a state of failure.

The Mohr envelope is commonly approximated as a straight line:

$$\tau_{ff} = \sigma'_{ff} tan\phi', \qquad \text{Eq. 3.7}$$

where τ_{ff} is the shear stress on the failure surface at failure, and ϕ' is the friction angle. At failure, the ratio of principal stresses is:

$$K_f = \frac{\sigma'_3}{\sigma'_1} = \frac{1 - sin\phi'}{1 + sin\phi'}. \qquad \text{Eq. 3.8}$$

The predicted failure surface is inclined at an angle θ_{cr} to the plane on which the major principal stress is acting:

$$\theta_{cr} = 45 + \frac{\phi'}{2}. \qquad \text{Eq. 3.9}$$

We plot stress paths to represent multiple stress states on a single diagram. For plane strain problems (such as an accretionary prism or normal-faulted system that does not change geometry significantly along strike), Roscoe and Burland (1968) suggest that average effective stress should be expressed as the average of the principal stresses in the plane of shearing (s'):

$$s' = \frac{(\sigma'_1 + \sigma'_3)}{2} \qquad \text{Eq. 3.10}$$

and that the shear stress should be expressed as the maximum shear stress in the plane of shearing:

$$t = \frac{(\sigma_1 - \sigma_3)}{2}. \qquad \text{Eq. 3.11}$$

The stress path for the material undergoing compressional failure in Figure 3.13a is shown in $s' - t$ space in Figure 3.13b. These stresses represent the brown dots at the peak of the Mohr circles (Fig. 3.13a). When we connect these points, we describe the stress path of the sample as it is loaded to failure (brown line, Fig. 3.13). In Figure 3.13b, the failure envelope intersects the average stress (s') at failure.

3.3.2 Stress State in the Earth

The Earth's surface is covered by fluid (air or water) that does not apply any shear tractions. Thus, the Earth's surface is a principal stress plane. With exceptions, such as near salt, and low friction faults, it is thought that one of the principal stresses is

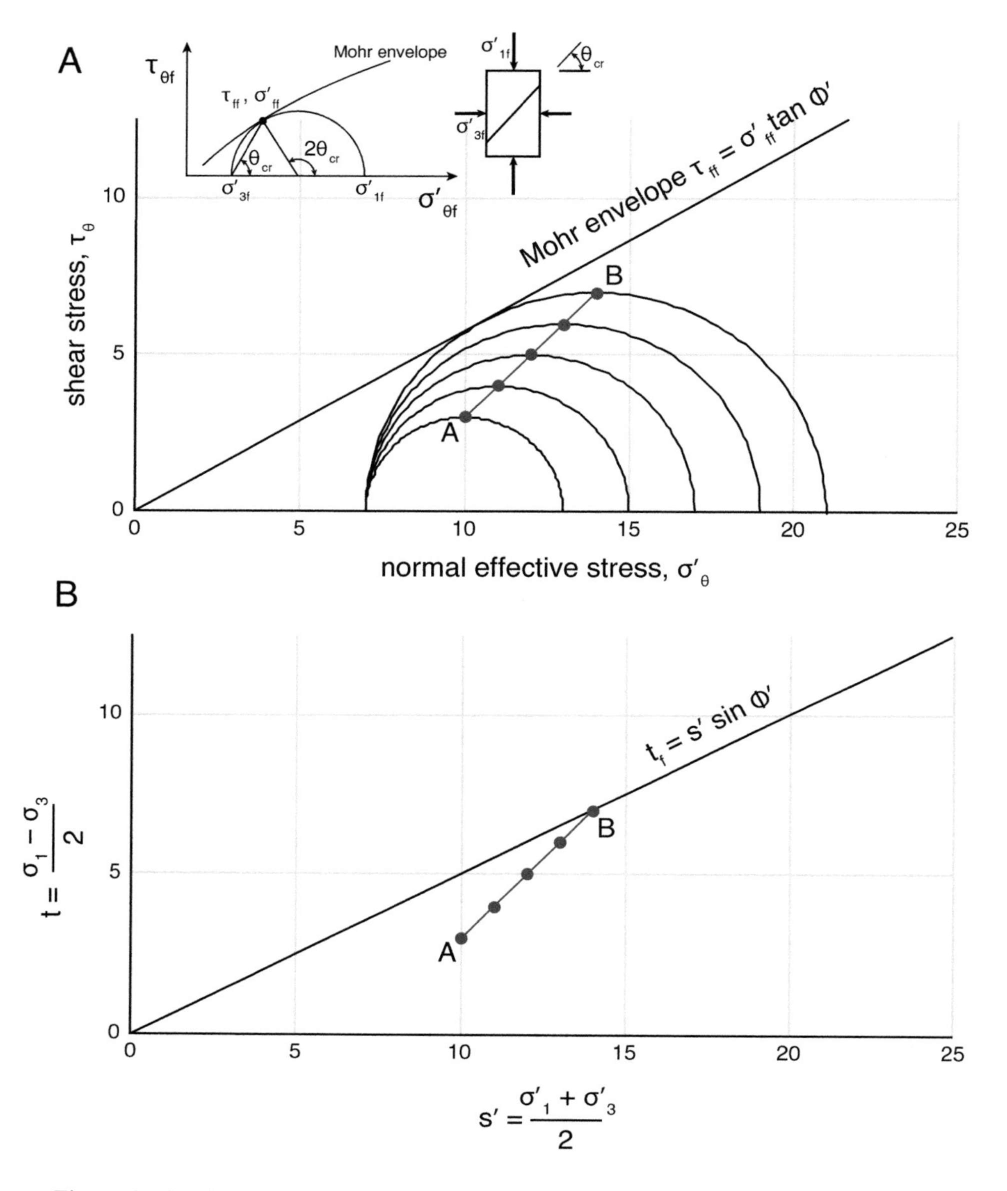

Figure 3.13 The stress evolution of a material undergoing compressional failure. (a) Mohr-diagram. Expanding Mohr circles record increase of the vertical effective stress (σ_1') while holding the horizontal effective stress (σ_3') constant. The limiting stress difference occurs when the Mohr circle becomes tangent to the Mohr envelope. Path A-B describes the maximum shear stress (peak of Mohr circle) versus the average effective stress (s') for each stress state. (b) Average stress (s') versus maximum shear stress (t). A friction angle (ϕ') of 30 degrees is assumed.

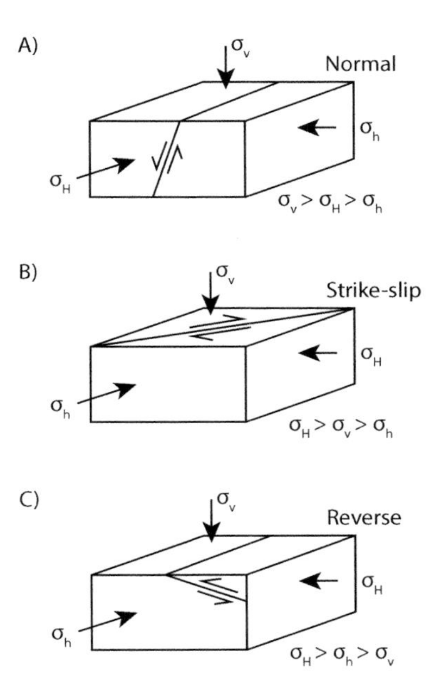

Figure 3.14 Stress states in three tectonic settings. Principal stresses are vertical (σ_v) and horizontal ($\sigma_h < \sigma_H$). (a) In an extensional setting, the overburden (σ_v) is the maximum principal stress, σ_H is the intermediate principal stress, and σ_h is the least principal stress. (b) Strike-slip system. (c) Reverse (or thrust) system. Modified from Zoback (2007). Reprinted with permission by Cambridge University Press.

vertical to the depth of the brittle–ductile transition (Zoback, 2007). This conceptual viewpoint provides a guiding approach to estimating Earth stresses. Anderson (1951) combined this viewpoint with geological observations to define three tectonic stress states: normal (extensional), strike slip, and reverse (compressional) (Fig. 3.14). Each of these systems are at failure, where the stress state is limited by friction along favorably oriented faults. An alternative viewpoint is that sediment is buried under vertical uniaxial strain (Fig. 3.15); in this case the vertical stress is the maximum principal stress and the two horizontal stresses are equal, and least principal, stresses (Fig. 3.15).

I describe the state of stress for normal-faulting, reverse-faulting, and uniaxial compression as a function of average effective stress (s') and maximum shear stress (t) (Fig. 3.16). σ_1' and σ_3' are replaced with σ_v' and σ_h' by assuming that one of the principal stresses is vertical (e.g., Figures 3.14 and 3.15). Thus,

$$s' = \frac{\sigma_v' + \sigma_h'}{2} \qquad \text{Eq. 3.12}$$

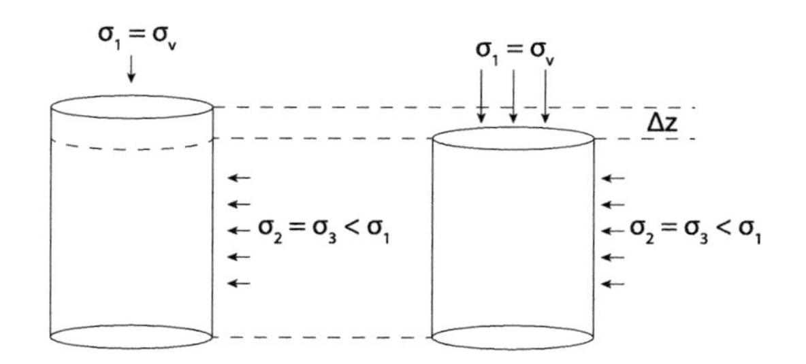

Figure 3.15 The stress state in sedimentary basins is often described with a uniaxial strain model. Because the cylinder of sediments is bounded on all sides by adjacent similar cylinders, the cylinder cannot expand or shrink laterally, and deformation thus occurs only in vertical direction (lateral deformation is zero). In this condition, in response to a vertical load, the lateral stress increases to maintain no lateral strain. Because the sediment has strength and can sustain differential stress, the lateral stress is less than the vertical stress. The maximum principal stress is vertical (σ_v). Because the system is radially symmetric, the other two principal stresses are horizontal and equal: $\sigma_v > \sigma_H = \sigma_h$.

and

$$t = \frac{\sigma_v - \sigma_h}{2}. \qquad \text{Eq. 3.13}$$

At any point above the x-axis ($t > 0$), the vertical stress is the maximum principal stress (e.g., a normal-faulting environment), whereas below the x-axis ($t < 0$), the horizontal stress is the maximum principal stress (e.g., a reverse faulting environment) (Fig. 3.16). In either case, we denote the horizontal in-plane stress as σ_h.

The slope (η) of an arbitrary line that originates at the origin on an $s' - t$ plot is:

$$\eta = \frac{t}{s'} = \frac{(\sigma_1 - \sigma_3)}{\sigma_1' + \sigma_3'}. \qquad \text{Eq. 3.14}$$

Different values of η record different values of the stress ratio K,

$$K = \frac{\sigma_3'}{\sigma_1'} = \frac{\sigma_3 - u}{\sigma_1 - u} \qquad \text{Eq. 3.15}$$

such that

$$\eta = \frac{1 - K}{1 + K}. \qquad \text{Eq. 3.16}$$

The slope of a material at Coulomb failure (η_τ) is determined from Equations 3.8 and 3.15:

$$\eta_\tau = \pm sin\phi'. \qquad \text{Eq. 3.17}$$

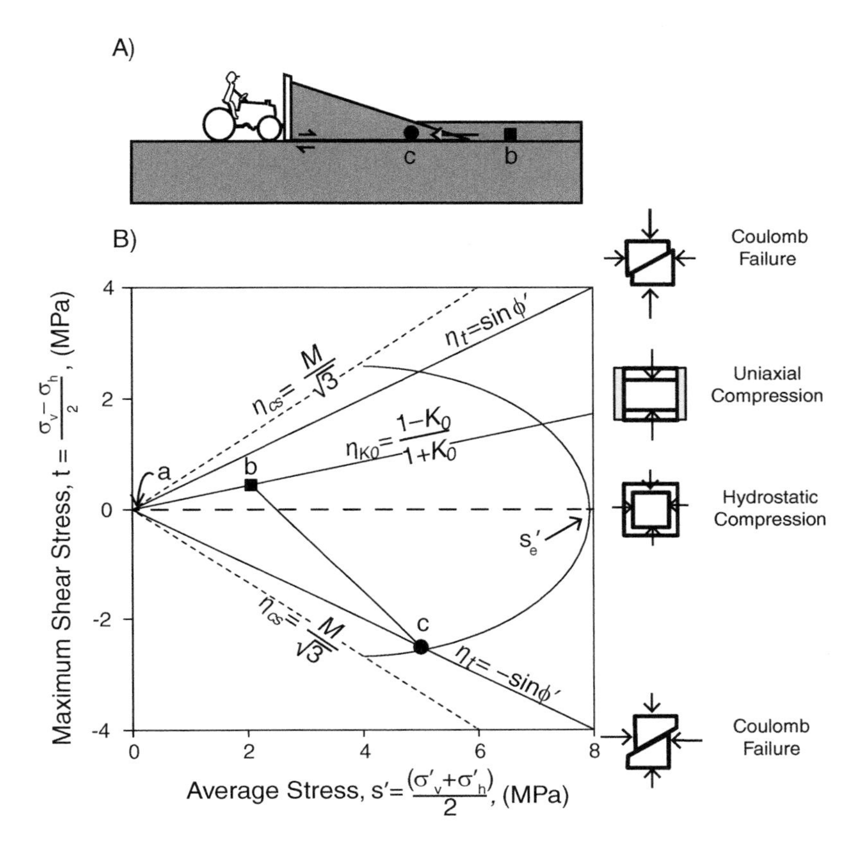

Figure 3.16 Representation of plane strain stress state. (a) To illustrate different stress states and stress paths, I describe material buried under uniaxial strain outside the prism (black square) that is subsequently incorporated into an advancing thrust belt (black circle). (b) Representation of stress with an average (s') versus maximum shear (t) stress plot. Hydrostatic compression occurs along the horizontal axis (dashed line): all three stresses increase equally and no shear stress is present ($t = 0$). Along the K_0 line (η_{K_0}), sediments undergo vertical uniaxial deformation. The Mohr Coulomb failure line (η_τ) records the stress state at Coulomb failure. Point B is outside the accretionary prism and underwent uniaxial compression from point A to point B. The sediment then enters the accretionary prism and horizontal stress increased until Coulomb failure occurs at point C. For this example, the system is assumed to be drained (no overpressure). Reprinted from Flemings and Saffer (2018) with permission by John Wiley and Sons.

In Figure 3.16, the normal-faulting stress regime (Fig. 3.14a) is recorded by the solid black line ($\eta_\tau = sin\phi'$) rising above the x-axis, whereas the reverse-faulting stress regime is recorded by the solid line that falls with increasing stress ($\eta_\tau = -sin\phi'$).

The slope of material at critical state (η_{cs}) is:

$$\eta_{cs} = \frac{M}{\sqrt{3}}, \qquad \text{Eq. 3.18}$$

where M is defined in Equation 3.4.

Under uniaxial loading (e.g., Figures 3.6 and 3.15), the horizontal effective stress scales with the vertical effective stress (Eq. 3.3). Its loading slope is:

$$\eta_{K_0} = \frac{1 - K_0}{1 + K_0}. \qquad \text{Eq. 3.19}$$

K_0 is the Earth coefficient at rest. We focus on materials where the present stress state acting on the material corresponds to the maximum stress state ever applied. Such materials are said to be in a state of normal compression. The uniaxial stress path for a material with $K_0 = 0.8$ is illustrated in Figure 3.16.

3.4 Generalized Mudrock Compaction Model

I present a conceptual and quantitative model of how rocks compact in response to variation in stress state. This sets the stage for understanding the causes of overpressure and how we interpret overpressure and stress.

3.4.1 Full Effective Stress Compaction Model

I use the modified Cam clay (MCC) model under plane strain conditions to illustrate how shear and mean stress drive compaction. In this model, compaction depends on average stress (s') and the maximum shear stress (t). To arrive at this solution, Roscoe and Burland (1968) assumed that the recoverable (elastic) components of deformation were negligible relative to the plastic (irrecoverable) components ($C_c \gg C_e$, Fig. 3.5) and made other modest engineering approximations. Their model demonstrated remarkable agreement between experiment and theory. These models are well known in the geotechnical community (Roscoe et al., 1958; Wood, 1990); however, their application to geological systems is only beginning (Flemings & Saffer, 2018; Hauser et al., 2014)

Yield surfaces describe the boundary between elastic (recoverable) and plastic (irrecoverable) deformation. In the MCC model, under plane strain conditions, these surfaces are approximated by:

$$\frac{s'}{s'_e} = \frac{\left(M/\sqrt{3}\right)^2}{\left(M/\sqrt{3}\right)^2 + \eta^2}. \qquad \text{Eq. 3.20}$$

Equation 3.20 describes an ellipse that parallels the s' axis and intercepts it at its 'equivalent' stress, s'_e (Fig. 3.16). A material undergoing isotropic loading ($t = 0$) supports approximately twice the average stress that it can support if it lies along the critical state line (Figures 3.16 and 3.17).

The yield surfaces described by Equation 3.20 also record constant pore volume (isoporosity) because it is assumed that elastic deformation is negligible. The void ratio (e) along any stress ratio (η) is described by:

$$e = e_\lambda - \lambda \ln(s'). \quad \text{Eq. 3.21}$$

e_λ is the void ratio at an average stress (s') of unity and λ is the slope on a plot of void ratio versus the natural log of average stress. While λ is independent of stress ratio (η), e_λ depends on the stress ratio. Compression along the isotropic axis ($t = 0$) is described by:

$$e = e_{\lambda_{iso}} - \lambda \ln(s'_e). \quad \text{Eq. 3.22}$$

$e_{\lambda_{iso}}$ is the void ratio for an average stress (s') of unity under isotropic loading. Equations 3.20, 3.21, and 3.22 are combined to solve for e_λ along any stress ratio (η):

$$e_\lambda = e_{\lambda_{iso}} - \lambda ln \left[\frac{\left({}^{M}/_{\sqrt{3}} \right)^2 + \eta^2}{\left({}^{M}/_{\sqrt{3}} \right)^2} \right]. \quad \text{Eq. 3.23}$$

Given λ and e_λ at a particular stress ratio (η), the compaction behavior along any stress ratio is determined by calculating e from Equation 3.21. The compaction for sediments under hydrostatic stress (η_{iso}), uniaxial strain (η_{K_0}), and at Coulomb failure (η_τ) are illustrated in Figure 3.17b. A key feature of the MCC model is that for a given average effective stress (s'), the higher the maximum shear stress (t), the greater the compaction.

I illustrate how to apply this approach with an example based on the compaction observed under uniaxial strain in Figure 3.5 ($e_0 = 0.9$ and $C_c = 0.44$), assuming $K_0 = 0.7$, and $\phi' = 30°$. The slope (λ) in a natural log system can be expressed as a function of the slope in a common log (C_c) as

$$\lambda = C_c log\,(2.7128). \quad \text{Eq. 3.24}$$

The void ratio at an average stress of unity ($s' = 1$ MPa) under uniaxial strain ($e_{\lambda_{K_0}}$) is found by combining Equations 3.1, 3.3, 3.12, and 3.23:

$$e_{\lambda_{K_0}} = e_0 - \lambda ln \left(\frac{2}{1 + K_0} \right). \quad \text{Eq. 3.25}$$

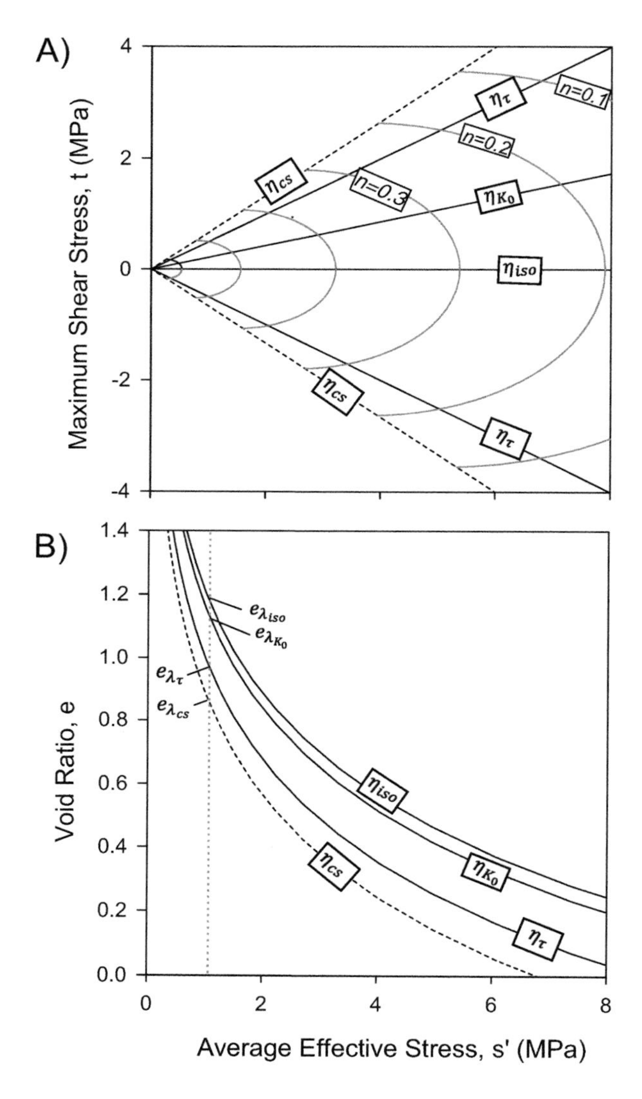

Figure 3.17 (a) Average effective stress (s') versus maximum shear (t). The plot is bounded above and below by the critical state line (η_{cs}). Equal pore volume contours are shown in porosity units (e.g., 'n = 0.3'). (b) Compression behavior. Void ratio versus mean effective stress for four stress ratios: (1) critical state (η_{cs}), (2) Coulomb failure (η_τ), (3) K_0 compression (η_{K_0}), and (4) isotropic ($\eta_{iso} = 0$). Modified from Flemings and Saffer (2018) with permission by John Wiley and Sons.

I set $e_{\lambda_{K_0}} = e_\lambda$ in Equation 3.24 and solve for $e_{\lambda_{iso}}$. Given $e_{\lambda_{iso}}$, e_λ for any stress ratio (η) can be calculated with Equation 3.23. The resultant values for isotropic, K_0, and uniaxial loading are listed in Table 3.1.

Table 3.1 *Compaction parameters calculated for different stress conditions.*

Void ratio at $\sigma'_v = 1$ MPa under uniaxial strain	e_0	0.9
Compression slope in common log space	C_c	0.44
Compression slope in natural log space	λ	0.19
Void ratio at $s' = 1$ MPa under hydrostatic stress ($t = 0$)	$e_{\lambda_{iso}}$	0.88
Void ratio at $s' = 1$ MPa under uniaxial loading	$e_{\lambda_{K_0}}$	0.87
Void ratio at $s' = 1$ MPa under Coulomb failure	$e_{\lambda\tau}$	0.80

3.4.2 Compaction Curves for Characteristic Stress States

I use the modified Cam clay (MCC) model to describe mudrock compaction as a function of vertical effective stress for any stress state where one of the principal stresses is vertical (Eq. 3.26). I apply this to the case of uniaxial strain, critically stressed normal-faulting, and critically stressed reverse-faulting (Figures 3.14 and 3.15). The generalized compaction equation for these conditions is:

$$e = \{e_0 + \Delta e_{s'} + \Delta e_q\} - \lambda ln\sigma'_v, \qquad \text{Eq. 3.26}$$

$$\Delta e_{s'} = \lambda \ln\left[\frac{1+K_0}{1+K}\right], \qquad \text{Eq. 3.27}$$

and

$$\Delta e_q = \lambda \ln\left[\frac{\left(\frac{M}{\sqrt{3}}\right)^2 + \left(\frac{1-K_0}{1+K_0}\right)^2}{\left(\frac{M}{\sqrt{3}}\right)^2 + \left(\frac{1-K}{1+K}\right)^2}\right]. \qquad \text{Eq. 3.28}$$

K is the ratio of the horizontal to vertical effective stress,

$$K = \frac{\sigma'_h}{\sigma'_v}, \qquad \text{Eq. 3.29}$$

and is dependent on the tectonic environment.

In Equation 3.26, the term in brackets describes the value of the void ratio at a vertical effective stress of unity (1 MPa). e_0 is the value under uniaxial strain. $\Delta e_{s'}$ accounts for the compaction that results from the difference in average effective stress between the tectonic condition (e.g., thrust or normal-faulting) and the uniaxial condition. For example, for a given vertical effective stress, the average effective stress is greater during thrust-faulting than under uniaxial strain and thus there is more compaction. Δe_q accounts for the effect of shear on compaction. For example, the shear stress is higher under both normal-faulting and thrust-faulting than it is under uniaxial strain. Thus, at a given average stress, the rock undergoing faulting will undergo more shear compaction than the rock under uniaxial strain.

Together, Equations 3.26, 3.27, and 3.28 have five unknowns: e_0, λ, ϕ', K, and K_0. The first two parameters are inferred by field observations of void ratio versus effective stress or through experiments (e.g., Fig. 3.5). As discussed, the friction angle is generally determined by experiment (e.g., Fig. 3.9). The uniaxial stress ratio (K_0) can be determined experimentally (e.g., Figures 3.7 and 3.8). However, it is also calculated directly in the MCC model or by other empirical approaches. I use the Jaky equation (Jaky, 1944):

$$K_0 = 1 - sin\phi'. \quad \text{Eq. 3.30}$$

In environments with no tectonic activity (uniaxial-strain conditions), $K = K_0$ (Eq. 3.3). In normal-faulted environments, Equation 3.8 is stated as:

$$K = \frac{1 - sin\phi}{1 + sin\phi}, \quad \text{Eq. 3.31}$$

whereas in thrust-faulted environments, Equation 3.8 yields:

$$K = \frac{1 + sin\phi}{1 - sin\phi}. \quad \text{Eq. 3.32}$$

Figure 3.18 illustrates the compaction curves for the three environments. All three curves have the same slope (λ), but they differ in their intercept (the void ratio at 1 MPa). For a given vertical effective stress, rocks compacted uniaxially (dashed line) are the least compacted and those compacted during reverse (solid line) are the most compacted. This is because for a given vertical effective stress, both the average effective stress $(\frac{\sigma'_H + \sigma'_v}{2})$ and the shear stress $(\frac{\sigma_H - \sigma_v}{2})$ are greatest during reverse-faulting. It is less intuitive that rocks deformed during normal-faulting are more compacted than those compacted uniaxially (dash-dot line versus dashed line, Fig. 3.18). For this case, the average stress (for a given vertical effective stress) during normal-faulting is less than during uniaxial conditions, but the shear stress is greater. The increased shear (Δe_q, Fig. 3.18, inset) overwhelms the decreased mean effective stress ($\Delta e_{s'}$, Fig. 3.18, inset) and as a result, rocks in the normal-faulting regime are slightly more compacted than those in the uniaxial regime (Fig. 3.18, dash-dot line).

I close by considering the role of shear and average stress in compaction by calculating the void ratio at 1 MPa for a range of friction angles for extensional,

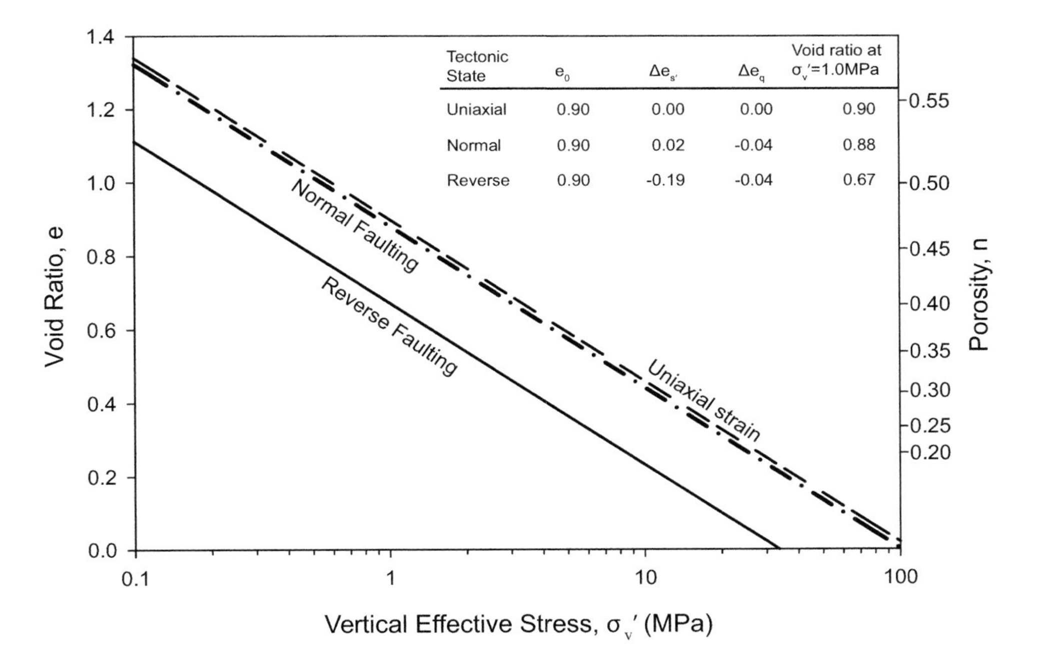

Figure 3.18 Compaction curves for three tectonic settings as a function of vertical effective stress as described by Equation 3.26: (1) uniaxial compaction (Fig. 3.15), (2) normal-faulting (e.g., Fig. 3.14a), and (3) reverse-faulting (e.g., Fig. 3.14c). The void ratio at 1 MPa is the sum of $e_0 + \Delta e_{s'} + \Delta e_q$ (Eq. 3.26), and these terms are shown in the inset. The uniaxial compaction behavior demonstrated in Figure 3.5 and a friction angle of 30 degrees is assumed. K_0 is calculated with Jaky's equation (Eq. 3.30).

compressional, and uniaxial strain conditions (Fig. 3.19). The void ratio at 1 MPa is the sum of the first three terms in the right hand side of Equation 3.26. The smaller the void ratio at 1 MPa, the greater the compaction. I first consider the case where shear does not affect compaction by setting the third term on the right hand side of Equation 3.26 to zero. This is analogous to poroelastic approaches, where volume change depends only on the mean effective stress (Wang, 2000). In this case, the void ratio at 1 MPa for normal-faulting conditions without shear compaction (solid line, Fig. 3.19) is greater than that for uniaxial burial (dotted line, Fig. 3.19). This is because, for a given vertical effective stress, the average effective stress is less for normal-faulting than for uniaxial stress. The line rises to the right because, for a given σ_v', the average stress (s') decreases as the friction angle increases (σ_h' reduces relative to σ_v'). In a similar fashion, sediment compacted under reverse-faulting without shear compaction (dashed line) is much more compacted than either the normal-faulting case (solid line) or the uniaxial case (dotted line). This is

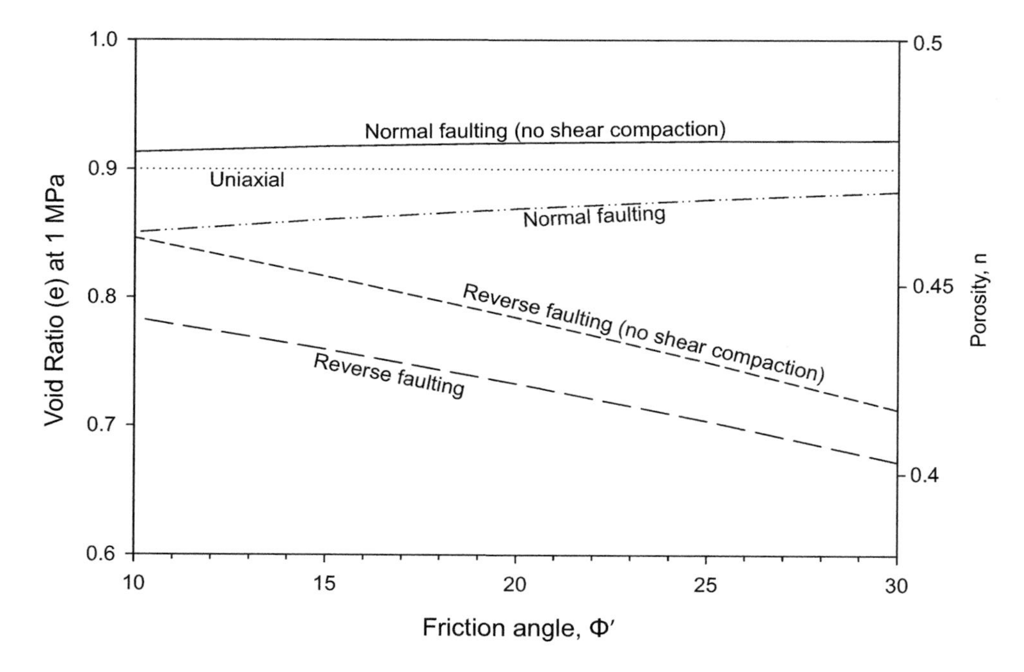

Figure 3.19 The void ratio (e) at a vertical effective stress of 1 MPa as a function of friction angle for normal-faulting, reverse-faulting, and uniaxial strain conditions. The compaction behavior demonstrated in Figure 3.5 is assumed. K_0 is calculated with Jaky's equation (Eq. 3.30).

because for a given vertical effective stress, the average effective stress is more for reverse-faulting than for either uniaxial stress or normal-faulting. The line decreases to the right because, for a given σ_v', s' increases as the friction angle increases (σ_h' increases relative to σ_v').

The full solution, which incorporates both average stress and shear stress, is calculated by summing all three terms on the right hand side of Equation 3.26. The vertical distance between the normal-fault case without shear (solid line) and with shear (dash-double dot line) is the shear-induced compaction. Shear drives the normal-faulting compaction curve beneath the uniaxial compaction curve. Similarly, for the case of reverse faulting, the addition of shear compaction decreases the void ratio (compare short dashed line with long dashed line).

3.5 Summary

Overpressures are generated when mudrocks are loaded by external stresses. If we can understand how rocks compact in response to these stresses, we will understand how overpressures are generated and be able to predict pore pressure from compaction state. In this chapter, I first described the compaction of mudrocks under uniaxial strain and how composition and texture control this compaction. I then reviewed the elastoplastic nature of mudrocks. In Chapter 6, I describe how elastoplastic behavior is incorporated into pore pressure prediction. I next reviewed the relationship between vertical and horizontal stress during one-dimensional compaction. In Chapter 8, I use these concepts to predict least principal stress. I then described how we measured strength and reviewed the concept of pressures induced by mean stress and shear stress. These insights drive both how we interpret least principal stress (Chapter 8) and how we predict pressure. Finally, I discussed creep behavior of mudrocks. I discuss in Chapter 6 how this impacts pore pressure prediction.

In closing, I presented a comprehensive model that incorporated the effect of mean and shear stress on compaction. I used this model to describe compaction in normal-faulting, thrust-faulting, and uniaxial compression environments. I apply these concepts for pore pressure prediction in Chapter 6 and Chapter 7. The key goal of this chapter was to develop insight into how we might expect compaction curves to systematically vary in different tectonic environments.

4

The Origins of Geopressure

4.1 Introduction

In this chapter, I review how overpressure is generated and preserved in sedimentary basins. I begin by describing how pore pressure is generated and how it dissipates under conditions of uniaxial strain. I then explore the overpressure that results during sedimentation under conditions of vertical uniaxial strain. I present several numerical solutions in one and two dimensions to extend these concepts. I illustrate how these types of basin models provide insight into the evolution of pressure and stress in basins. I also use these results to explore the role of thermal expansion and smectite-illite diagenesis in generating pore pressure. I then discuss hydromechanical models that couple the full stress state with pore fluid flow under three-dimensional strain. I use these models to illustrate the role of non-uniaxial loading in salt systems and thrust belts. Finally, I explore the potential role of viscous compaction in pore pressure generation.

4.2 Pore Pressure Generation and Dissipation under Vertical Loading

Terzaghi (1923) used a simple analogy to describe pore pressure generation in a cylinder of saturated soil under a vertical load (Fig. 4.1a). The plate above the soil is porous, allowing the load to be applied to the soil and yet permitting escape of the fluid from the soil. The analogue model represents the soil solid structure as a spring and the soil pore structure as a valve (Fig. 4.1b).

When the load is applied to the soil sample, the soil (spring, Fig. 4.1b) tends to compress under the load. However, water in the soil does not have time to escape from the tiny pores of the soil (closed valve, Fig. 4.1c). Because water is almost incompressible, the stone plate barely moves, the spring cannot compress and thus cannot bear any load, and the water supports the entire load. This results in water pressure greater than hydrostatic pressure (Fig. 4.1f, early time). When the valve is opened, the fluid pressure forces water through (Fig. 4.1d). As the water escapes,

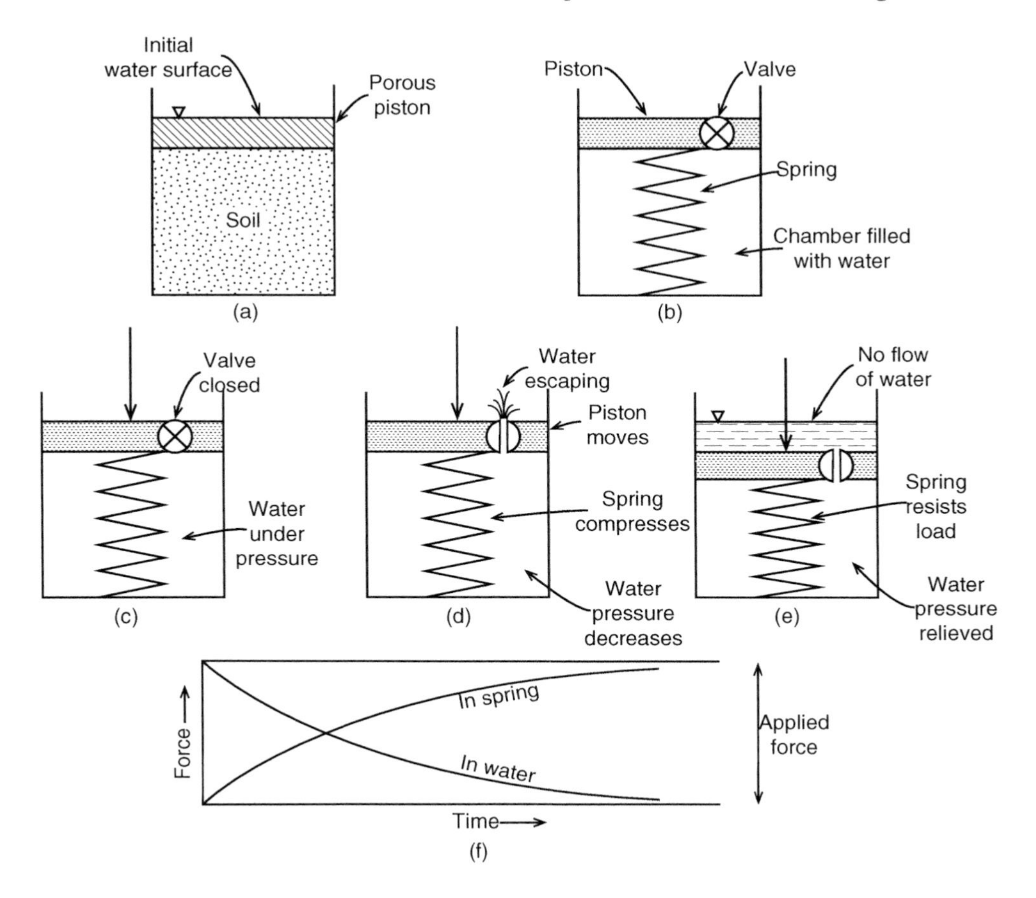

Figure 4.1 Hydromechanical analogy for uniaxial consolidation under a vertical load after Terzaghi (1923). (a) Physical sample. (b) Hydromechanical analogue. (c) Load applied with valve closed. Piston does not move because water cannot escape and water supports the entire load. (d) Load applied with valve open. Piston moves down as water escapes. Load gradually transfers from water to spring. (e) Hydraulic equilibrium (hydrostatic pressure, no water flow). Spring supports the entire load. (f) Gradual transfer of load from water to spring as consolidation occurs. From Lambe and Whitman (1979), reprinted with permission by John Wiley and Sons.

the spring shortens and begins to carry a greater fraction of the applied load (Fig. 4.1f). As the load on the spring increases, the fluid pressure drops (Fig. 4.1e, f).

Only a limited amount of water can escape from the pores of the soil (valve, Fig. 4.1d) during any interval of time. Thus, the overpressure dissipation and soil consolidation is a time-dependent process. The duration of this process scales with the stiffness of the solid structure (m_v), the viscosity of the pore fluid (μ), the thickness of the soil (d) (representing the drainage length), and the permeability (k) (representing the size of pores).

This simple hydromechanical analogy lies at the core of understanding the origin of overpressure, how to predict it from compaction state, and the timescales over which overpressure can be preserved. It can be used to understand basin overpressure where stress is the primary source of overpressure. Overpressure in these basins is generated by loading (e.g., weight of overburden or tectonic loads), and its dissipation is controlled by the geometry and the properties of the rock and pore fluid.

I now explore the process illustrated in Figure 4.1. I begin with a more detailed discussion of effective stress, which was introduced in Chapter 2. I then describe undrained loading, which describes the early state of the soil immediately after the load is applied (Fig. 4.1c), and drained behavior, which describes the final state of the soil long after the load is applied and all overpressure is dissipated (Fig. 4.1e). Finally, I discuss the time dependent process of consolidation, which describes the intermediate state of the soil when overpressure is dissipating and the load is being transferred from water to the sediment structure (Fig. 4.1d).

4.2.1 Effective Stress

Craig (2004) presented a simple physical model to illustrate the concept of effective stress in a saturated granular material under a vertical load (Fig. 4.2). Consider

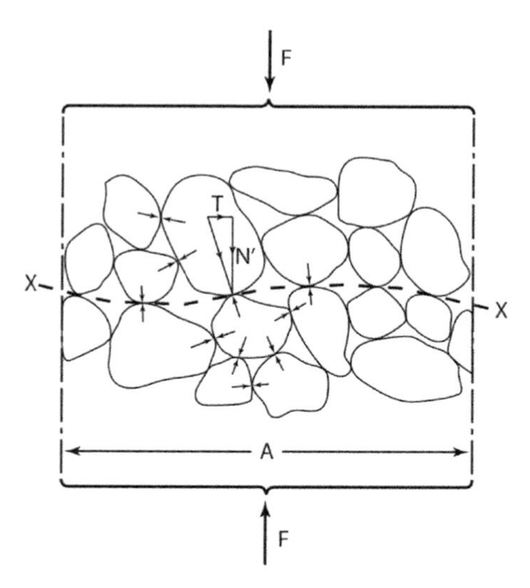

Figure 4.2 Concept of effective stress. From Craig (2004). Reproduced with permission of Taylor and Francis (Books) Limited UK through PLS Clear.

a horizontal surface that crosses the material along the contact points between the grains (Fig. 4.2). At every grain contact, the interparticle forces can be divided into a vertical (N') and a horizontal component (T) (Fig. 4.2). The effective normal stress is defined as the sum of normal interparticle forces divided by the total area across which these forces act:

$$\sigma' = \frac{\sum N'}{A}. \quad \text{Eq. 4.1}$$

The total normal stress is the applied load (F) divided by the area:

$$\sigma = \frac{F}{A}. \quad \text{Eq. 4.2}$$

If we assume the contact areas are infinitesimally small, then pore pressure (u) acts along entire the area (A) and the force balance can be written as:

$$F = \sum N' + (u \times A). \quad \text{Eq. 4.3}$$

This can be rewritten as:

$$\frac{F}{A} = \frac{\sum N'}{A} + u, \quad \text{Eq. 4.4}$$

which reduces to a simplified form of the effective stress equation (Eq. 2.7):

$$\sigma = \sigma' + u. \quad \text{Eq. 4.5}$$

The total stress (σ) applied is balanced by the effective stress felt by the grains (σ') and the pore pressure (u). In fact, as discussed in Chapter 2 (Eq. 2.7), the effective stress equation is more precisely stated as:

$$\sigma = \sigma' + \alpha u \quad \text{Eq. 4.6}$$

where the pore pressure coefficient (α) is a function of the relative compressibility of solid grains (c_s) and bulk structure (c_b) $(\alpha = 1 - c_s/c_b)$. In soft rocks, the compressibility of grains is much less than the bulk compressibility and α is commonly assumed to equal 1. In contrast, in low porosity mudrocks, where bulk compressibility can also be very small, the pore pressure coefficient can be much smaller. Ma and Zoback (2017) found the pore pressure coefficient to lie between 0.3 and 0.9 and to decrease with increasing stress in the Bakken play in the Williston Basin. Suarez-Rivera and Fjær (2013) found the pore pressure coefficient to be approximately 0.5 in samples from the Haynesville Formation.

4.2.2 Undrained Loading

I next consider the pore pressure response to loading immediately after a load is applied to the material. Pore water does not have time to escape: the soil is in an undrained condition. This represents the initial state in Terzaghi's geomechanical analogue, where pore pressure increases abruptly in response to the load (Fig. 4.1c; time = 0, Fig. 4.1f).

For the case of an isotropic loading, the pore pressure increases (du) in proportion to the increase in the load $(d\sigma_m)$:

$$du = Bd\sigma_m. \qquad \text{Eq. 4.7}$$

B is Skempton's coefficient and a function of the rock porosity (n) and the bulk, solid, and fluid compressibilities:

$$B = \frac{c_b - c_s}{(c_b - c_s) + n(c_f - c_s)}. \qquad \text{Eq. 4.8}$$

Green and Wang (1986) present Equation 4.8 and describe its relationship to earlier work (Biot, 1941; Rice & Cleary, 1976; Skempton, 1954). Equation 4.8 assumes the porous media is uniform and all of the pore space interconnected (Green & Wang, 1986). If the behavior of the bulk structure is poroelastic, Equations 4.7 and 4.8 describe the pore pressure response to change in mean stress in non-isotropic loadings (Wang, 2000).

For the case of uniaxial-strain loading, pore pressure is obtained as:

$$du = Cd\sigma_v. \qquad \text{Eq. 4.9}$$

Where C is termed the loading efficiency (Wang, 2000) and is a function of the individual compressibilities:

$$C = \frac{m_v - c_s}{(m_v - c_s) + n(c_f - c_s)}. \qquad \text{Eq. 4.10}$$

m_v is the compressibility of the bulk structure under uniaxial strain:

$$m_v = \frac{dV}{V}\frac{1}{d\sigma'_v} = \frac{dH}{H}\frac{1}{d\sigma'_v} = \frac{1}{1+e}\frac{de}{d\sigma'_v}, \qquad \text{Eq. 4.11}$$

where e is void ratio.

I evaluate Skempton's coefficient (B) and the loading efficiency (C) for several rock types and porosities listed in Table 4.1. For the Gulf of Mexico mudrocks, which typically are not significantly lithified and have relatively high porosities, the bulk compressibility is much greater than the solid and the water compressibilities

Table 4.1 *Loading efficiencies (C) and Skempton's coefficients (B) for different rock types and porosities. To calculate C (Eq. 4.10) and B (Eq. 4.8), a fluid compressibility of water of* $5 \times 10^{-10} Pa^{-1}$ *and a solid compressibility of quartz* $(2 \times 10^{-11} Pa^{-1})$ *was assumed (de Marsily, 1986). References: (1) Casey (2014), (2) Ma and Zoback (2017), (3) Suarez-Rivera and Fjær (2013).*

Material	Uniaxial Compressibility (m_v) Pa^{-1}	Porosity (n)	C (Eq. 4.10)	Reference
Resedimented Gulf of Mexico Mudstone (RGoM-EI) ($\sigma_v' = 0.1$ MPa)	1×10^{-7}	0.82	1.00	1
Resedimented Gulf of Mexico Mudstone (RGoM-EI) ($\sigma_v' = 35$ MPa)	2×10^{-9}	0.25	0.94	1
Material	**Isotropic Compressibility (C_b) Pa^{-1}**	**Porosity (n)**	**B (Eq. 4.8)**	**Reference**
Bakken Fm. (silicate rich) ($\sigma_v' = 24.4$ MPa)	5×10^{-11}	0.07	0.47	2
Bakken Fm. (carbonate rich) ($\sigma_v' = 24.4$ MPa)	3.3×10^{-11}	0.1	0.21	2
Haynesville Shale (depth of ~4 300 m)	7.25×10^{-11}	0.082	0.57	3
Bossier Shale (depth of ~4 000 m)	4.35×10^{-11}	0.045	0.52	3

and as a result both coefficients (B and C) are very close to 1. This explains why both coefficients are commonly approximated as 1.0 for soils in geotechnical engineering (Terzaghi, 1923).

In contrast, for unconventional shales, which are highly indurated and have low porosity, the solid, fluid, and bulk compressibilities are of the same order. In these cases, Skempton's coefficient (B) is much less than 1 (Table 4.1). In these calculations, water was assumed as the pore fluid. If a more compressible fluid is present (e.g., gas), then the values of B and C will be much lower.

Figure 4.3 illustrates undrained loading in a well in the Nankai Trough, offshore Japan. Pore pressure was monitored at the seafloor and multiple depths at ODP Site 1173. Sawyer et al. (2008) compare the tidal load as recorded by the water pressure at the seafloor to the pore pressure within the sediments. For example, at 359 meters below seafloor (mbsf), the pore pressure (as referenced to the seafloor) records 85% of the water pressure at the seafloor: the loading efficiency is 0.85. In this location, pressure is measured in compressible and high porosity sediments. In contrast, at

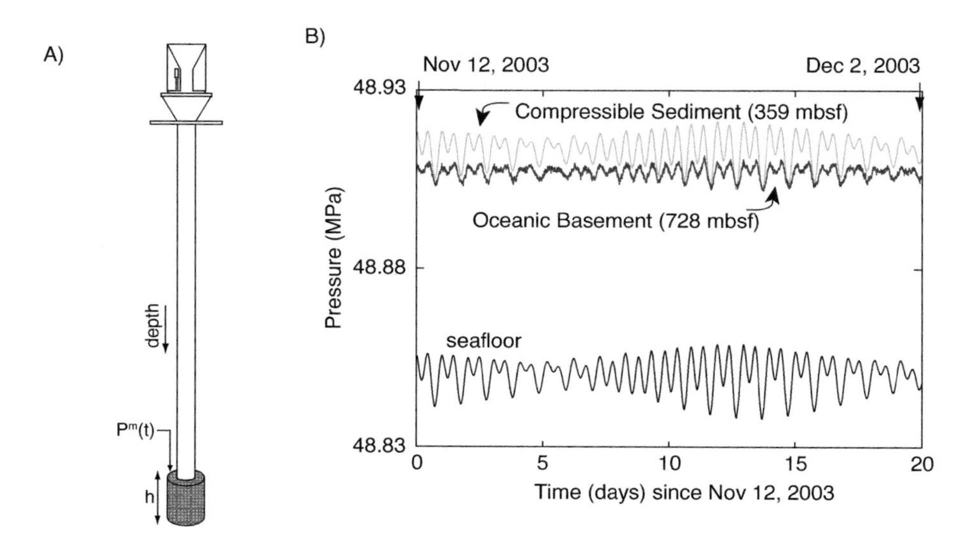

Figure 4.3 Pore pressure and seafloor pressure recorded at ODP Site 1173, offshore Japan. (a) Schematic illustration of pressure monitoring system. Pressure is recorded over a screen of height h. A thin tube (not shown) extends from the monitoring screen to the seafloor where pressure is measured. Simultaneously, pore pressure is measured at the seafloor. (b) The pressure record at the seafloor, and at two screened intervals at 359 mbsf and 728 mbsf. Reprinted from Sawyer et al. (2008) with permission by John Wiley and Sons.

728 mbsf, the loading efficiency is 0.4. In this location, pore pressure is measured in stiff volcaniclastic sediment and perhaps the underlying oceanic basement. The lower loading efficiency at 728 mbsf relative to 359 mbsf may reflect that stiffer material is present at the deeper location. This demonstrates that loading instantaneously elevates pore pressure and that the response is proportional to the rock and fluid compressibilities.

4.2.3 Drained Compression

I next consider the end member of drained loading, which describes the state of the soil sample long after loading, when overpressure has dissipated completely (hydrostatic pressure) and no water escape from the sample occurs (Fig. 4.1e). In this condition, the rock structure (spring, Fig. 4.1e) supports the entire load and the rock compression is completed. Laboratory examples of drained loading were presented in Chapter 3 (e.g., Eq. 3.1).

There is no overpressure under drained conditions. Thus, the vertical effective stress is the total stress, and the compression due to the applied vertical load can be directly calculated. The volumetric strain due to compression can be expressed as a function of the difference between initial and final void ratios:

$$\frac{1}{V_0}\Delta V = \frac{e_0 - e_1}{1 + e_0} = \frac{n_0 - n_1}{1 - n_1}. \quad \text{Eq. 4.12}$$

Under vertical uniaxial strain, the volumetric strain equals the vertical strain. Therefore,

$$\frac{1}{V_0}\Delta V = \frac{1}{h_0}\Delta h = \frac{e_0 - e_1}{1 + e_0} = \frac{n_0 - n_1}{1 - n_1}. \quad \text{Eq. 4.13}$$

We describe compression with either the coefficient of compressibility (m_v) or the coefficient of consolidation (C_c):

$$m_v = \frac{1}{h_0}\left(\frac{h_0 - h_1}{\sigma_1' - \sigma_0'}\right) = \frac{1}{1 + e_0}\frac{e_0 - e_1}{\sigma_1' - \sigma_0'} \quad \text{Eq. 4.14}$$

and

$$C_c = \frac{e_0 - e_1}{log\left(\sigma_1'/\sigma_0'\right)}. \quad \text{Eq. 4.15}$$

Thus, the total settlement due to a change in effective stress is:

$$\Delta h = m_v d\sigma' h_0 \quad \text{Eq. 4.16}$$

or

$$\Delta h = \frac{C_c log\left(\sigma_1'/\sigma_0'\right)}{1 + e_0} h_0. \quad \text{Eq. 4.17}$$

Of course, if there are several layers of different properties over depth, the total settlement would be the sum of the settlement of each layer calculated individually using Equation 4.16 or Equation 4.17.

I apply Equation 4.17 to estimate the compaction of the Gulf of Mexico mudrock described in Chapter 3 (Figure 3.5) with $C_c = 0.44$ and e_0 (at 1 MPa) $= 0.9$. I assume this sediment package is initially at an effective stress of 1 MPa and then is loaded to 10 MPa. Void ratio decreases from 0.9 $(n = 0.47)$ to $e = 0.46$ $(n = 0.31)$ (Fig. 4.4), resulting in a vertical strain of 23%. For the rock to compact due to this load, a pre-fluid volume equivalent to 23% of the total volume must be expelled (Fig. 4.4b).

4.2.4 Consolidation

I next consider uniaxial vertical consolidation, which is the intermediate time-dependent process that occurs between overpressure generation (undrained condition) due to an instantaneous load through complete overpressure dissipation

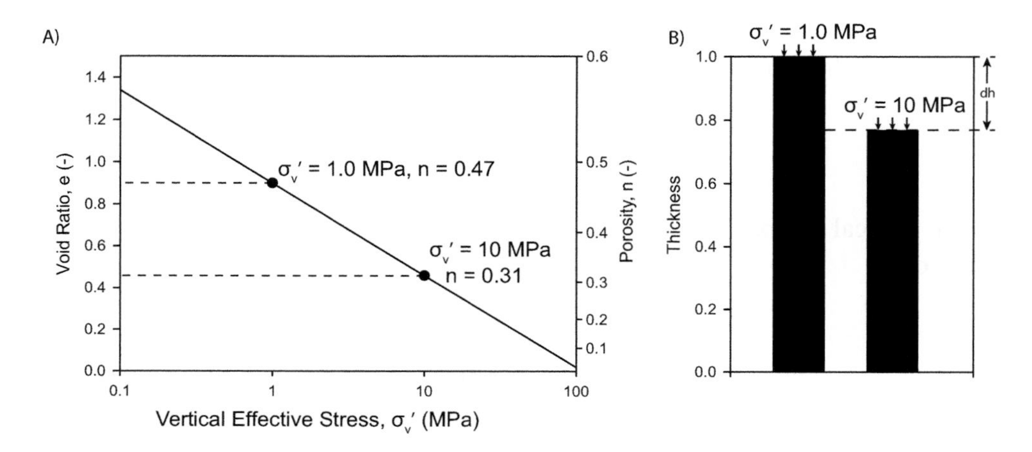

Figure 4.4 (a) Compression curve for Gulf of Mexico sediment (see Fig. 3.5). (b) Resultant compaction due to increase in surface load.

(drained condition) (Fig. 4.1f). In this process, overpressure causes pore fluid to flow vertically toward the surface and escape the medium. This is accompanied by the compression of the bulk structure. I present the equation of consolidation, which is derived by combining relationships that describe the flow of pore water through the porous medium and the compression of bulk structure. The following are assumed: (1) the sediment is homogenous; (2) the sediment is fully saturated; (3) the solid particles and the water are incompressible; (4) compression and flow are uniaxial and vertical; (5) strains are small; (6) pore water flow follows Darcy's law; (7) there is a unique relationship between void ratio and effective stress that does not change with time; and (8) the permeability, fluid viscosity, and compressibility do not change through the process. I closely follow the presentation of Craig (2004) below. Eloquent descriptions are also presented by de Marsily (1986) and Ingebritsen et al. (2006).

Consider a differential element of the consolidating material (Fig. 4.5). According to Darcy's law, the velocity (q) of pore fluid flowing vertically through any cross-section of the element is proportional to the overpressure vertical gradient at that section:

$$q = -\frac{k}{\mu}\frac{\partial u^*}{\partial z}. \qquad \text{Eq. 4.18}$$

μ is the fluid viscosity, k is the intrinsic permeability, and u^* is the overpressure. It must be noted that q is an apparent fluid velocity because it represents fluid velocity

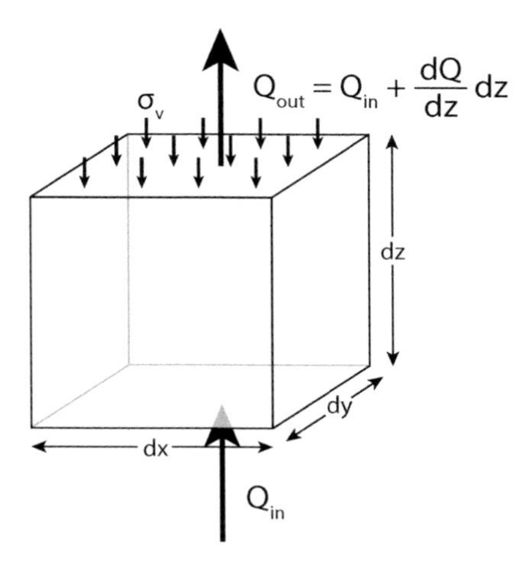

Figure 4.5 Differential element of sediment with uniaxial vertical flow (Q) and a constant vertical load (σ_v).

averaged over the total area of the cross-section, which includes solid structure. The actual pore fluid velocity is greater than q.

The net volume of fluid that escapes the differential element per unit time $\left(\frac{dV}{dt}\right)$ is the difference between the flux exiting the element at the top $\left(Q_{out} = q_{top}dxdy\right)$ and the flux entering the element at the bottom $(Q_{in} = q_{in}dxdy)$ (Fig. 4.5):

$$\frac{dV}{dt} = Q_{out} - Q_{in} = -\frac{d}{dz}\left[\left(\frac{k}{\mu}\right)\frac{\partial u^*}{\partial z}\right]dx\,dy\,dz. \qquad \text{Eq. 4.19}$$

The escape of pore fluid from the element allows for compression of the element, which increases effective stress by $d\sigma_v{}'$:

$$\frac{dV}{dt} = m_v \frac{d\sigma_v{}'}{dt} dx\,dy\,dz, \qquad \text{Eq. 4.20}$$

where m_v is the uniaxial compressibility of the bulk structure (Eq. 4.11).

Because the total load on the element is constant through the consolidation process $\left(d\sigma = d\sigma_v^{'} + du = 0\right)$, the increase in effective stress causes an equal decrease in overpressure:

$$d\sigma_v^{'} = -du^*. \qquad \text{Eq. 4.21}$$

Equations 4.20 and 4.21 are combined to express the volume change as a function of the change in overpressure:

$$\frac{dV}{dt} = -m_v \frac{du^*}{dt} dxdydz. \qquad \text{Eq. 4.22}$$

Equations 4.22 and 4.19 are combined to produce the consolidation equation:

$$\frac{\partial u^*}{\partial t} = c_v \frac{\partial^2 u^*}{\partial z^2}, \qquad \text{Eq. 4.23}$$

where c_v is the coefficient of consolidation:

$$c_v = \frac{k}{\mu m_v}. \qquad \text{Eq. 4.24}$$

Equation 4.23 has many simplifying assumptions. Nonetheless, this simple form provides extraordinary insight into the dissipation of overpressures over time.

4.2.5 Pore Pressure Dissipation

I now use the consolidation equation (Eq. 4.23) to quantitatively describe the process of consolidation that was qualitatively presented in Figure 4.1. Consider a water-saturated sediment column of thickness d that is loaded instantaneously at the surface with a unit load (Fig. 4.6a). Pore water flow is zero at the bottom boundary, and overpressure is zero at the surface. Because the load is applied instantaneously, pore water does not have time to drain (undrained condition), causing pore pressure to increase instantaneously (Eq. 4.9). Because the fluid and solid are assumed incompressible, the loading efficiency (C) is equal to 1, and the pore pressure increases to the value of the applied load (a value of 1.0).

If the compressibility (m_v) is constant and compression is small, the consolidation equation (Eq. 4.23) yields overpressure as a function of time and depth as,

$$u^* = \sum_{m=0}^{m=\infty} \frac{2u_i}{M} \left(sin \frac{Mz}{d} \right) \exp(-M^2 T_v), \qquad \text{Eq. 4.25}$$

where

$$M = \frac{\pi}{2}(2m+1), \qquad \text{Eq. 4.26}$$

and

$$T_v = \frac{c_v t}{d^2}. \qquad \text{Eq. 4.27}$$

The overpressure given by Equation 4.25 is illustrated in Figure 4.6b over depth at different times. Initially ($T_v = 0$), the overpressure is equal to 1.0 everywhere along the column (except the very surface). With time, as fluid is expelled, overpressure

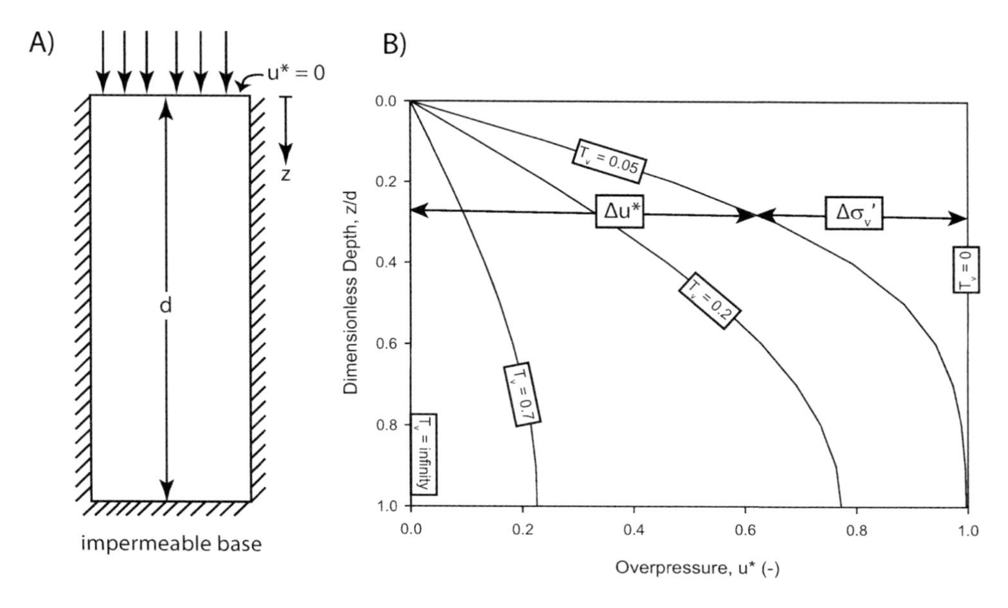

Figure 4.6 Overpressure dissipation during consolidation. (a) A sediment column of height d is loaded instantaneously by a unit stress at the top surface (arrows). The top surface is held open ($u^* = 0$). The base is impermeable. (b) The resultant overpressure profile over depth at different times. Overpressure decreases nonuniformly along the column from an initial value equal to the applied load ($T_v = 0$) to zero long after the load is applied ($T_v = \infty$). Modified from Craig (2004). Reproduced with permission of Taylor and Francis (Books) Limited UK through PLS Clear.

(u^*) dissipates and the column compacts particularly near the upper surface. The decrease in overpressure is equal to the increase in effective stress $\left(\Delta\sigma'\right)$. At infinite time ($T_v = \infty$), the overpressure is zero, and the effective stress equals the applied load.

The average degree of consolidation at any given time (U) is the average overpressure along the column:

$$U = 1 - \sum_{m=0}^{m=\infty} \frac{2}{M^2} e^{-M^2 T_v}. \qquad \text{Eq. 4.28}$$

As illustrated in Figure 4.7, when the time factor (T_v) equals 1.0, the consolidation is approximately 95% complete and the overpressure is approximately 95% dissipated.

The time factor (T_v), which determines the degree of consolidation, is inversely proportional to the square of the column thickness (d, Eq. 4.27). This thickness thus has a substantial control on the rate of consolidation. For example, the consolidation of a column of 1 000 m thickness is 10^4 longer than the consolidation of the

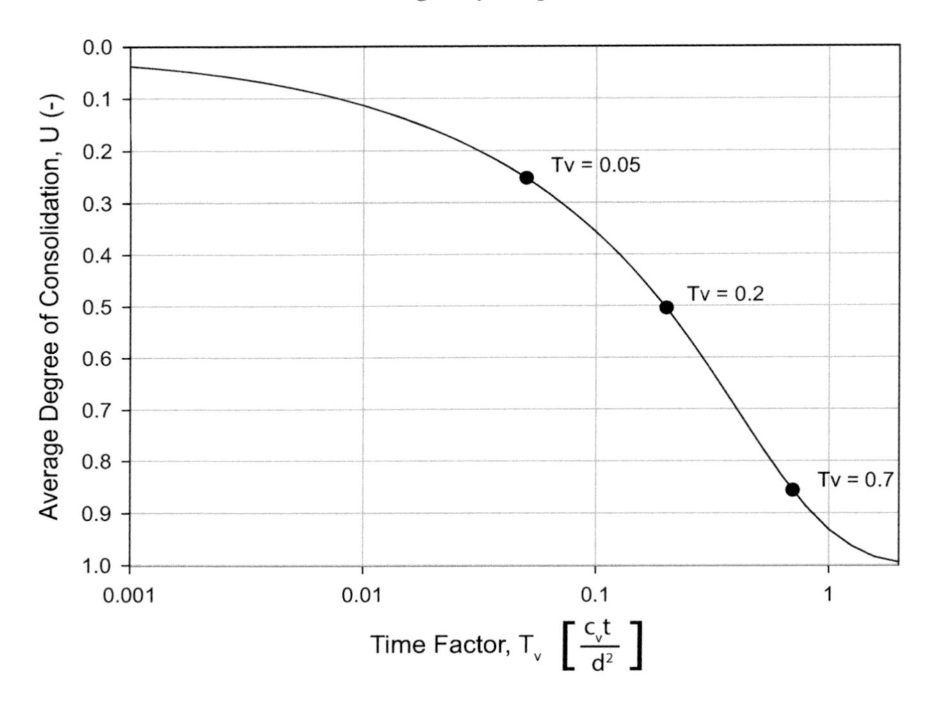

Figure 4.7 The average degree of consolidation (Eq. 4.28), which is equivalent to the average overpressure, during one-dimensional consolidation. Modified from Craig (2004). Reproduced with permission of Taylor and Francis (Books) Limited UK through PLS Clear.

same column with a thickness of 10 m. Figure 4.8 compares the times needed for 95% consolidation ($T_v = 1$) of different thicknesses of mudrock for typical values of c_v. While it takes years for the 10 m thick column to reach this degree of consolidation, it takes millions of years for the 1 000 m thick column to reach the same degree of consolidation. This is why thick columns of mudstone retain overpressure for very long timescales.

4.2.6 Hydromechanical Properties

Compressibility (m_v), permeability (k), and the resulting coefficient of consolidation (c_v) are the rock hydromechanical properties that control consolidation (Equation 4.23 and 4.24). In Chapter 3, I explored the compaction behavior of resedimented smectite-rich Gulf of Mexico mudrock (RGoM-EI) and illite-rich Boston Blue Clay (RBBC) (Fig. 3.2). Here, I explore the hydromechanical properties of these rocks (Fig. 4.9). The compressibility (m_v) is calculated from the compaction curve of the rocks (Eq. 4.14). m_v varies significantly with effective

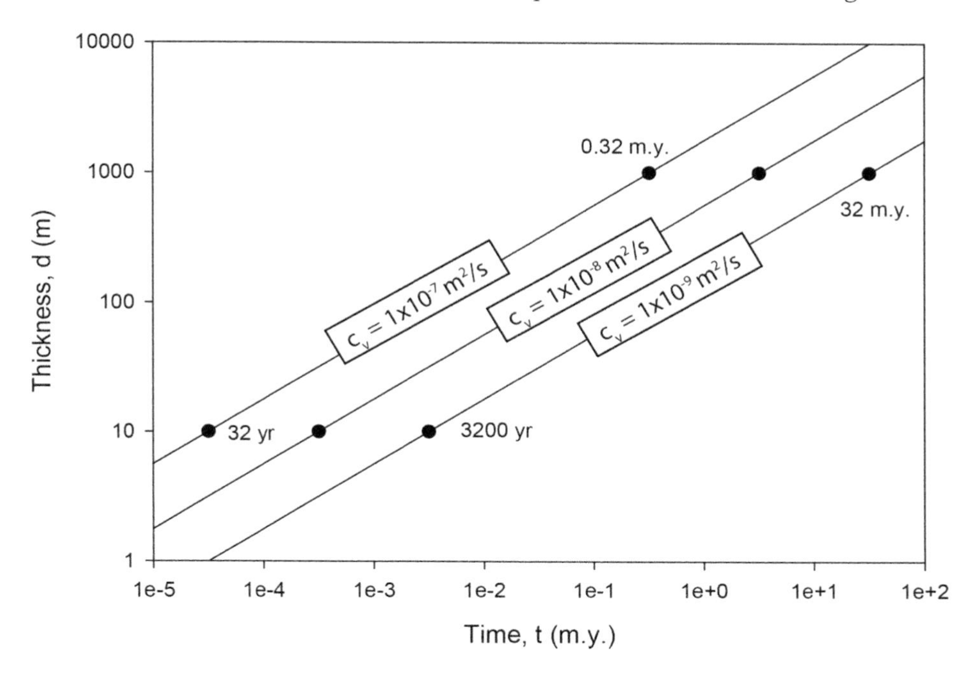

Figure 4.8 Time for 95% consolidation ($T_v = 1$) as a function of sediment column thickness.

stress; it is high at low stresses and declines orders of magnitude as stress increases to 40 MPa (Fig. 4.9a). Permeability (k) also varies significantly with stress. As stress increases, the rocks compact and their porosity decreases, causing their permeability to drop orders of magnitude (Fig. 4.9b). The coefficient of consolidation (c_v), which is obtained from the ratio of the permeability to the compressibility (Eq. 4.24), varies with stress (Fig. 4.9c). However, because both the permeability and compressibility decline with stress, the stress variation of the coefficient of consolidation is not large: c_v is almost constant for the illite-rich mudrock (RBBC) and declines approximately one order of magnitude for Gulf of Mexico mudrock (RGoM-EI) (Fig. 4.9c). The decline in c_v with stress for the Gulf of Mexico mudrock (Fig. 4.9c) derives from the large drop in the permeability of these rocks with stress (Fig. 4.9b).

The behavior of c_v with stress is important for understanding at what depths overpressure will be generated and where it will persist. c_v defines the timescale that it will take for dissipation for a given thickness of sediment (Fig. 4.8). A common misconception is that low permeability alone will determine whether overpressure is generated and preserved over large timescales. In fact, dissipation is driven by two factors: the permeability controls how rapidly the water can

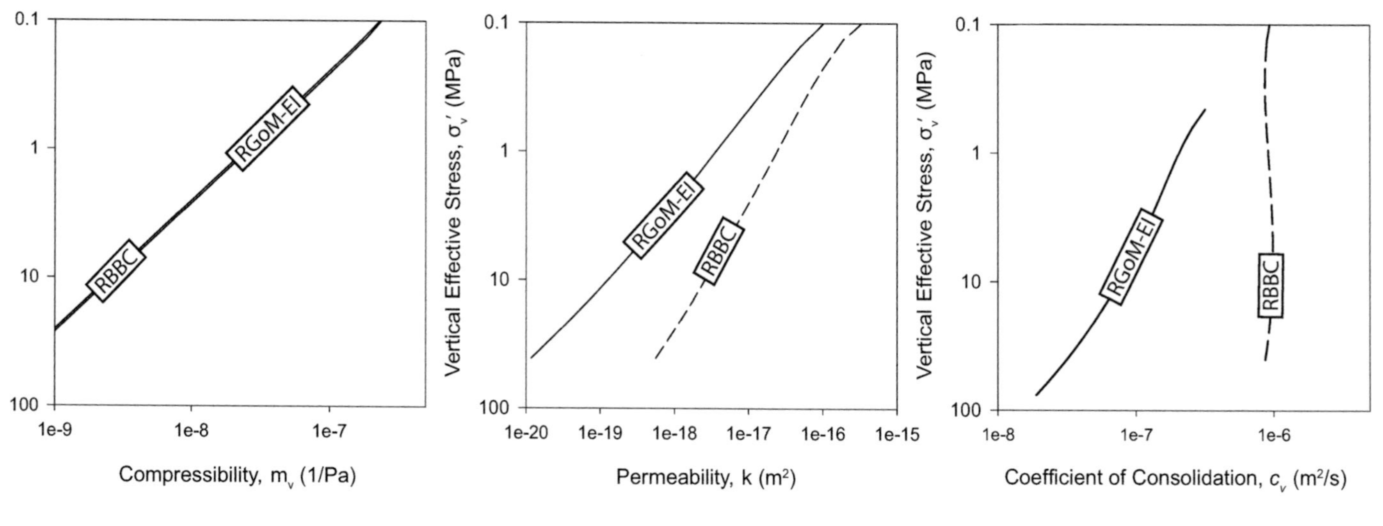

Figure 4.9 Variation in (a) compressibility, (b) permeability, and (c) coefficient of consolidation with effective stress for resedimented Gulf of Mexico Mudrock (RGoM-EI, solid line) and resedimented Boston Blue Clay (RBBC, dashed line). To calculate c_v (Eq. 4.24), a constant viscosity (μ) of $1 \times 10^{-3} PaS$ was assumed.

escape, and the compressibility controls how much water must escape. This analysis shows that shallow (low effective stress), poorly compacted mudrocks often have similar coefficients of consolidation as deeper, more compacted mudrocks.

4.3 Sedimentation and Overpressure

4.3.1 Overpressure During Sedimentation

In the previous section, I described the overpressure that develops from a surface load. However, overpressure in geologic settings develops from the weight of overburden being deposited. Gibson (1958) derived the governing equation for overpressure that develops from sedimentation over an impermeable base (Fig. 4.10):

$$\frac{\partial u^*}{\partial t} = c_v \frac{\partial^2 u^*}{\partial z^2} + \left(\rho_b - \rho_f\right) g \frac{dh}{dt}. \qquad \text{Eq. 4.29}$$

Equation 4.29 is the consolidation equation (Eq. 4.23) with an additional term for sediment loading. The terms on the right hand side of Equation 4.29 describe the sources for overpressure change. The first term is generally negative and represents the decrease in overpressure due to fluid escape. The second term is positive and represents the overpressure increase due to load added by sedimentation $\left(\frac{dh}{dt}\right)$. This increase is equal to the pressure increase under undrained conditions for a system with a loading efficiency (C) of 1.0 (Eq. 4.9). Gibson (1958) derived an analytical solution to Equation 4.29 for the case of a constant sedimentation rate and constant bulk density. This solution depends on a new dimensionless time factor, T:

$$T = \frac{m^2 t}{c_v}, \qquad \text{Eq. 4.30}$$

where m is the sedimentation rate. This solution is illustrated in nondimensional form in Figure 4.10b. The horizontal axis is the nondimensional overpressure, the overpressure divided by the 'reduced weight' of the entire sediment column (the overburden stress less the hydrostatic pressure, e.g., Fig. 2.1c). The vertical axis is the nondimensional elevation, the elevation of any given point along the column divided by the total thickness of the column. If there is no pressure dissipation, then the overpressure equals the reduced overburden stress (diagonal line, Fig. 4.10b) and the pressure equals the overburden stress.

The sedimentation rate is a key parameter in the time factor (T), which controls the overall degree of overpressure (Eq. 4.30). When a column of height (H) forms under a high sedimentation rate (m), pore water has little time to escape, thus, most of the overburden stress is balanced by overpressure (e.g., $T > 64$). In contrast,

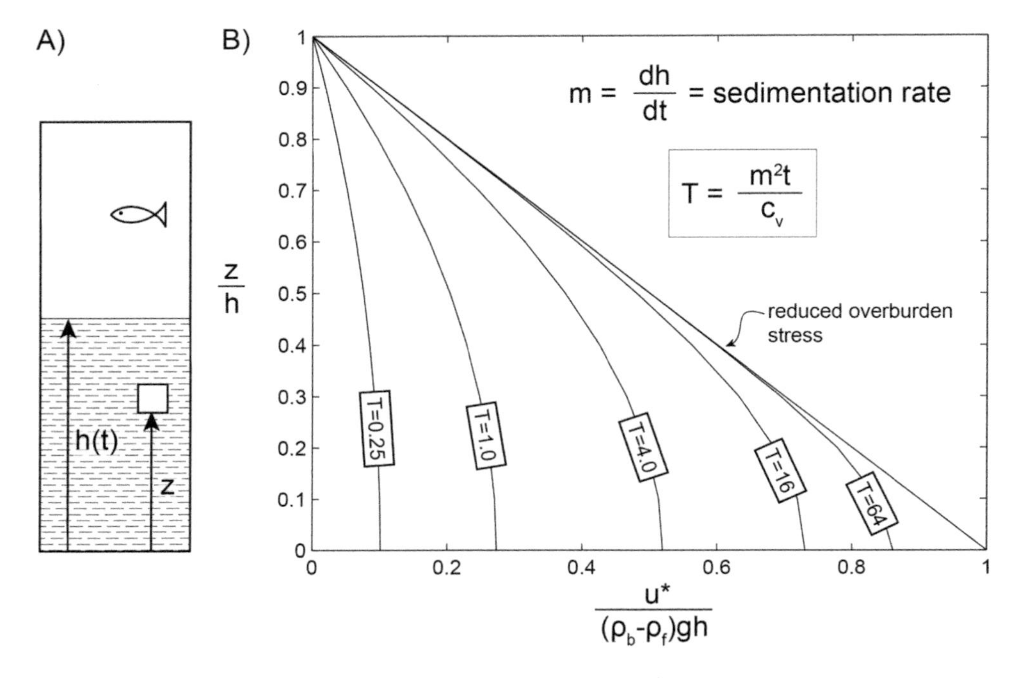

Figure 4.10 Overpressure in a column under sedimentation at a constant rate. (a) Schematics of the column, after Gibson (1958, p. 174, Fig. 1). (b) Overpressure along the column at different times/sedimentation rates, after Gibson (1958, p. 174, Fig. 3). Republished with permission of ICE Publishing, from Gibson (1958); permission conveyed through Copyright Clearance Center, Inc.

when the same column forms under a low sedimentation rate, pore water has time to escape, thus, little overpressure develops (e.g., $T<0.25$).

Flemings et al. (2008) measured pore pressures directly with a penetrometer in mudrocks at IODP Site U1324 in the Ursa Basin, Gulf of Mexico (Fig. 4.11). These measurements are one of the few examples of direct pore pressure measurements in mudrocks. They show that overpressure in the shallow sediments is high, approximately 70% of the reduced lithostatic pressure (λ^*~0.7). In contrast, overpressure in deeper sediments is only about 30% of the reduced overburden stress (λ^*~0.3). I map the measured overpressures to the plot used in Gibson's solution (Fig. 4.12); I divide the measured overpressures (symbols, Fig. 4.11f) by the reduced overburden stress at the base of the profile ($\sigma_v - u_h$ at 600 mbsf, Fig. 4.11f) to calculate the nondimensional pressures in the Gibson's plot (Fig. 4.12). A profile thickness of 600 m is assumed to calculate the normalized depths in this plot.

Ursa mudrocks form most of the shallow section of the basin (up to 300 mbsf, Fig. 4.11b) and have a coefficient of consolidation (c_v) of ~$2 \times 10^{-8} \text{m}^2/\text{s}$. Beneath these mudrocks, siltstones have c_v values greater than ~$2 \times 10^{-7} \text{m}^2/\text{s}$ (Long et al.,

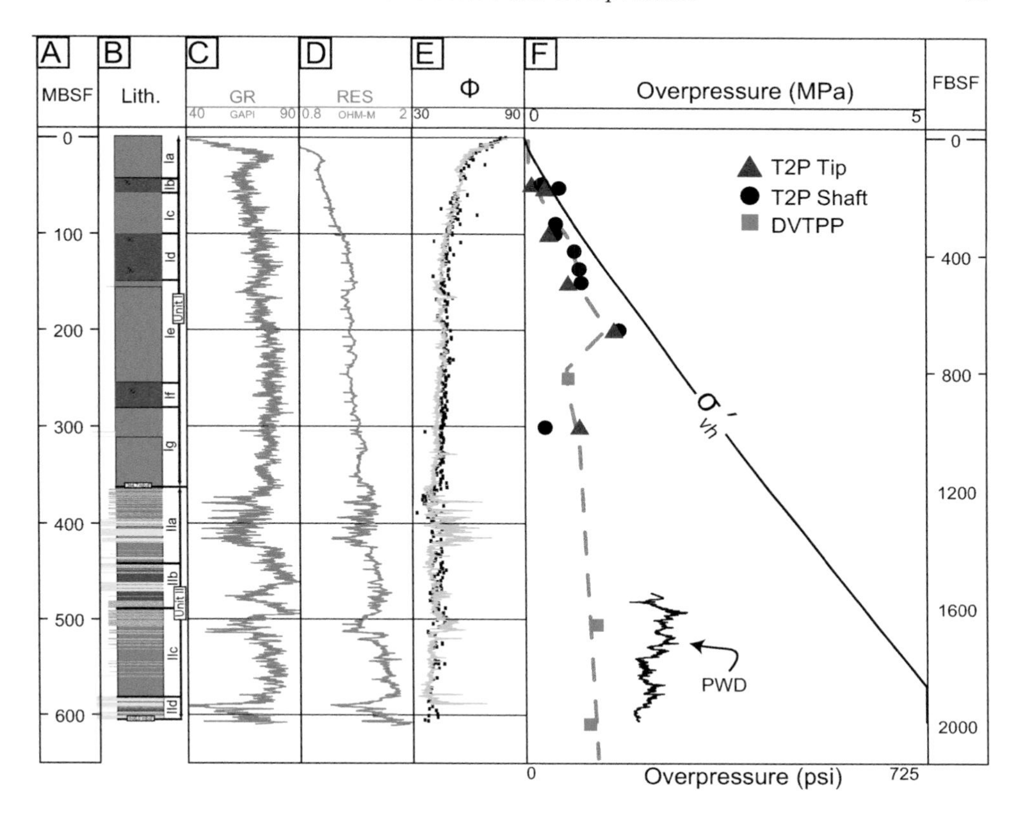

Figure 4.11 IODP Site U1324. (a) Depth in meters below seafloor (mbsf). (b) Lithology. Mass Transport Deposits (MTDs) are delineated in purple, siltstones in orange, sandstones in yellow, and mudstone in green. (c) Gamma Ray (GR) from logging while drilling (LWD) data. (d) Resistivity log (RES) log. (e) Porosity log interpreted from shipboard moisture and density (MAD) measurements (solid symbols). Porosity interpreted from logging while drilling (LWD) bulk density log. (f) Overpressure (u^*) determined from pore pressure penetrometer. Overpressure measured during drilling ('PWD') is shown with solid line. The grey dashed line is our best estimate of the in situ pressure. σ'_{vh} is the hydrostatic effective stress $\left(\sigma'_{vh} = \sigma_v - u_h\right)$. The overburden stress σ_v is calculated by integrating shipboard-derived core density measurements. Reprinted from Flemings et al. (2008) with permission of Elsevier.

2011). The average sedimentation rate (m) at this location is approximately 12 mm/year (Flemings et al., 2008), and thus the duration (t) of deposition for the 600 m thick section is 50 000 years. These values result in time factors $(T$, Eq. 4.30) of 11 for the mudrock and 1.1 for the siltstone. These pressures (symbols, Fig. 4.12) are compatible with those predicted by the Gibson's solution (Fig. 4.12). This example illustrates that it is quite possible to generate high pore pressures very near the seafloor. Such conditions occur when the coefficient of consolidation is low and/or the sedimentation rate is high.

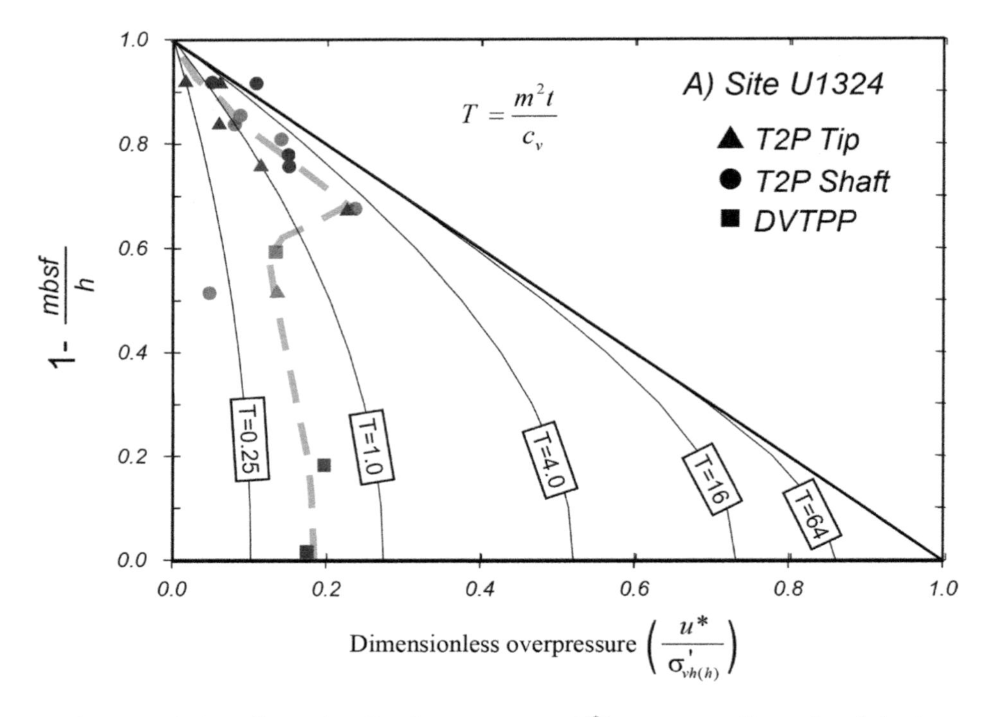

Figure 4.12 Nondimensionalized overpressure ($\frac{u^*}{\sigma'_{vh}}$) versus nondimensional depth ($1 - \frac{mbsf}{h}$). σ'_{vh} is the reduced overburden stress ($\sigma'_{vh(h)} = \sigma_v - u_h$) at the base of the profile ($h = 600\,mbsf$) at IODP Site U1324. Gibson (1958) one-dimensional solutions for overpressure during constant sedimentation rate with no flow at the base are shown for different time factors (T) (solid black lines). Symbols record direct pressure measurements plotted in nondimensional form. I divide the measured overpressures (symbols, Fig. 4.11f) by the reduced overburden stress at the base of the profile (600 mbsf, Fig. 4.11f) to calculate the nondimensional pressures. Grey dashed line records the interpreted in situ pressure profile (see grey dashed lines in Figure 4.11). The shallow pressure data are compatible with a time factor (T) of approximately 4, whereas the deeper data are compatible with T between 0.25 and 1.0. As discussed in text, T equals 11 for mudrocks and 1.0 for siltstones in this location. Reprinted from Flemings et al. (2008) with permission of Elsevier.

4.3.2 Numerical Uniaxial Basin Models

Analytical models such as those provided by Gibson (1958) provide extraordinary insight, but they apply only to specific conditions. Numerical approaches are necessary to describe more complicated geological systems. These models can consider large strain and nonuniform rock and fluid properties. They can also include two- and three-dimensional pore water flow, and the evolution of stratigraphy and structure. Hantschel and Kauerauf (2009) provide a detailed discussion of these approaches. The Basin2 model (e.g., Bethke et al., 1988) and PetroMod® (e.g., Burwicz et al., 2017) are two of the many codes that are available.

Most of these models assume uniaxial strain and the equations that are solved are quite similar to the sedimentation equation (Eq. 4.29), but often include additional source terms. For example, Gordon and Flemings (1998) described the one-dimensional evolution of pressure during sedimentation with a numerical solution to the following equation:

$$\frac{Du}{Dt} = \frac{1}{S_t}\left[\frac{k}{\mu}\right]\frac{\partial^2 u^*}{\partial z^2} + \frac{1}{S_t}\left[\frac{n}{(1-n)}\right]\beta\frac{D\sigma_v}{Dt} + \frac{\alpha_m}{S_t}\frac{DT}{Dt}, \qquad \text{Eq. 4.31}$$

where

$$S_t = \frac{n}{(1-n)}\beta + nc_f - nc_s, \qquad \text{Eq. 4.32}$$

and,

$$\alpha_m = n\alpha_f - \frac{n}{(1-n)}\alpha_b - n\alpha_s. \qquad \text{Eq. 4.33}$$

Gordon and Flemings (1998) developed Equation 4.31 to describe the one-dimensional evolution of pressure due to sedimentation, illite-smectite diagenesis, and thermal expansion. Equation 4.31 assumes that porosity is an exponential function of effective stress through the term β (Eq. 5.1). Equation 4.31 is developed in substantial coordinates (e.g., $\frac{Du}{Dt}$), where the differential element contains a constant solid mass. As a result, large strain is considered. Given the large strains that occur during mudrock compaction, this is a considerable advance over the consolidation equation (Eq. 4.29). Gordon and Flemings (1998) assumed that permeability and porosity vary as a function of effective stress and that viscosity decreases with temperature in their solutions to Equation 4.31.

4.3.2.1 Overpressure During Constant Sedimentation

I begin by simulating a constant rate of sedimentation using the permeability and compressibility behavior measured for resedimented Gulf of Mexico mudrock (RGoM-EI) and resedimented Boston Blue Clay (RBBC) (Fig. 4.9). I solve numerically Equation 4.31 and consider only the effect of sediment loading. I assume a sedimentation rate of 2 mm/year for 5 million years, no overpressure at the top, and no flow at the base. The RGoM-EI mudrock generates a pore pressure profile where overpressure begins within a few hundred meters of the seafloor and then increases at a rate sub-parallel to the overburden stress (Fig. 4.13a). With increasing time, the pore pressure shifts toward the overburden stress as predicted by the Gibson approach (Fig. 4.10). For example, after 1.3 million years of deposition (t_1, Figure 4.13a), the pore pressure at the base lies halfway between the hydrostatic

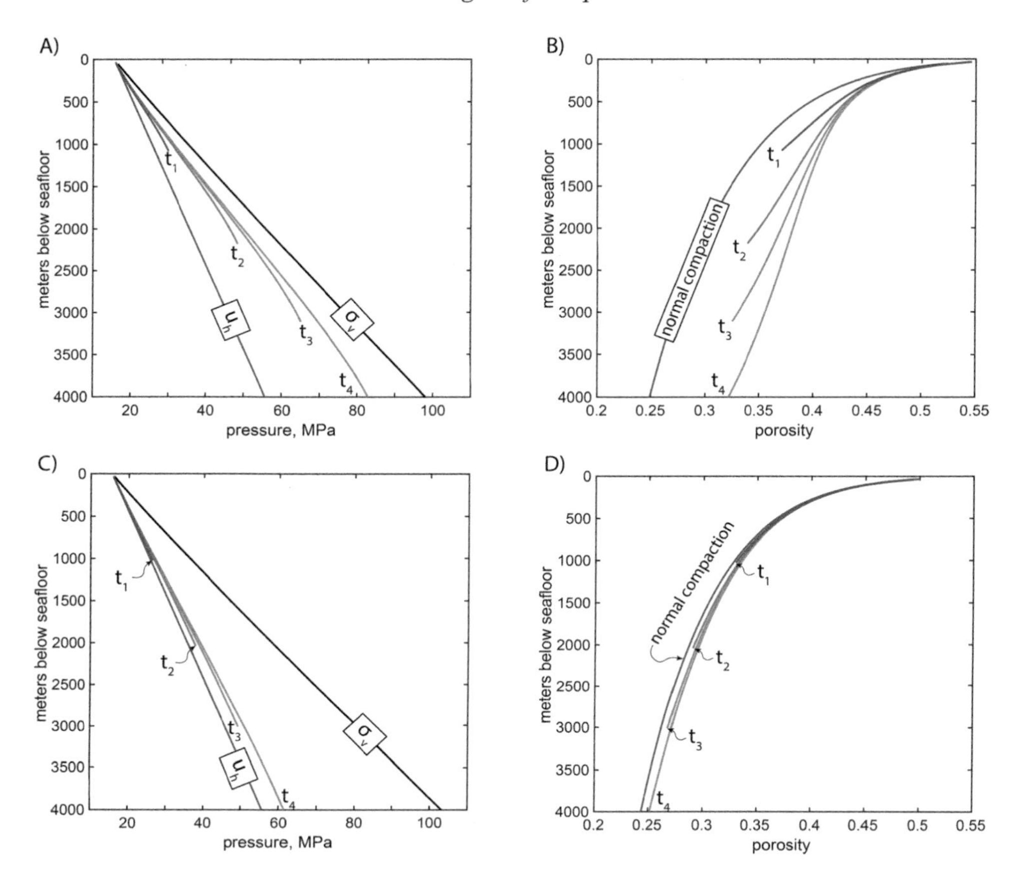

Figure 4.13 Pore pressure and porosity evolution during constant sedimentation rate of 2 mm/year. (a) Pressure versus depth for RGoM-EI mudrock illustrated in Figure 4.9. (b) Porosity versus depth for RGoM-EI mudrock. (c) Pressure versus depth for RBBC mudrock. (d) porosity versus depth for RBBC mudrock. Four timesteps of the simulation are shown. t_1 = 1.3 million years, t_2 =2.8 million years, t_3 = 4.3 million years, t_4 = 5.0 million years.

pressure and the overburden stress whereas after 4.3 million years (t_4, Fig. 4.13a) of deposition, the pore pressure lies two-thirds of the way between the hydrostatic pressure and the lithostatic stress. The simulated pressure for the RBBC mudrock is much lower (Fig. 4.13c and Fig 4.13d). This is not surprising because the coefficient of consolidation is much lower for this material (Fig. 4.9).

The pressure profile after 4 km of deposition has broad similarity to the overpressure field observed at the Macondo well (Fig. 1.3 and 6.12). In the very shallow section (<200 mbsf), the porosity follows the normal compaction trend (the porosity present at hydrostatic pressure), and in this zone the pore pressure is hydrostatic

(Fig. 4.13). However, beneath this, the porosity is less than that predicted by the normal compaction trend, because the overpressure is bearing some of the applied load (Fig. 4.13b).

4.3.2.2 Pressure Retention Depth and Unloading

In Figure 4.13a, the pore pressure is hydrostatic for approximately 200 m and then the pore pressure is subparallel to the lithostatic stress. A common conceptual view is that this type of pore pressure profile is caused by initial burial in a drained fashion followed by continued burial in an undrained fashion. We can envision that this might be due to either the increasing burial depth, which increases the drainage length, or a decrease in the coefficient of consolidation with effective stress (e.g., Fig. 4.9c). Either of these would result in a higher value for the time factor, T (Eq. 4.30), at deeper depths and hence greater overpressure (Fig. 4.10).

I illustrate this process in a cartoon fashion (Fig. 4.14). The mudrock is buried to a certain depth under hydrostatic conditions (t_2). At this point, the system becomes undrained. Any further load applied by sedimentation is borne by the pore fluid pressure for the case where the loading coefficient (C) is equal to 1.0 (Eq. 4.10), and the pore pressure profile then parallels the overburden stress (dash-dot line, Fig. 4.14a). In this process, the rock compacts to t_2 and then the porosity stays constant thereafter (t_{3a}) (Fig. 4.14b). Thus, if sedimentation is the only source of overpressure and there is no drainage, the pore pressure will increase at the overburden gradient and further burial occurs at a constant effective stress (Fig. 4.14).

In contrast, if there is an additional source of overpressure (e.g., thermal expansion, or smectite-illite diagenesis), then, under undrained conditions, pore pressure will rise more rapidly with depth than the overburden stress ($t_2 - t_{3b}$, Fig. 4.14). In this case, the vertical effective stress decreases with further burial. This decrease in effective stress is termed unloading. For most mudrocks, when the effective stress decreases, the porosity deviates from the normal compaction trend (Fig. 3.5, Fig. 4.14b, $t_2 - t_{3b}$). This shows that to generate unloading during burial, there must be a pressure source other than sediment loading. Furthermore, this pressure source must exceed any pressure lost due to fluid flow. Osborne and Swarbrick (1997) discuss this behavior further, and I describe how to account for this unloading in pressure prediction in Chapter 6.

4.3.2.3 The Relative Role of Sedimentation, Thermal Expansion, and Smectite-Illite Diagenesis in Overpressure Generation

Thermal expansion and smectite-illite diagenesis are commonly invoked as possible sources of overpressure (Osborne & Swarbrick, 1997). As the sediment

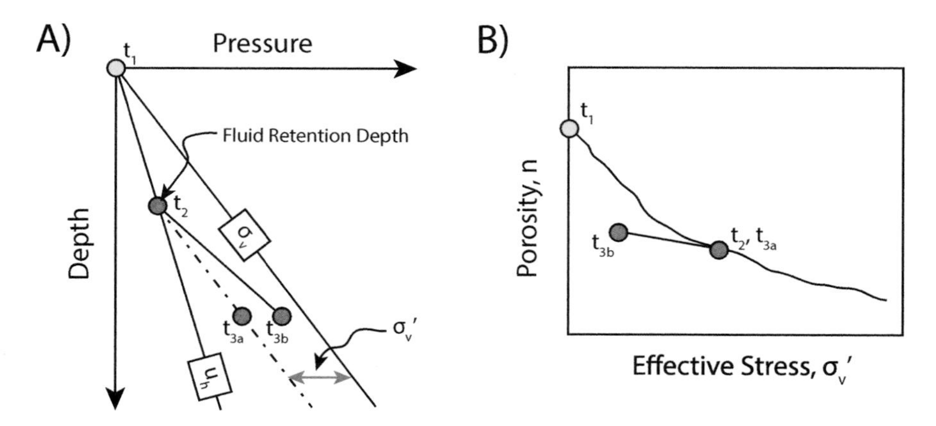

Figure 4.14 Evolution of pressure and porosity for a sediment volume buried first under drained conditions (t_1-t_2), and then undrained conditions (t_2-t_3). (a) Pressure profile. If sedimentation is the only source of pressure, then under undrained conditions, the pore pressure will parallel the overburden stress (t_2-t_{3a}). If there is an additional source of overpressure, then the pore pressure can rise toward the lithostatic stress with further burial (t_2-t_{3b}). (b) Porosity evolution. If no unloading occurs, the porosity is constant during burial under undrained conditions (t_2-t_{3a}). If unloading occurs due to an extra pressure source, then effective stress will decrease and porosity will also decrease, but not along the same trend as occurred during burial (t_2-t_{3b}).

volume is buried, it is heated and as a result it will expand under undrained conditions, resulting in overpressure generation (Eq. 4.31). In addition, as heating occurs, any smectite present may convert to illite driving off water in the process. This is sometimes called clay dehydration. The water that is expelled is commonly treated as a fluid source resulting in overpressure (Gordon & Flemings, 1998).

Gordon and Flemings (1998) explored the role of these processes in overpressure generation for a column under sedimentation at a constant rate (Fig. 4.15). The simulation is broadly similar to the example in Fig. 4.13. Overpressure begins soon after burial and rises with depth. Figure 4.16 illustrates the relative contributions of sedimentation, heating, and smectite-illite diagenesis to the overpressure at the base of the section during the burial. The amount of pressure generated by sedimentation (solid line, Fig. 4.16) is orders-of-magnitude larger than that generated by either smectite-illite diagenesis (dash-dot) or thermal expansion (dashed line, Fig. 4.16). This is a common result for basins undergoing active deposition, and it has been demonstrated by various authors (Harrison & Summa, 1991; Osborne & Swarbrick, 1997). Osborne and Swarbrick (1997) discuss the possible role of other overpressure sources such as biogenic or thermogenic processes. They emphasize that

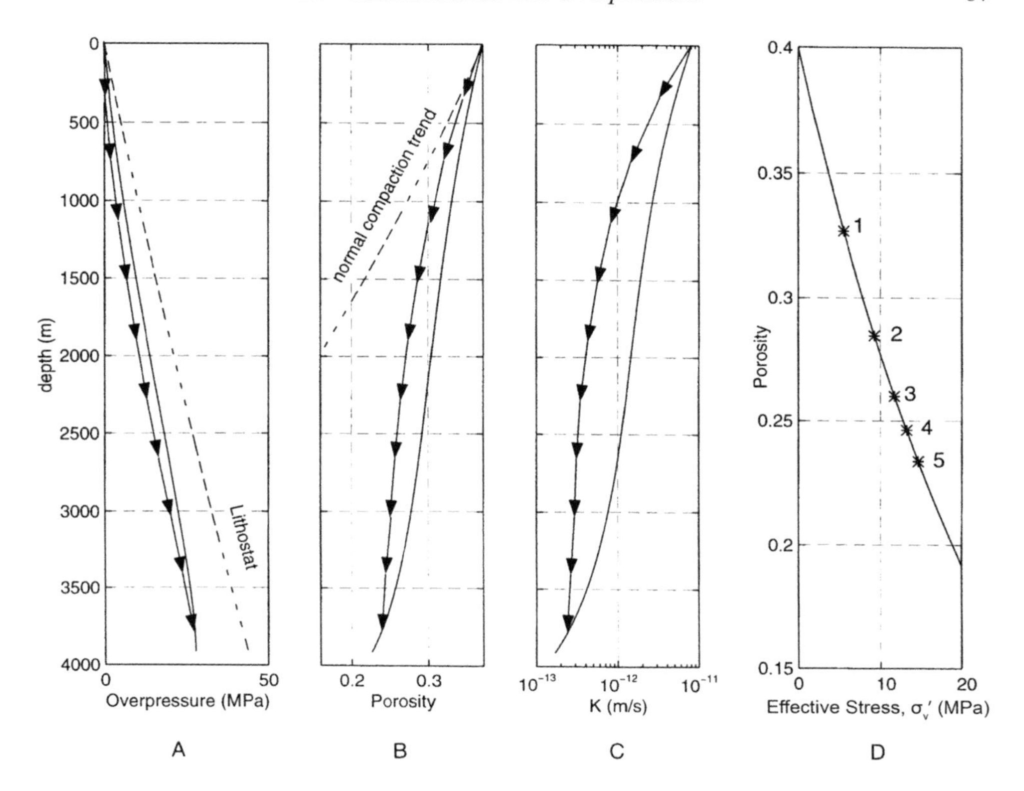

Figure 4.15 One-dimensional pressure evolution for a constant sedimentation rate of 1mm/year for a total thickness of 3 750 m. The line without arrows represents results along the column after it reaches the maximum thickness of 3 750 m. The line with arrows represents results at the base of the column as it thickens. (a) Overpressure, (b) porosity, (c) hydraulic conductivity ($K = \frac{k\rho_f g}{\mu}$), (d) vertical effective stress. Reprinted from Gordon and Flemings (1998) with permission by John Wiley and Sons.

the relative role of these sources is hard to estimate because it is hard to estimate the total volume change that these transformations cause.

4.3.2.4 Stratigraphic Evolution and Pore Pressure Profiles

Basin models are an excellent way to visualize how pore pressures develop in basins with complex stratigraphy. As a simple example, I illustrate the overpressure evolution that occurs when permeable rocks (sandstone) bury low permeability rocks (mudrocks) (Fig. 4.17). The mudrock deposited early in the simulation (delineated with a '*', Fig. 4.17a) is overpressured from the start of deposition until it reaches its final depth at the end of the simulation (follow arrows, Fig. 4.17b). Throughout this burial, the effective stress in the mudrock continues

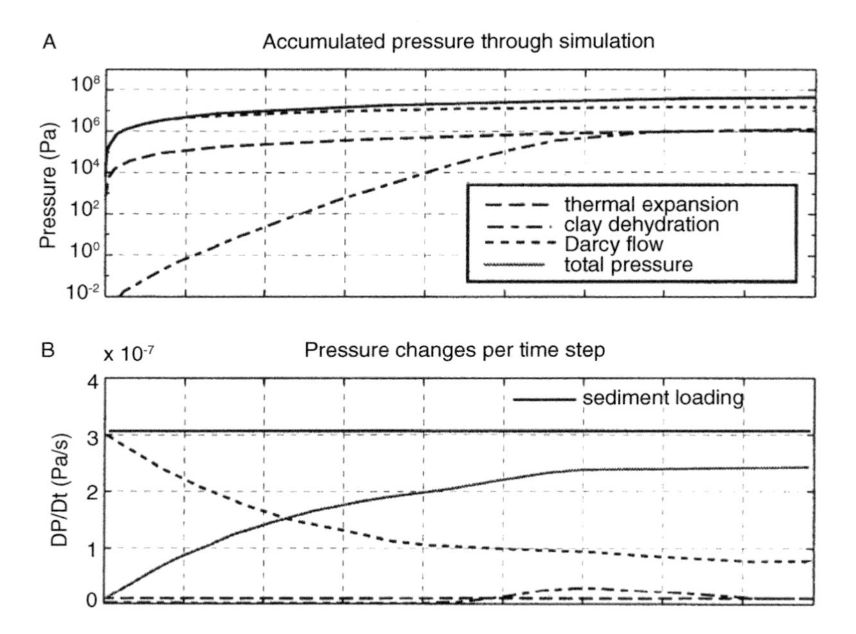

Figure 4.16 Overpressures produced by sedimentation, clay dehydration (smectite-illite diagenesis), thermal expansion, and fluid dissipation at the base of the column during sedimentation. The distance between each vertical grid line is 0.5 million years. (a) Generated overpressure. (b) Rate of overpressure generation. The rate of thermal pressuring is constant (line is horizontal) because it is proportional to the rate of temperature increase, which increases approximately at a constant rate with depth. The constant sedimentation rate also causes a constant rate of pressuring from sedimentation loading (line is horizontal). Reprinted from Gordon and Flemings (1998) with permission by John Wiley and Sons.

to increase and it further compacts (Figure 4.17c). No overpressure is generated in the overlying sand-rich strata (Fig. 4.17b). As a result, the final pore pressure profile is hydrostatic until the mudrock is reached, and thereafter there is an abrupt increase in overpressure (solid line, Fig. 4.17b). In this final profile, the effective stress $(\sigma_v^{'})$ first increases with depth in the hydrostatic section and then decreases dramatically (Fig. 4.17b).

Figure 4.17 illustrates that deposition of low permeability sediments followed by deposition of high permeability sediments generates a pore pressure profile that is hydrostatic in the shallow section and overpressured in the deep section. It is a different mechanism from what I showed in Figure 4.14, where a similar profile was generated by burying a single homogenous sediment first in a drained fashion and then in an undrained fashion. The distinction is important because in the first example (Fig. 4.14), the effective stress history of the deep overpressured mudrock

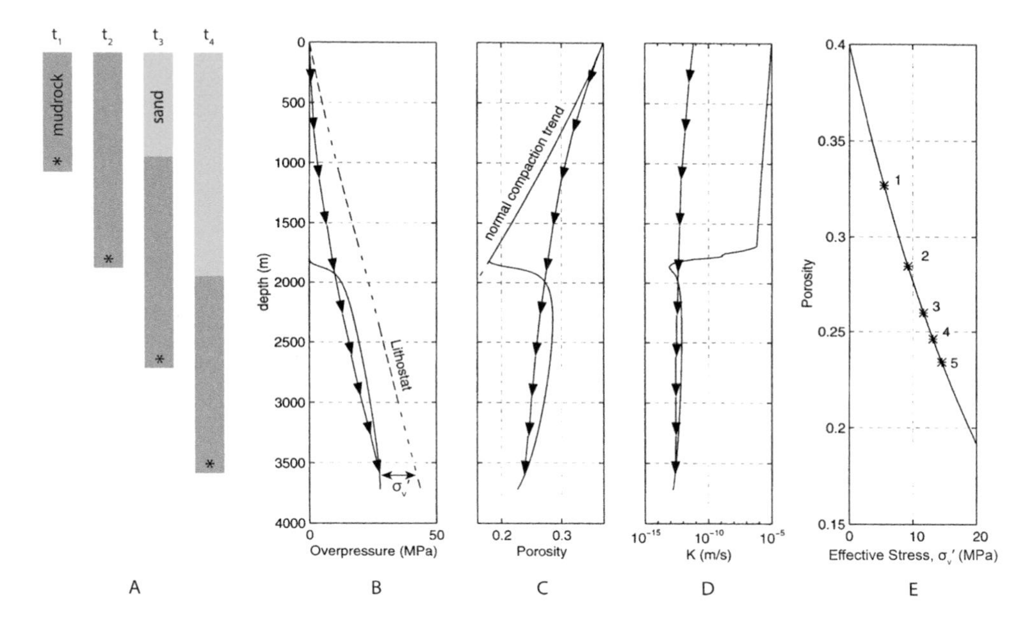

Figure 4.17 Simulation of pressure that results by sedimentation of first mudrock and then sandstone. (a) Sketch of burial history. Mudrock is deposited for 2.2 million years and this is followed by deposition of sandstone for 2.2 million years. A deposition rate of 1mm/year is applied. (b) Overpressure. The line without arrows records the pressure profile at the end of the simulation. The line with arrows records pressure path of mudrock deposited early in the simulation as it is buried. (c) Porosity. (c) Hydraulic conductivity ($K = \frac{k\rho_f g}{\mu}$), (d) Vertical effective stress. Reprinted from Gordon and Flemings (1998) with permission by John Wiley and Sons.

is similar to the variation in effective stress with increasing depth. In contrast, in the second example (Fig. 4.17), the effective stress history of the deep mudrock has nothing to do with the present-day pressure profile. The significant difference between the final effective stress profile and the effective stress history of the deep mudrock (Fig. 4.17) shows that it is a mistake to use the present-day effective stress profile to infer the effective stress history for a particular rock within it. Nonetheless, this is often done. I discuss in Chapter 6 how application of this assumption overestimates the amount of unloading that can occur.

The burial of low permeability mudrock by high permeability sandstone is a common geological feature. For example, the shelf and upper continental slope of the northern Gulf of Mexico record progradation of sand-rich sediment deposited in a shallow depositional environment over mud-rich sediment deposited on the continental slope (Fig. 4.18a). Harrison and Summa (1991) used a basin model to

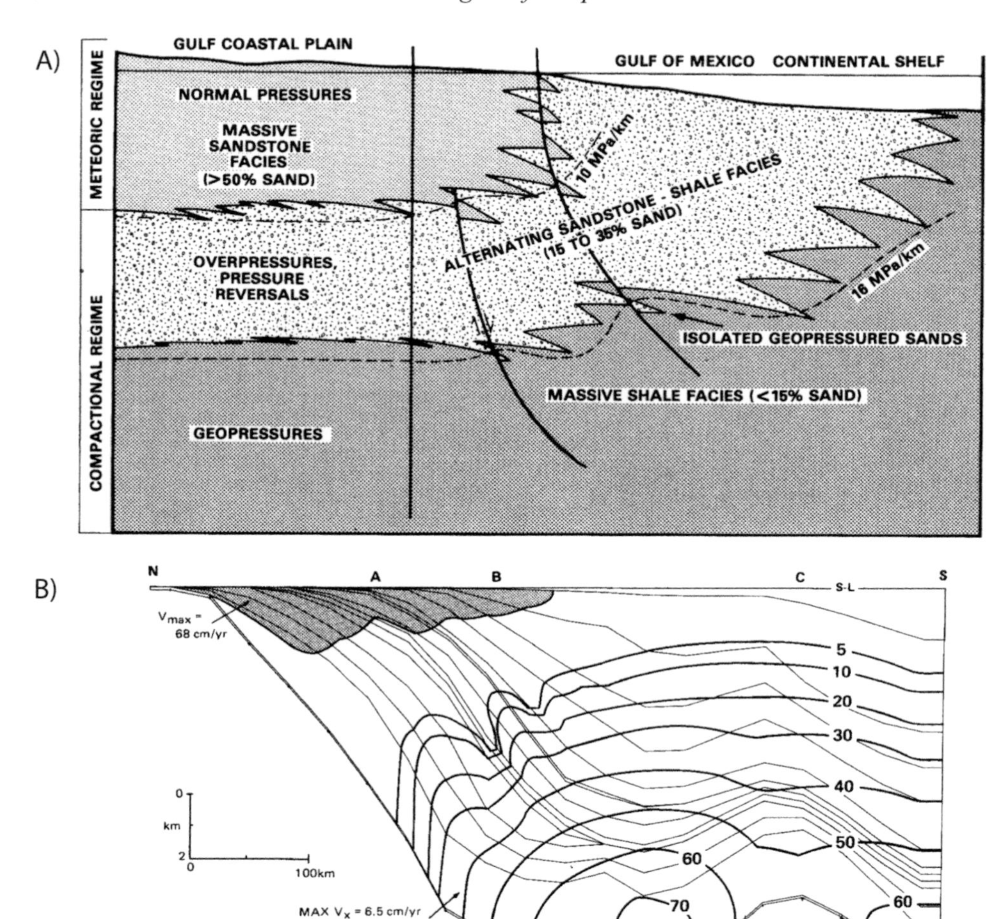

Figure 4.18 (a) Generalized lithofacies distribution and interpreted controls on geopressure for Gulf of Mexico continental margin. (b) Simulated overpressure in MPa. The shaded regime delineates the meteoric zone where water is flowing toward the basin from land. The white region is the compactional regime where flow is directed upward. The model illustrates the shallowing of the overpressure zone in the offshore where rapid deposition of low permeability material is occurring. Locations A, B, and C are shown in Figure 4.19. Republished with permission of the American Journal of Science, from Harrison and Summa (1991); permission conveyed through Copyright Clearance Center, Inc.

simulate the two-dimensional evolution of this margin (Fig. 4.18b). They simulated hydrostatic pressures in the landward zone where only permeable sediments were deposited (left) (Fig. 4.18b, 4.19, location 'A'). In contrast, on the present day upper

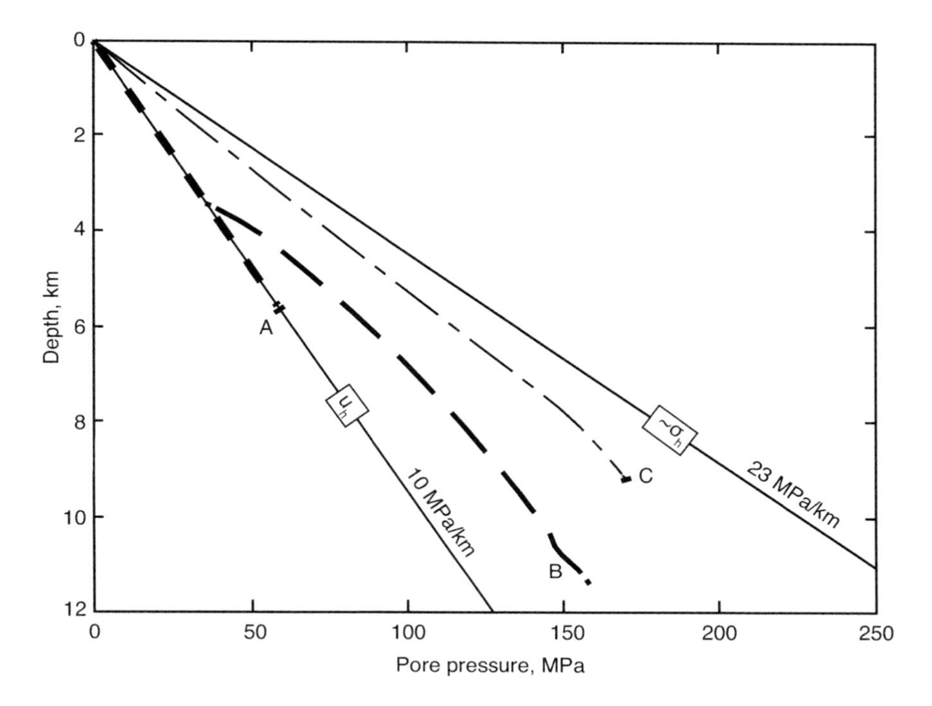

Figure 4.19 Simulated pore pressure profiles for three locations across the Gulf of Mexico continental shelf (located in Figure 4.18b). (a) represents an onshore well that only penetrates normally pressured strata. (b) represents a well on the continental shelf. Finally, (c) represents the outboard location where mudrock is present from the seafloor. Republished with permission of the American Journal of Science, from Harrison and Summa (1991); permission conveyed through Copyright Clearance Center, Inc.

continental shelf, a zone of hydrostatic pressure overlies an abrupt pressure increase (Fig. 4.18b, 4.19, location 'B'). Finally, in the deepwater, where mud deposition dominates, overpressure is present from the seafloor downwards (Fig. 4.18b, 4.19, location 'C'). Harrison and Summa (1991) compared their model results to field observation and found strong similarity.

This example demonstrates that the stratigraphic evolution is perhaps the most critical control on the development of overpressure. To understand the present-day distribution of overpressure, it is necessary to understand the stratigraphic evolution, which includes understanding the rates of formation, the geometries, and the hydrodynamic properties of the stratigraphic layers. The progradation of permeable strata over low permeability strata is one example of a characteristic stratigraphic succession that results in a characteristic pore pressure response. However, there are many other possible stratigraphic scenarios. For example, in the deepwater Gulf of Mexico, it is common to find permeable sands that are themselves buried by low permeability mudrocks. In this environment, it is common to find lower overpressures overlain by higher overpressures. This is discussed more in Chapters 10 and 11.

4.4 Pore Pressure in Complex Stress States

4.4.1 Hydromechanical Models

Up until this point, I have considered how pore pressure is generated and dissipated under conditions of vertical uniaxial strain. Under these conditions, the loading term can be described by the change in overburden stress due to erosion or deposition. Of course, under uniaxial loading, the other principal stresses also change. However, they change in proportion to the overburden stress: a change in vertical effective stress results in a proportional change in horizontal effective stress through the stress ratio K_0 (Eq. 3.3). However, as discussed in Chapter 3, many geologic systems undergo more complex strain histories (e.g., Fig. 3.14). In these cases, principal stresses are not necessarily vertical and there is no simple scaling between the principal stresses such as the K_0 relationship. Thrust belts or regions adjacent to salt bodies are obvious locations where we might expect complex stress fields. In fact, there are probably very few geological environments that undergo purely vertical uniaxial strain.

A new generation of coupled hydromechanical models has been developed to address the challenge of simulating overpressure under complex stress fields. These models incorporate the effect of the full stress tensor on compaction and pore pressure response. Examples of such modeling approaches include work by Morency et al. (2007), who explored the evolution of pressure and stress in an evolving delta, and Albertz et al. (2011), Gradmann et al. (2012), and Gradmann and Beaumont (2012), who explored deformation and overpressure along passive margins. In thrust belts, Ellis et al. (2019), Obradors-Prats et al. (2017), and Gao (2018) have used hydromechanical models to understand pore pressure development.

4.4.2 The Role of Mean- and Shear-Induced Pore Pressure

An important component of these models is how they include the pore pressure response to changes in both mean stress and shear stress. As illustrated in Figure 3.9, the pore pressure generated by loading can be decomposed into a shear-induced component and a mean stress-induced component. The amount of shear-induced pore pressure is a strong function of the initial stress state. For example, at isotropic stresses, the effective stress path is nearly horizontal and there is little shear-induced pressure (Fig. 3.9). However, at stresses near the critical state (near the failure line), the effective stress path is nearly horizontal and the shear-induced pressure is large.

M. Nikolinakou (personal communication, March 2020), incorporated both shear and mean stress to describe pore pressure response:

$$\frac{du}{dt} = \underbrace{\frac{\xi_m}{S}\frac{d\sigma_m}{dt}}_{du^m/dt} + \underbrace{\frac{\xi_q}{S}\frac{dq}{dt}}_{du^q/dt} + \underbrace{\frac{1}{S}\frac{k}{\mu}\nabla^2 u^*}_{du_e^{diss}/dt}, \qquad \text{Eq. 4.34}$$

where

$$S = c_f n + \xi_m. \qquad \text{Eq. 4.35}$$

ξ_m is the volume compressibility for a change in mean stress (σ_m) and ξ_q is the volume compressibility for a change in shear stress (q), where the mean total stress is:

$$\sigma_m = \frac{\sigma_1 + \sigma_2 + \sigma_3}{3}, \qquad \text{Eq. 4.36}$$

and the differential (shear) stress is:

$$\text{q} = \sqrt{\frac{(\sigma_1 - \sigma_2)^2 + (\sigma_2 - \sigma_3)^2 + (\sigma_3 - \sigma_1)^2}{2}}. \qquad \text{Eq. 4.37}$$

The loading terms in Equation 4.34 are expressed as the pore pressure induced by a change in mean stress (du^m) and the pore pressure induced by a change in differential (shear) stress (du^q). Equation 4.34 is similar to that presented by Neuzil (1995) to describe the generation and maintenance of anomalous fluid pressure due to a range of geological driving mechanisms. However, unlike the formulation of Neuzil (1995), Equation 4.34 includes the effect of shear stress (q) on pore pressure. The ξ_m/S ratio is approximately equal to 1 for compressible saturated sediments and relatively insensitive to the level of shear; hence mean-induced pressures are essentially equal to the mean-stress change. On the other hand, for a critical state model such as modified Cam clay, ξ_q/S is highly dependent on the initial stress state: shear-induced overpressures are minimized at or near the uniform loading state $\left(\frac{\xi_q}{S} = 0\right)$, and increase nonlinearly as the stress state approaches critical state (e.g., Fig. 3.9). ξ_q/S can reach values up to 10 for typical modified Cam clay inputs.

While the effect of both mean and shear stress on pore pressure generation is included in the modified Cam clay, it is not always considered. For example, the work of Morency et al. (2007), Albertz et al. (2011), Gradmann et al. (2012), and

Gradmann and Beaumont (2012) all assume that only mean stress controls compaction and thus pore pressure response.

4.4.3 Full Effective Stress Hydromechanical Model Examples

4.4.3.1 Pore Pressure and Stress around an Evolving Salt Diapir

I first describe the pore pressure and stress in an evolving salt diapir with a transient evolutionary model that was built with Elfen software (Rockfield, 2017). The model is described in detail in Nikolinakou et al. (2018). The approach is based on a finite-strain, quasi-static, finite-element formulation (Perić & Crook, 2004). Lagrangian and Eulerian reference frames are used for the mechanical and fluid phases, respectively. The salt is modeled as a solid viscoplastic material using a reduced form of the Munson and Dawson formulation (Munson & Dawson, 1979). Basin mudrocks are assumed homogenous and isotropic and broadly similar to the resedimented Gulf of Mexico mudrock (RGoM-EI) presented in Chapter 3. They are modeled as porous elastoplastic using the SR3 constitutive model (Crook et al., 2006). SR3 is broadly similar to the modified Cam clay model presented in Chapter 3. The model is initiated with a flat salt layer (Fig. 4.20). Sediment

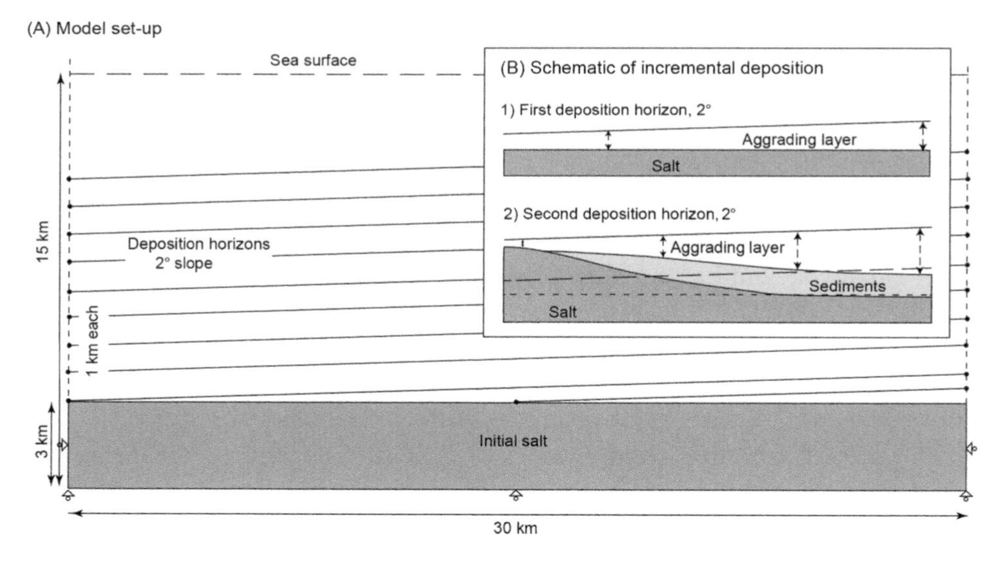

Figure 4.20 (a) Model set-up for simulation of a rising salt wall. Initially, a flat salt layer 3 km thick and 30 km wide is present. Sediment aggrades onto this salt with a 2 degree slope. There is no displacement and no friction at sides or base. (b) Sediment aggrades vertically such that the upper surface maintains a 2 degree slope. Reprinted from Nikolinakou et al. (2018) with permission of Elsevier.

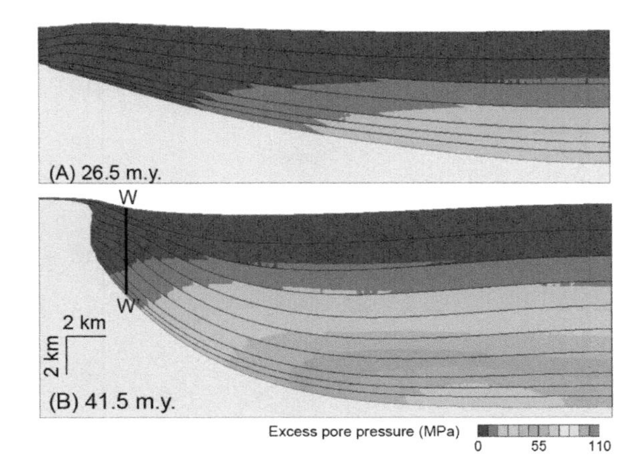

Figure 4.21 Transient evolutionary model of rising salt wall developing into salt sheet. Color contours show overpressure. (a) After 26.5 million years of deposition, differential sediment loading (e.g., Fig. 4.20b) drives rise of salt at left edge of model. (b) After 41.5 million years of deposition, the salt wall upbuilds to basin surface. Reprinted from Nikolinakou et al. (2018) with permission of Elsevier.

aggrades through time at a constant slope of 2 degrees (Fig. 4.20a). The base-level surface moves upward at a rate of 200 m/million years.

I illustrate the first 41.5 million years of the simulation (Fig. 4.21) and I focus on the impact of the salt diapir on pore pressure. I compare horizontal stress and overpressure adjacent to the diapir (Fig. 4.21, location WW') with the corresponding stress and pressure simulated from uniaxial deposition of the same material (solid versus dashed lines, Fig. 4.22). The pore pressure from the fully coupled simulation is twice the overpressure from the uniaxial simulation (compare dark blue solid line to blue dashed line, Fig. 4.22). This is because the growing salt diapir is laterally expanding and increasing the horizontal stress (compare dark green to dashed green line) relative to the case of uniaxial deposition. This extra loading is contributing to the pore pressure generation. This type of effect cannot be captured in models that assume uniaxial strain.

4.4.3.2 Pore Pressure and Stress in an Evolving Accretionary Prism

Thrust belts or accretionary prisms are a dramatic example of non-uniaxial loading. I illustrate pressure and stress in an evolving accretionary prism following the modeling results of Gao (2018). Transient loading, consolidation, fluid flow, and the evolution of pressure and stress are simulated as sediments enter a subduction zone (Fig. 4.23). A plane-strain numerical finite-element model (Elfen) is used. A 5 km thick low-permeability, high-compressibility mudrock layer represents the

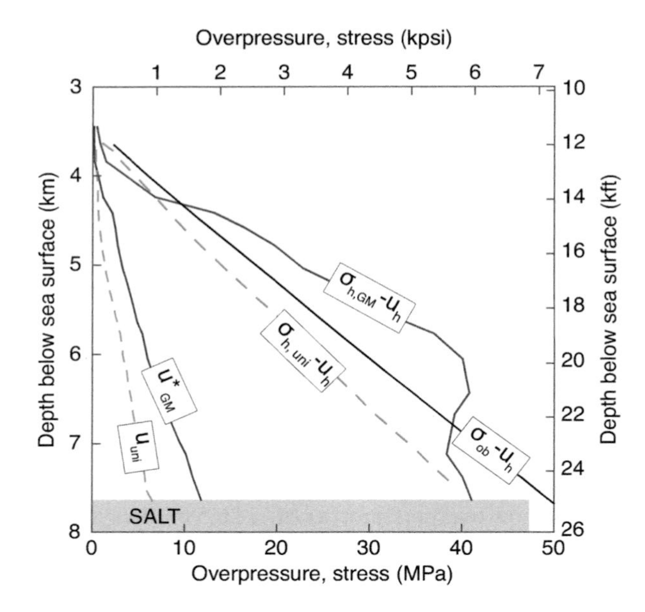

Figure 4.22 Overpressure and stress profiles illustrate the effect of sedimentation and salt load on pore pressure. Overpressure predicted by the geomechanical model near salt (solid dark-blue line) is higher than that predicted by a uniaxial model for the same average sedimentation rate (dashed light-blue line). This is because the vertical salt face (41.5 million years, Fig. 4.21b) imposes a horizontal load on the sediments. The horizontal stress predicted by the geomechanical model (solid green line) is much higher than horizontal stress predicted by the uniaxial model (dashed green line). Reduced overburden (solid black line) is shown for reference. Reprinted from Nikolinakou et al. (2018) with permission of Elsevier.

sediments entering the subduction zone. A proto-décollement is located 3 km beneath the seafloor (Fig. 4.23a). The accretionary wedge is formed by displacing the left edge of the sediment above the proto-décollement at a constant rate of 5 mm/year (Fig. 4.23a).

Overpressures are present within and beneath the prism (Fig. 4.23b). Near the trench, overpressures immediately above the décollement are higher than beneath it; landward (>30 km), overpressures increase monotonically with depth (Fig. 4.23b). At any given location, there is a discontinuity in stress across the décollement. The mean total stress (σ_m) is higher in the prism than the footwall (Fig. 4.23c) due to the lateral compression imposed on the wedge sediments; this lateral compression also leads to significant increase in differential (shear) stress. The relative level of shear is expressed as the ratio q/q_f (Fig. 4.23d). This ratio illustrates the proximity of the sediment element to shear failure.

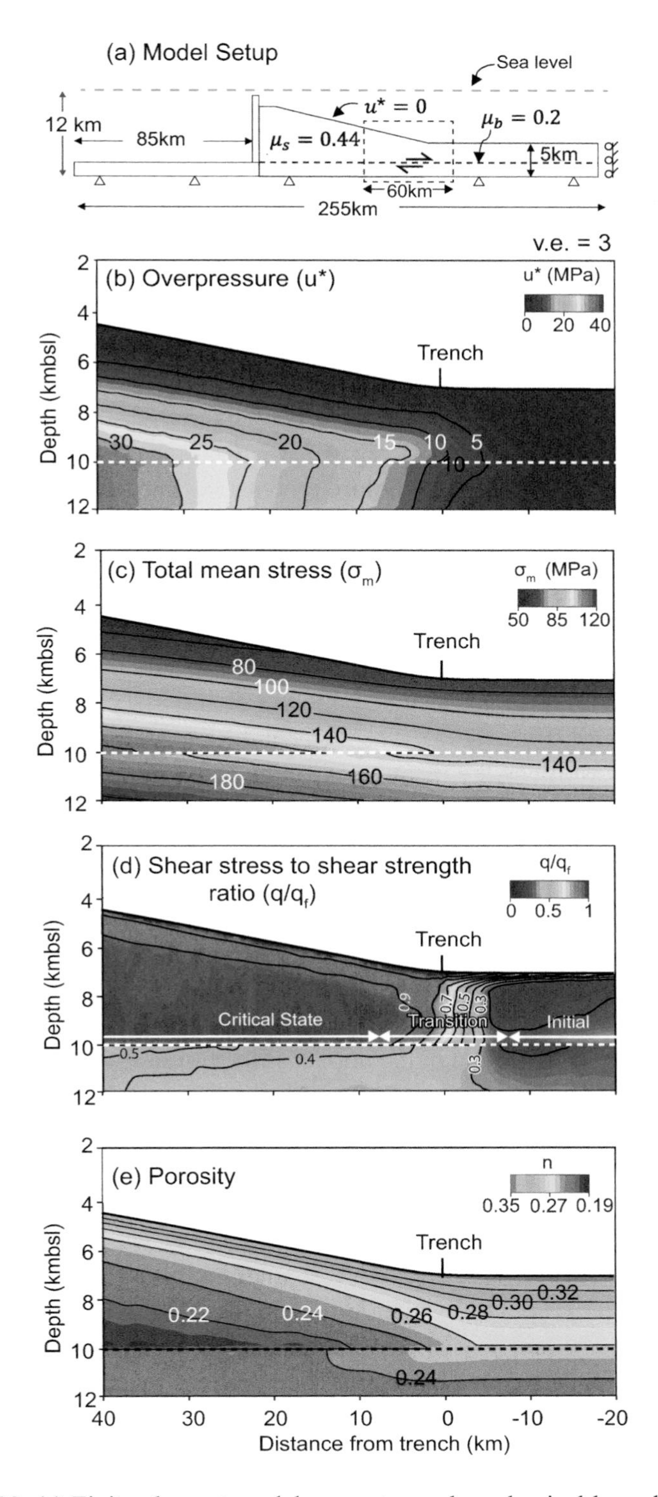

Figure 4.23 (a) Finite element model geometry and mechanical boundary conditions for simulating the evolution of an accretionary prism. The studied area expands from 20 km seaward of the trench to 40 km landward of the trench (dashed rectangle). (b) Overpressure (u^*). (c) Total mean stress (σ_m). (d) Shear stress to shear strength ratio (q/q_f). A ratio of 0 indicates no shear and a ratio of 1 corresponds to frictional failure (critical state). The ratio for the uniaxial stress state is 0.25. (e) Porosity (n). Modified from Gao (2018). Reprinted with permission by Baiyuan Gao.

Seaward, away from the accretionary prism on the subducting plate, the stress state is uniaxial, with a vertical maximum principal stress, resulting from sediment burial ($q/q_f = 0.25$, Fig. 4.23d). There is a transition in the stress state that begins 10 km in front of the trench, as sediments experience the effects of tectonic compression imposed by the approaching wedge ('Transition' zone, Fig. 4.23d). Above the décollement, the maximum principal stress rotates from vertical to approximately horizontal across this transition zone. Once incorporated into the accretionary prism, the material is at pervasive Coulomb failure (critical state) due to tectonic compression ($q/q_f = 1$; Fig. 4.23d). The transition in stress state from vertical uniaxial to compressional failure increases shear stress inside the wedge. The stress state below the décollement does not change significantly from the uniaxial state because of the low friction along the décollement surface. The mean stress is notably lower than in the hanging wall (Fig. 4.23c) and the relative shear stress remains low (Fig. 4.23d).

I describe the different contributions to the pore pressure through time (Eq. 4.34) for a sediment volume that is consumed into the accretionary prism above the plate interface ($a \rightarrow b \rightarrow c$, Fig. 4.24a). The shear-induced overpressure (u^{*q}, red line; Fig. 4.24c) is almost twice the mean stress-induced overpressure (u^{*m}, green line; Fig. 4.24c). The dissipated overpressure (u^{*diss}, black line) is a significant fraction of the combined stress-induced pressures; this dissipation is manifested by both porosity loss and dewatering across the transition in stress regimes near the trench (Fig. 4.24b). This dewatering reflects partial drainage of the loading-driven pressures; however, the dissipation does not keep pace with the loading, and thus there is a progressive increase in overpressure as the volume passes into the subduction zone (u^*, Fig. 4.24c).

This is a dramatic illustration of the impact of the full stress field on pore pressure in settings of non-uniaxial strain. High pore pressures are present in the hanging wall of the accretionary prism (Fig. 4.23b), yet the porosity above the décollement is less than that below it (Fig. 4.23e). These effects are caused by the lateral load, which increases both the mean and shear stress, driving both compaction and elevated pressure. Discontinuities in porosity, such as we simulate, are present across subduction décollements at several margins either observed directly or inferred from negative polarity seismic reflections (Bangs et al., 2004; Screaton et al., 2002; Stoffa et al., 1992). In Chapter 6, I discuss how to invert the pore pressure from the porosity profile in thrust belt systems by accounting for the high lateral stresses present in these systems.

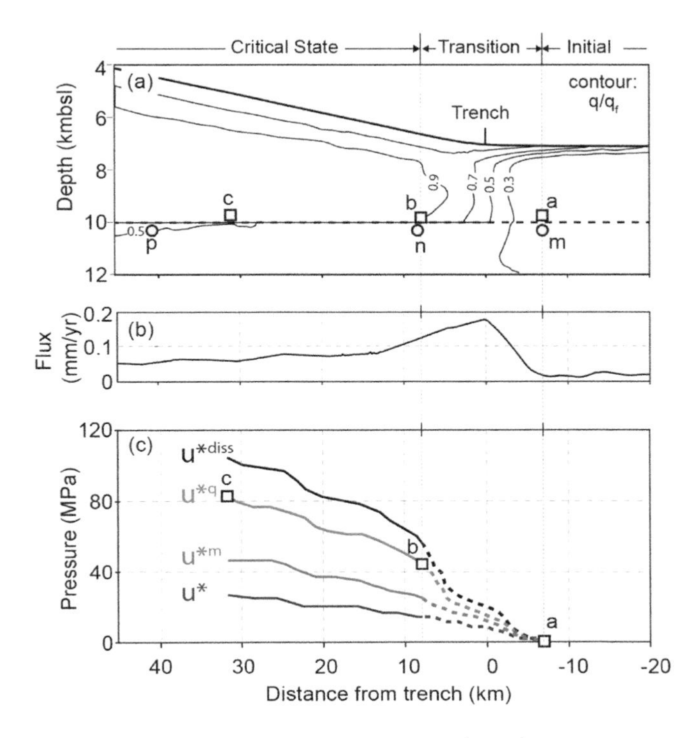

Figure 4.24 (a) Shear stress to shear strength ratio (q/q_f) for simulation shown in Figure 4.23. Points 'a,' 'b,' and 'c' record the path of a sediment volume entering the subduction zone immediately above the décollement, with 'b' marking the beginning of the critically stressed wedge. Points 'm,' 'n,' and 'p' record the path of a subducted sediment volume beneath décollement at equivalent simulation time of points 'a,' 'b,' and 'c.' (b) Simulated fluid flux across the seafloor. (c) Overpressure is generated by both increased mean stress (u^{*m}, green line) and shear stress (u^{*q}, red line). Much of the overpressure generated is dissipated (u^{*diss}, black line), leaving the resultant overpressure (u^{*}, blue line). Modified from Gao (2018). Reprinted with permission by Baiyuan Gao.

4.5 Viscous Compaction

Up to this point, I have only considered the case where there is a unique relationship between porosity and stress for a given mudrock. However, if creep occurs, then the same mudrock will have a lower porosity at the same effective stress with increasing time (Section 3.2.5). In Chapters 5 and 6, I discuss that it is common to encounter deeper, older, and hotter mudrocks that are more compacted than younger and cooler shallow mudrocks at the same effective stress. One interpretation for this behavior is that creep has occurred. Pore pressure prediction has strived for years to account for this behavior, as is discussed in Chapter 6.

Modeling studies have also incorporated this effect and it has broadly been termed viscous compaction. In these cases, these models generally infer that porosity loss occurs by grain dissolution at intergranular contacts (e.g., Birchwood & Turcotte, 1994; Fowler & Yang, 1999; Schneider et al., 1996). This approach is also used to describe creep-controlled viscous compaction laws used in studies dealing with magma transport in the Earth's mantle (McKenzie, 1984).

Morency et al. (2007) incorporate viscous compaction into a fully coupled hydromechanical model. The compaction is modeled as:

$$\frac{1}{1-n}\frac{Dn}{Dt} = -c_b\frac{D\sigma'_m}{Dt} - \frac{\sigma'_m}{\eta(n)}. \qquad \text{Eq. 4.38}$$

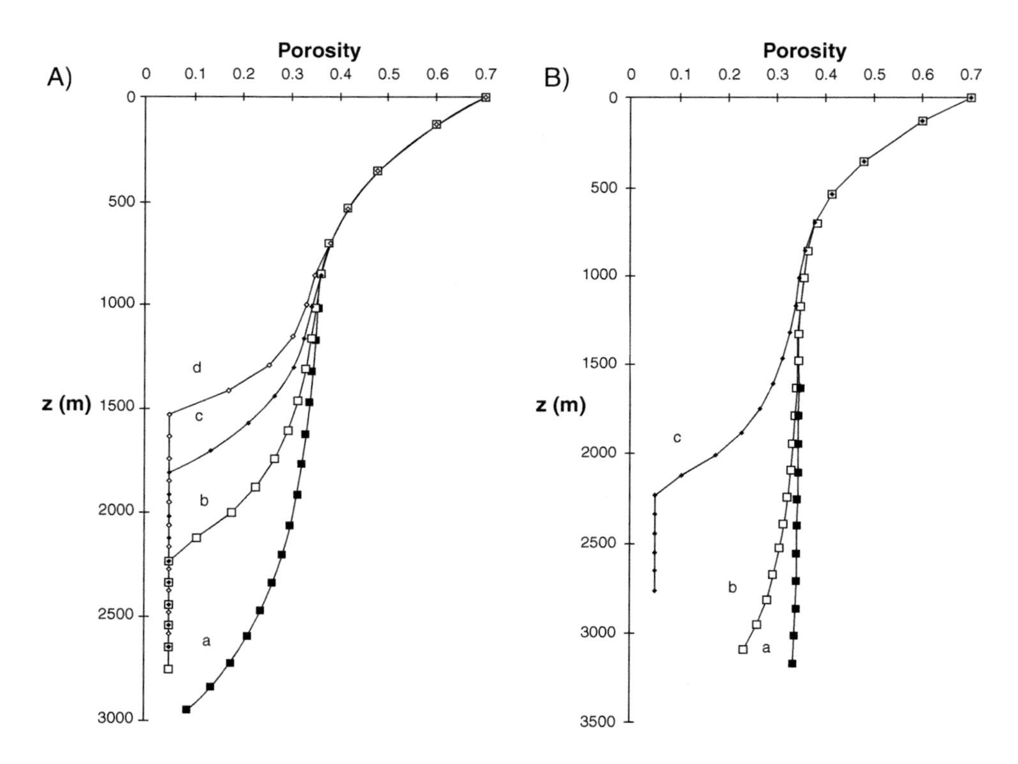

Figure 4.25 (a) Evaluation of temperature effect on porosity due to viscous compaction. Each simulation was performed at a different thermal gradient (a: 15 deg C/km; b: 30 deg C/km, c: 45 deg C/km; d: 60 deg C/km). Porosity dramatically decreases with increasing temperature. (b) The effect of time on viscous compaction. All of the parameters were the same except the total simulation time was 1 Ma for curve 'a,' 10 Ma for curve 'b,' and 100 Ma for curve 'c.' The porosity declines with increasing time. Reprinted from Schneider et al. (1996) with permission of Elsevier.

In Equation 4.38, the first term on the right hand side is the mechanical compaction term that I have discussed previously. The second term is the viscous compaction term, where η is the sediment viscosity. η is a function of porosity and has been modeled both as a function of time alone and also as a function of temperature (Schneider et al., 1996). In Figure 4.25, the effect of time and temperature due to viscous creep are illustrated on the porosity profile for hydrostatic pore pressure. At a given depth and effective stress, hotter basins will have a lower porosity (Fig. 4.25a). In a similar fashion, at a given depth, but with a fixed temperature, older sediments will be more compacted (Fig. 4.25b).

The viscous term is important because it represents an additional source of overpressure. When the porosity declines due to viscous creep, the framework collapses, and the pore pressure will increase in order to take on a greater fraction of the applied load (e.g., Fig. 6.9). For viscous compaction to be an efficient source of overpressure, either low viscosity (reflecting an ability to alter pore volume quickly) or a low sedimentation rate is required (Morency et al., 2007). This is illustrated in Figure 4.26. In this simulation, Morency et al. (2007) have assumed both a low sedimentation rate and a low permeability. During the early stages of sedimentation, the fluid pressure is hydrostatic. However, with time overpressures

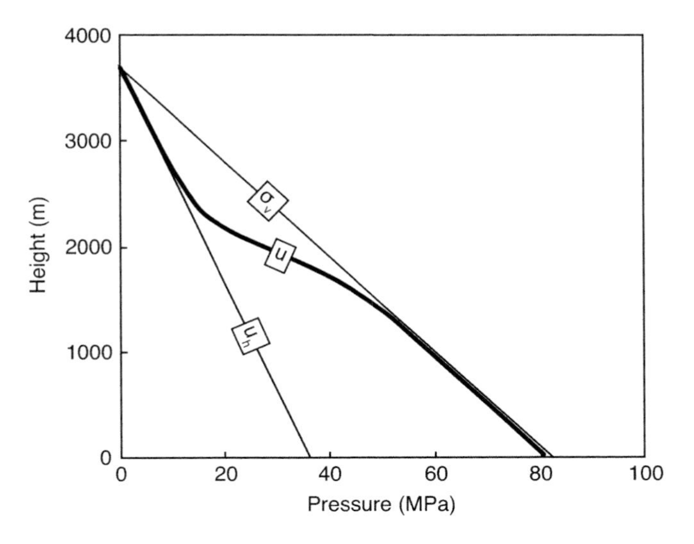

Figure 4.26 One-dimensional simulation of overpressure during constant sedimentation including both mechanical compaction applied by sediment loading and viscous compaction. The sedimentation rate is relatively low and the mechanical loading does not produce significant overpressure. However, the viscous compaction is significant and in the older, deeper sediment it generates significant overpressure. Reprinted from Morency et al. (2007) with permission by John Wiley and Sons.

develop in a progressively thicker layer at depth (Fig. 4.26). In this case, any fluid pressure generated by mechanical compaction dissipates. However, the deeper and older sediments undergo viscous compaction and generate overpressure. This is a third way to generate a pressure profile with hydrostatic pressure in the shallow section and overpressure in the deep section (compare Figures 4.14, 4.17, and 4.26). It is an intriguing result and provides a mechanism for how extremely old basins, not presently undergoing loading, maintain high overpressures. A better understanding of creep behavior in mudrocks is an outstanding challenge for pore pressure prediction.

4.6 Summary

At this point, the reader may be overwhelmed by the many possible sources of overpressure in porous media. I have emphasized that early work in geotechnical engineering on undrained pore pressure response, drained compression, and pore pressure dissipation provides a strong foundation to understand pore pressure in basins. These contributions provide us with a sense of what drives pore pressure and what the timescales of pore pressure dissipation are.

Advanced numerical basin models are used to extend these concepts to explore the evolution in pore pressure in complex geological environments. Most of these models assume vertical uniaxial strain. These models illustrate the relative role of different mechanisms in generating pore pressure, and they show how the geological evolution of the system strongly controls the pore pressure. Most recently, more advanced models have incorporated solution of the full stress field simultaneously with calculation of the pore pressure field. These models illustrate that in settings of non-uniaxial strain, the full stress field must be calculated in order to estimate the pore pressure. The effect is particularly dramatic around salt bodies and in thrust belts. Finally, I have also reviewed viscous compaction. During viscous compaction (or creep), porosity is lost as a function of time and temperature at a fixed effective stress. This is an additional source of overpressure.

We have debated for decades the relative role and importance of different mechanisms for generating overpressure. I have emphasized that in a growing sedimentary basin, the roles of sediment loading and tectonic forces are dominant, while the overpressure generated by either diagenesis or thermal expansion are small. I do, however, think that viscous compaction may play a significant role in pore pressure generation in many basins around the world and that a better understanding of this behavior will improve our ability to predict pressure.

5

Pore Pressure Prediction in Mudrocks

5.1 Introduction

In this chapter I describe how to predict pressure from the compaction state of mudrocks. The compaction state is either measured directly (e.g., from cuttings and core) or indirectly (e.g., from resistivity, velocity, or density). I present a detailed example from the Eugene Island 330 (EI-330) oil field, Gulf of Mexico. I then review two common challenges: (1) how to pick mudrocks and (2) how we establish normal compaction trends in overpressured basins. I close with a review and comparison of five different compaction models.

In Chapter 4, I described the pore pressure response to loading and I emphasized that there is a quantitative relationship between compaction, applied load, soil strength, and pore pressure. To predict pressure, we invert this process. We use the degree of compaction to determine the pressure. For example, if we examine the porosity profiles in two sedimentary basins and find that one is less compacted than the other (Fig. 5.1b), yet all material properties are the same, then the basin that is less compacted will have a higher pore pressure (Fig. 5.1c).

I give an example of how we predict pore pressure from these concepts, which I term the Hubbert approach. Hubbert and Rubey (1959) suggested that compaction was an exponential function of effective stress:

$$n = n_0 e^{-\beta\sigma'_v}. \qquad \text{Eq. 5.1}$$

Equation 5.1 can be rearranged to solve directly for the pore pressure:

$$u = \sigma_v - \sigma'_v = \sigma_v - \frac{\ln(n_0/n)}{\beta}. \qquad \text{Eq. 5.2}$$

If we know the total vertical stress and can describe the compaction (n_o and β) of the mudrock, then pore pressure can be calculated.

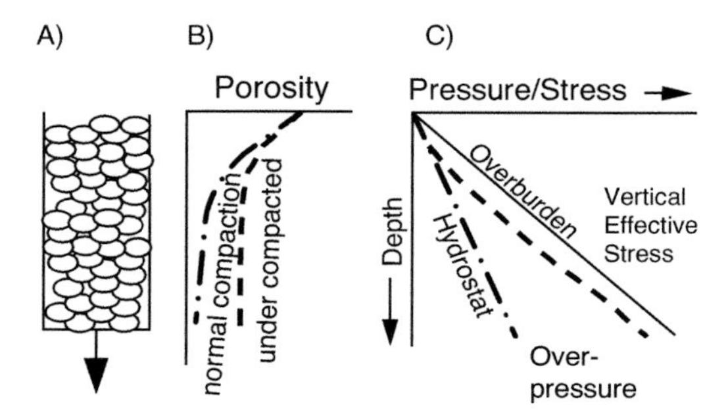

Figure 5.1 Concept of pore pressure prediction. (a) Sedimentary basin. (b) Porosity versus depth profiles extracted from two sedimentary basins. One basin (dash-dot line) is more compacted than the other (dashed line). (c) The less compacted basin is interpreted to have higher pressures: the pore fluid is supporting a greater fraction of the load and is inhibiting compaction.

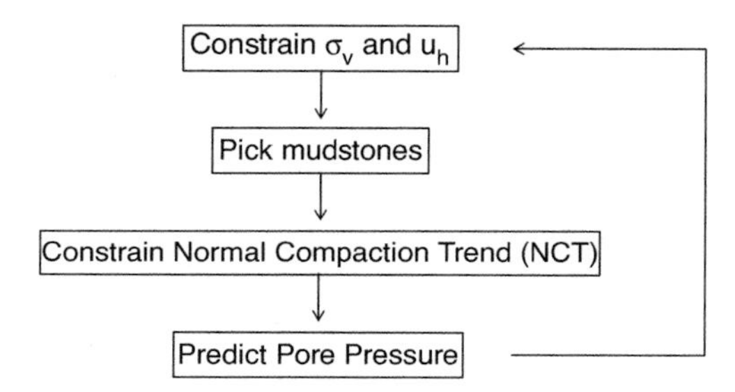

Figure 5.2 To predict pore pressure, the overburden stress (σ_v) and hydrostatic pressure must first be determined. Then mudrocks must be selected from the complex lithologies present. Next, a normal compaction trend must be established that relates effective stress to some measurable parameter (e.g., porosity, velocity, or resistivity). Finally, pore pressure is predicted and the results are compared to the data available. Often the process is cyclic and the multiple pore pressure predictions are made as components of the pore pressure prediction are refined.

In summary, the practice of pore pressure prediction (Fig. 5.2) can be understood in the context of Terzaghi's spring model (Fig. 4.1). First, the overburden stress (σ_v) must be determined. Next, a relationship between stress and strain must be developed (compaction). This describes the deformation of the spring in response to a particular effective stress and is described in this example by n_o and β (Eq. 5.1). Finally, given a measure of the strain in the rock through measurement of porosity (n), the pressure can be predicted (Eq. 5.2). This approach assumes there is a unique relationship between porosity and vertical effective stress. I relax this requirement in Chapter 6. These steps are conceptually simple, but complex in practice and there are often multiple approaches.

5.2 Application to the Eugene Island 330 Oil Field and the Pathfinder Well

5.2.1 Observations in the Pathfinder Well

The EI-330 field of the Gulf of Mexico consists of progradational shelf-margin sediments bound to the north by a prominent growth fault and its associated splays (Fig. 5.3). The field has been extensively studied (Alexander & Flemings, 1995; Hart et al., 1995). A striking characteristic of wells drilled in this field is that beneath 6 800 ft velocity, density, and resistivity decrease with depth (Fig. 5.4). Core-based measurements of the mudrock porosity show that the decrease in velocity, density, and resistivity beneath 6 800 ft records an increase in porosity. Porosity was also estimated from log velocities within the mudrocks, and these values are within 5% of the core-derived measurements (Fig. 5.4b). Finally, direct pressure measurements in the Pathfinder well (Fig. 5.4c) show that pore pressures are hydrostatic to a depth of ~5 000 ft and below this overpressures increase dramatically (Fig. 5.4).

Viewed in the context of the spring model (Fig. 4.1), the pore pressure in the zone beneath 6 800 ft is supporting a greater fraction of the overburden causing the rock to be less compacted. A central concept is that because we know that the shallow sedimentary section (above 5 000 ft) has no overpressure, we can examine the loss of porosity in this section to describe the compaction behavior. Then, once we have characterized the porosity-effective stress behavior in this region, we can use that relationship to predict pressure at deeper depths.

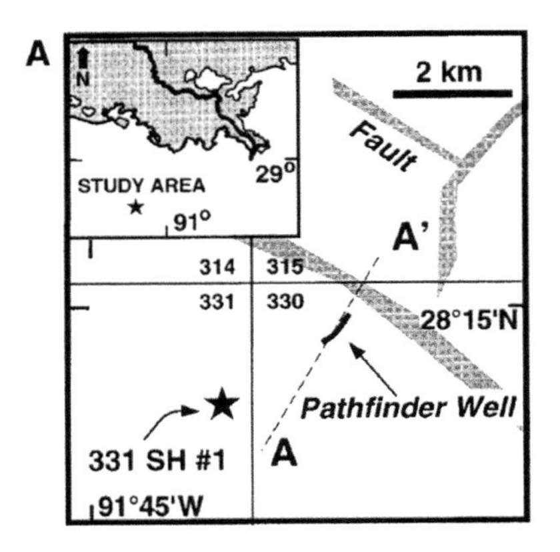

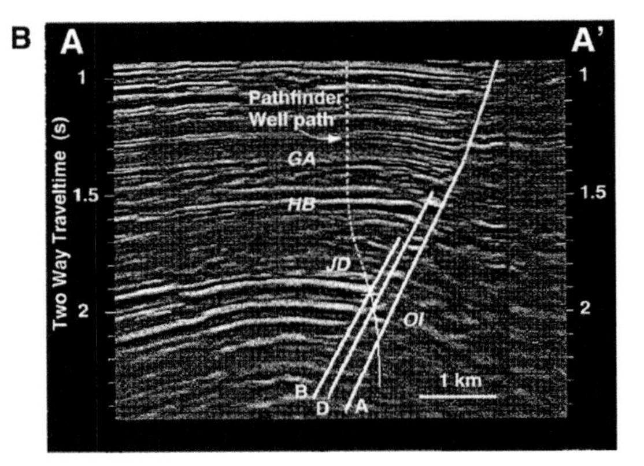

Figure 5.3 (a) Location of Eugene Island 330 oil field and EI-331 SH#1 well. (b) Seismic line showing Pathfinder well path with respect to location of principal stratigraphic (GA, HB, JD, OI) sands and structural elements (fault splays B, D, and A). Republished with permission of the Geological Society of America, from Hart et al. (1995); permission conveyed through Copyright Clearance Center, Inc.

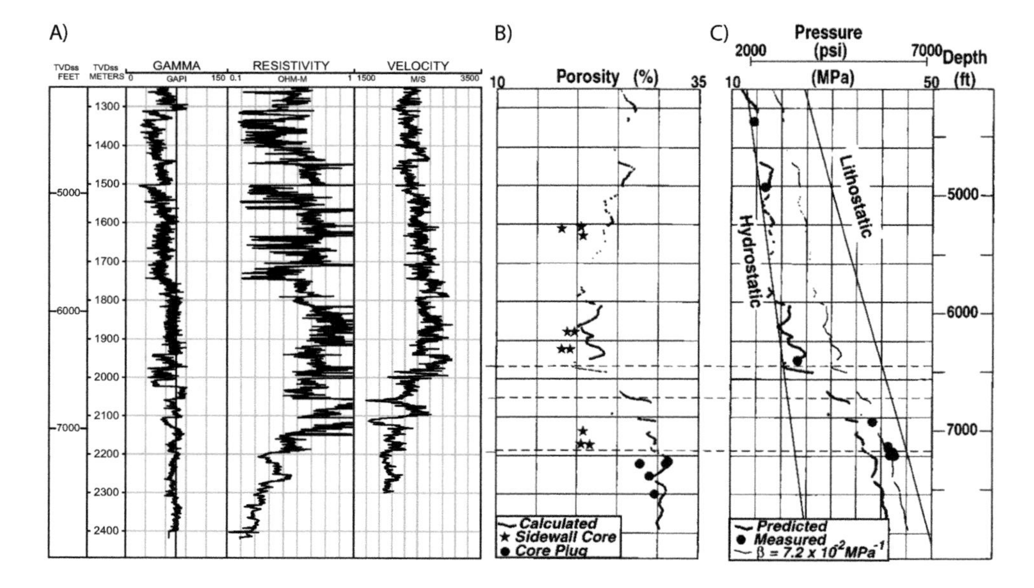

Figure 5.4 (a) Log values in the Pathfinder well (EI-330 A20 ST2). At depths greater than 6 800 ft, the velocity and resistivity decrease with depth. (b) Porosity measured in core (symbols) and calculated from wireline velocity with Equation 5.3. (c) Predicted pressure. Direct pressure measurements are shown with solid circles. Republished with permission of the Geological Society of America, from Hart et al. (1995); permission conveyed through Copyright Clearance Center, Inc.

5.2.2 Determination of Normal Compaction Trend

I determine the relationship between porosity and effective stress in the shallow section (Eq. 5.1). This is commonly termed the normal compaction trend (NCT). I estimate the mudrock porosity from the velocity log using the expression suggested by Issler (1992):

$$n = 1 - \left(\frac{dt_{ma}}{dt}\right)^{\frac{1}{f}}, \qquad \text{Eq. 5.3}$$

where dt and dt_{ma} are the log-derived sonic travel time and the matrix travel time (travel time is the inverse of velocity $\left(dt = \frac{1}{V}\right)$), respectively, and f is a constant. I assumed $dt_{ma} = 220\frac{\mu sec}{m}$ and $f = 2.19$ as proposed by Issler (1992) for low total organic carbon, non-calcareous mudrocks.

I use the EI-331 SH #1 well where there is log data in the shallow sedimentary section. I use the Gamma Ray log data to pick zones where only mudrocks are present. I then calculate mudrock porosity in the EI-331 SH#1 well with Equation 5.3. I smooth these values using a moving average to eliminate noise due to borehole effects or minor lithologic changes.

I calculate the overburden stress as a function of depth (Eq. 2.6). I use the bulk density log and, where there is no data in the very shallow section, I assume the density varies exponentially from the seafloor to the first log-derived density measurement (see Chapter 8).

There is no overpressure from the surface to approximately 5 000 ft. Thus, I assume that in the shallow section (0 ft to 5 300 ft), the effective stress is equal to the overburden stress less the hydrostatic pressure $\left(\sigma'_{vh} = \sigma_v - u_h\right)$. The hydrostatic pressure (Eq. 2.1) is assumed to have a gradient of 10.5 MPa/km (0.465 psi/ft). Next, I plot the porosity versus the hydrostatic effective stress $\left(\sigma'_{vh}\right)$ (black circles, Fig. 5.5). Finally, I use linear regression on the data between 1 000 and 5 300 ft and calculate the values of β and n_o. I find $\beta = 2.15x10^{-4}psi^{-1}$, and $n_o = 0.38$. I excluded the data above 1 000 ft because porosity changes very rapidly in this interval and because these data lie outside of the range in which we wish to predict pore pressure.

5.2.3 Pore Pressure Prediction in the EI-330 A20 ST2 Well

I next calculate the overburden stress and the mudrock porosity in the well where I wish to predict pressure. A density log was not run above 1 250 m in this well, so I combined the density log from the A20-ST2 well with the density log from shallow parts (<1 250 m) of the 331 SH #1 well to compute the overburden stress (σ_v).

I next calculate the pressure (Eq. 5.2) based on the values of β and n_o constrained in Figure 5.5. The calculated pressures are nearly hydrostatic down to approximately 6 000 ft (1 829 m). Between 6 000 ft (1 829 m) and 7 300 ft (2 226 m), predicted pressures rise abruptly toward the lithostatic pressure reaching a maximum overpressure at ~7 500 ft (2 286 m) (Fig. 5.6). The predicted pressures agree with the measured values down to approximately 6 500 ft (1 982 m). However the predicted pressures are 3–5 MPa (435–725 psi) less than the measured pressures at the deepest depths (~7 500 ft, 2 286 m).

An alternative way to look at this misfit is through an examination of the porosity versus effective stress behavior (Fig. 5.5). The porosities at the most overpressured location at 7 300 ft (2 226 m) are approximately 0.31 (yellow squares, Fig. 5.6a). However, the effective stresses recorded by direct pressure measurements at these locations are less than would be expected from the normal compaction trend (yellow squares are to the left of the normal compaction trend at a porosity of 0.31 in Figure 5.5). The mudrock in the overpressured section is more compacted than predicted by the normal compaction trend. The difference in effective stress between the yellow squares and the normal compaction trend is the magnitude of the misfit in the predicted pore pressure (Fig. 5.5).

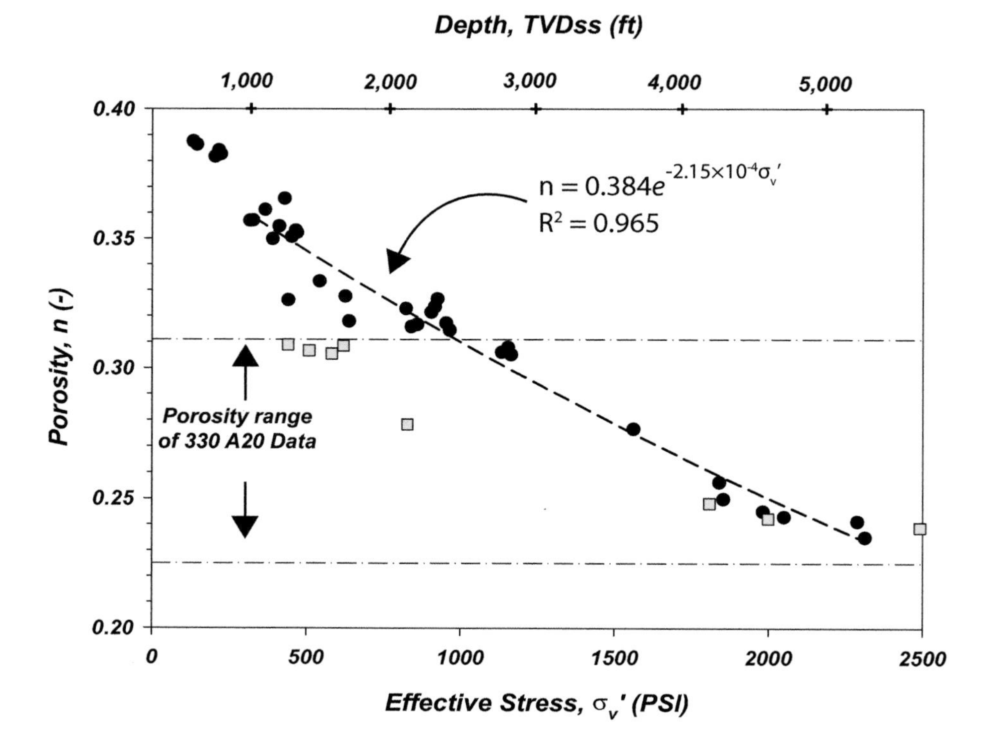

Figure 5.5 Mudrock porosity versus hydrostatic effective stress for mudrocks within the hydrostatic section (1 000 ft to 5 300 ft) in the Eugene Island 331 #1 well (black circles). Porosity data are derived from the velocity log using Equation 5.3. Dashed line shows the line calculated by least squares regression for Equation 5.1 for depths between 1 000 ft and 5 300 ft. Yellow squares mark the effective stresses calculated from direct pressure measurements deeper in the well. The porosities for these yellow squares are the mudrock porosities that are immediately adjacent to where the pressures were measured (see Fig. 5.6a, yellow circles). Porosities for the 330 A20 well (where we wish to predict pressure) lie between the dash-dot lines. On the top axis, the actual depths below sea surface (TVD_{ss}) for the shallow hydrostatically pressured data are shown.

A variety of explanations are possible for this misfit. First, changing depositional conditions might have resulted in different mineralogy or texture for the deeper mudrocks relative to the shallow mudrocks with the result that the deep and shallow mudrocks cannot be described by the same coefficients β and n_o. Second, it is possible that the deeper mudrocks, which are older and hotter than the shallow mudrocks, may have undergone creep and or diagenesis and as a result they do not lie on the original compaction trend (e.g., Fig. 3.11). Third, under some situations, the mudrock pressure is not exactly equal to the adjacent reservoir pressure (Chapter 10). Fourth, if the present day effective

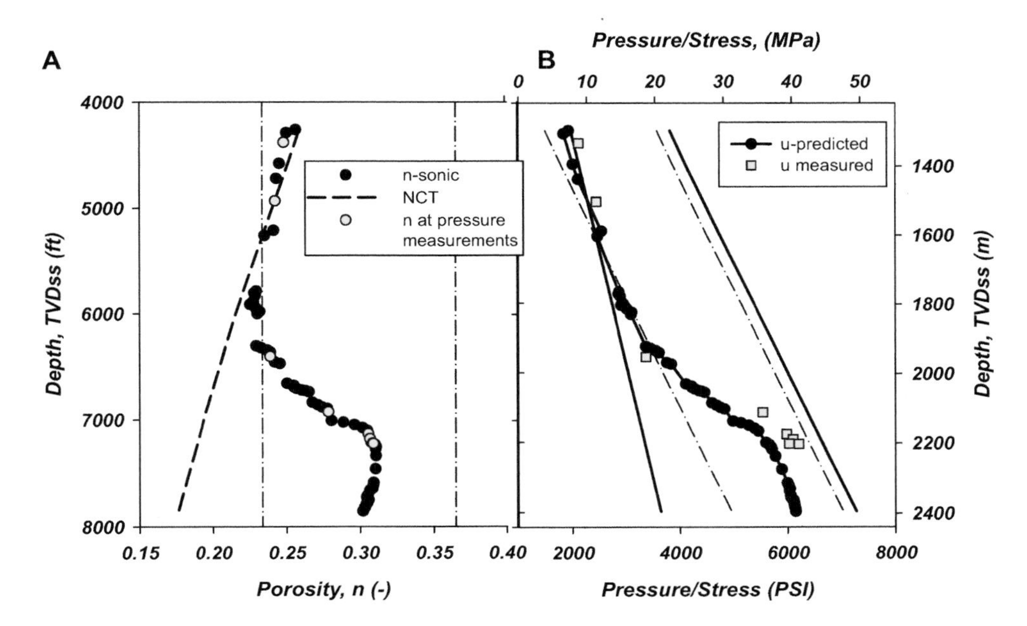

Figure 5.6 (a) Porosity calculated from velocity log (Eq. 5.3) versus depth in the EI-330 A20 well (black circles). Mudrock porosity at depth of pressure measurements (yellow circles). Dashed line (NCT) is the normal compaction trend, which is the porosity expected if pore pressure was hydrostatic throughout the interval. (b) Pressure calculated from Equation 5.2 (black circles) and measured water-phase pressures (i.e., corrected for hydrocarbon column height) (yellow squares). The predicted and observed pressures rise with depth. However, the predicted pressures are lower than the observed pressures. Dash-dot lines bound the lower and upper limits of the regression done on the EI-331 #1 well (between 1 000 ft and 5 300 ft in Figure 5.5). See Hart et al. (1995) for further discussion.

stress is less than the preconsolidation stress, it is likely that they will not lie on the same compaction trend (Chapters 3 and 6). We will discuss these issues further in the next chapter and explore mechanisms to explain why we underestimate the pore pressure at depth.

5.3 Details of Pore Pressure Prediction

Pressure prediction using the normal compaction trend approach is conceptually simple (Figures 5.1 and 5.2) but challenging in practice. In my experience, there are three particular challenges: (1) selecting a consistent lithology for pressure prediction, (2) determining a normal compaction trend within overpressure, and (3) determining total vertical stress. In the following sections I discuss approaches to address the first two of these issues and in Chapter 8, I discuss approaches to determine the total vertical stress.

5.3.1 Picking Lithologies for Pore Pressure Prediction

Pore pressure prediction is practiced on mudrocks because mudrocks undergo dramatic compaction as effective stress increases (Chapter 3). Other lithologies (e.g., sandstones, limestones) in addition to being less compressible, can suffer from diagenesis that affects their compaction history. Thus, a key step is to select mudrocks of a similar composition to use as the foundation for pore pressure prediction.

Perhaps the most common approach is to use the Gamma Ray (GR) log to pick mudrocks of consistent properties. Mudrocks have a higher clay content than siltstones or sandstones and as a result are commonly more radioactive; this is recorded with higher Gamma Ray (GR) values. There are various approaches to filtering log data. I compare auto-picking and hand-picking with an example from Well 826–1 in the Mad Dog field (Fig. 5.7).

In auto-picking, rocks with a GR value above a threshold value are assumed to be mudrocks. Often, over the depth of a well, multiple constant gamma-ray cutoffs are employed, each spanning several thousand feet. In Well 826–1, GR values greater than 110 API are assumed to record mudrocks (to the right of the vertical blue dashed lines, Fig. 5.7a). Blue dots on Figure 5.7a record the GR values in these mudrocks, and the velocities and resistivities at these locations are shown with blue dots in Figures 5.7b and 5.7c. The continuous blue lines (Fig. 5.7b, c) result from an 11-point moving average filter to the resistivity (Fig. 5.7b) and the velocity (Fig. 5.7c) values.

In hand-picking, mudrocks are manually chosen approximately every 30–40 ft (9–12 m). In this approach, mudrocks are picked (red dots, Fig. 5.7a-c) where GR values are high and where the resistivity and sonic logs also record mudrock (in this case, a higher GR, higher resistivity and lower velocity) (Fig. 5.7). An 11-point moving average filter is applied to reduce noise caused by borehole effects or small lithology changes (continuous red lines, Fig. 5.7b, c). The typical width of the averaging window is 400 ft (122 m).

In the shallow section (7 000 ft to 11 000 ft), the hand-picked approach results in slightly higher pressures than the auto-picked approach (Fig. 5.7e). This is most likely because the hand-picked values have a slightly higher GR value and slower velocity than the auto-picked values. Near 12 000 ft (3 658 m), the auto-picked values produce distinctly lower pressures than the hand-picked values: a thin layer of high GR material has high velocity, and this results in low pressure. The two approaches differ most in the middle Miocene, where the pressure profile for the auto-picked example has high frequency oscillations above and below sand bodies with direct pressure measurements. In contrast, the pore pressure curve for the hand-picked example does not have large fluctuations. From the middle to lower

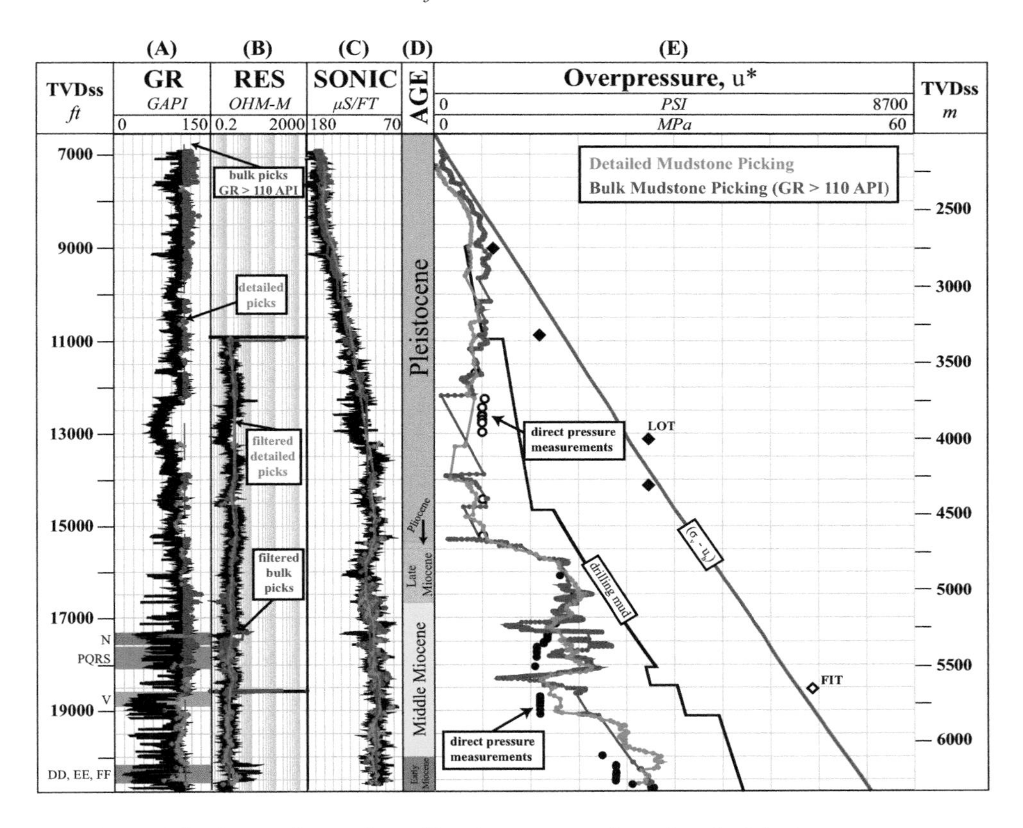

Figure 5.7 Comparison of predicted pressures from hand-picked versus auto-picked approach in Mad Dog well 826–1. (a) Gamma Ray log. Vertical blue dashed line marks auto-picking cutoff: GR values greater than 110 API are assumed to record mudrock and are marked with blue dots. Red dots record hand-picked mudrocks. (b) Resistivity log. Blue dots are auto-picked and red dots are hand-picked. The blue and red continuous curves are 11-point moving average filters of the auto- and hand-picked mudrocks, respectively. (c) Velocity log. Auto-picked shown in blue and hand-picked shown in red. (d) Geologic age. (e) Predicted mudrock overpressures. Auto-picked mudrocks recorded in blue; hand-picked mudrocks recorded in red. The overpressures for the auto-picked approach exhibit high-frequency oscillations that are unlikely to record pore pressure variation. From Merrell et al. (2014). Reprinted by permission of the AAPG whose permission is required for further use. AAPG© 2014.

Miocene, both pressure profiles build at similar rates and ultimately converge on the deepest set of direct pressure measurements.

The high-frequency pressure oscillations are present in intervals of interbedded sandstone and mudrock. I interpret that changes in the composition and fabric of the mudrock occur that cannot be resolved with a bulk GR cut-off. Compositional changes are driven by changes in silt fraction near sand bodies and, often, by the presence of flooding surfaces that cap sandstone bodies (very clay-rich deposits formed during transgression). These changes in fabric and composition result in very different pressure predictions. Careful hand-picking by comparing a number of different logs allows the user to avoid these intervals. In practice, high frequency pressure oscillations are characteristic of pore pressures computed from auto-picking. Practitioners of this approach will commonly smooth these oscillations out, either through moving average filters, or by hand.

Ultimately, as is the case with choosing the appropriate normal compaction trend function, what is important is to understand what is driving the apparent oscillations in predicted pore pressure. More elegant auto-picking procedures may improve auto-picking results. For example, Yang and Aplin (2004) used neural networks with a standard input wireline data set (caliper, GR, density, sonic, resistivity) to calculate the clay fraction, cementation factor, total organic content, and density.

5.3.2 Calibrating Compaction Behavior in Overpressure

Early pore pressure prediction approaches were developed on the continental shelf. In this environment, progradation of sand-rich shelf systems over mud-rich slope systems resulted in hydrostatic pore pressure in the shallow section overlying overpressured, mudrock-rich systems such as in the EI-330 A20 ST2 well (Fig. 5.4). However, in deeper water and other more complex environments, there is not necessarily a shallow section that is hydrostatically pressured. For example, in numerous locations, it has been demonstrated that overpressures start at or near the seafloor (Flemings et al., 2008; Ostermeier et al., 2001; Ostermeier et al., 2002).

In this environment, a different approach must be used to establish the normal compaction trend. Direct measurements of reservoir or aquifer pressure must be correlated to mudrock properties (Fig. 5.8). For example, mudrock velocities are determined above and below a sandstone body (black circles, Fig. 5.8c) and the pressure is directly measured in the sandstone (or some other permeable unit) (open square, Fig. 5.8e). The water phase overpressure in the mudrock is assumed to equal the water phase overpressure in the sandstone (vertical dashed line, Fig. 5.8e). I emphasize the water-phase pressure because the effects of buoyancy due to hydrocarbons should be removed (e.g., Fig. 2.6). The assumption that the overpressure in the mudrock and the reservoir are equal should be viewed with caution. In Chapter 10, I discuss why there are sometimes systematic differences between the water pressures in the reservoir and the bounding mudrock. The effective stress (σ_v') is then calculated by subtracting the water phase pore pressure from the overburden stress. The porosity and effective stresses of these mudrock points are then plotted on a velocity (or porosity) versus effective stress plot (Fig. 5.9).

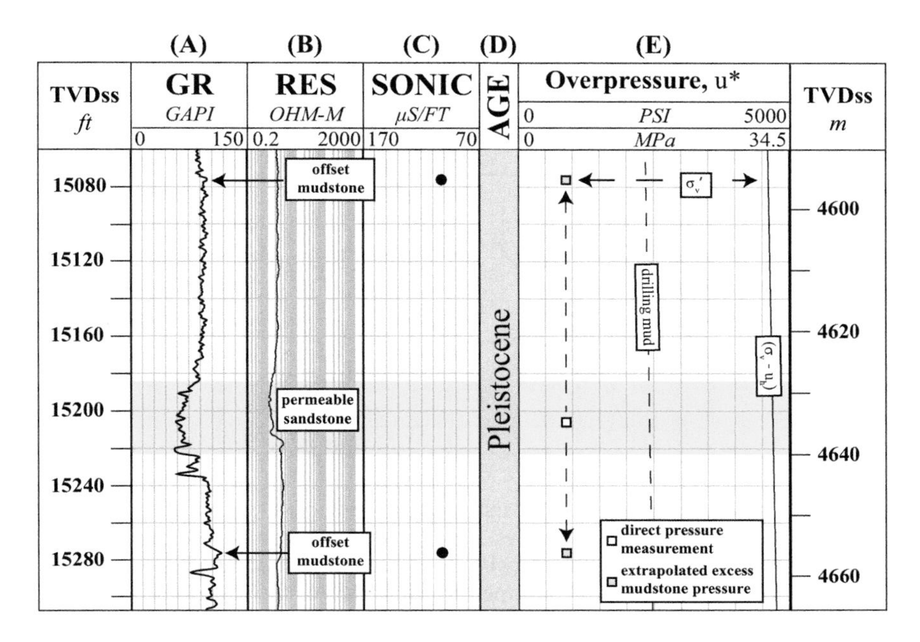

Figure 5.8 Approach used to determine mudrock pressure from measured pressure in nearby sandstone. The pressure is directly measured in the sandstone (open square). The solid black circle marks the mudrock velocity pick (see Figure 5.7 for approach). The water phase overpressure (yellow squares) in the mudrock is assumed to equal the water phase overpressure measured in the nearby sandstone. The extrapolated pore pressure is subtracted from the total vertical stress (σ_v) to determine the vertical effective stress (σ_v'). From Merrell et al. (2014). Reprinted by permission of the AAPG whose permission is required for further use. AAPG© 2014.

The approach is illustrated for the reservoirs in the Bullwinkle field in the offshore Gulf of Mexico (Flemings & Lupa, 2004) (Fig. 5.9). The normal compaction trend is very different when based on the reservoir pressures versus when based on the shallow compaction data if hydrostatic pressure is assumed (grey circles versus black diamonds, Fig. 5.9): the effective stresses are higher at a given porosity for the deeper pressure measurements than for the shallower measurements. As shown by Flemings and Lupa (2004), predicted pore pressures based on the shallow compaction data overpredict the observed pressures at depth. This is the opposite behavior from what was observed in the Eugene Island example (Fig. 5.6) and is a common result when the shallow section is overpressured. Examples of the approach of using direct pressure measurements to constrain the normal compaction trend are common (Flemings & Lupa, 2004; Hauser et al., 2013; Merrell et al., 2014).

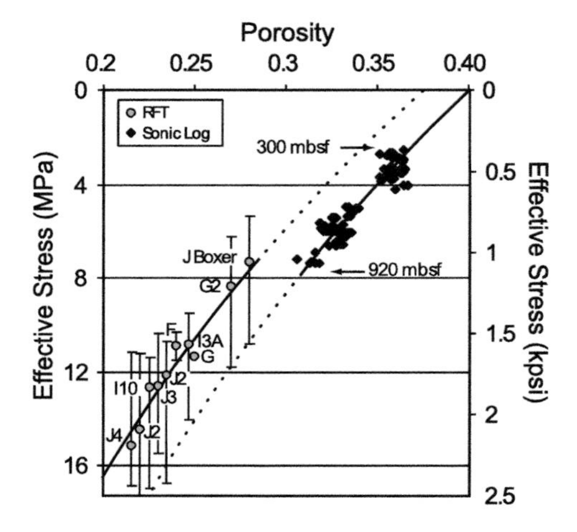

Figure 5.9 Mudrock porosity (n) versus vertical effective stress (σ_v'). The n-σ_v' relationship is determined in two ways. First, velocities from mudrocks adjacent to sandstones are selected and their porosity is calculated from Equation 5.3. Next, the overpressure in the mudrock is assumed to equal that of the sandstone (e.g., Fig. 5.8) (squares). Given the pressure present, the effective stress is plotted (grey circles). The vertical error bars represent the range of water phase vertical effective stresses from the crest of the reservoir to the base. The black diamonds record the porosity versus effective stress behavior between 300 and 920 mbsf assuming the shallow section is hydrostatically pressured. There is an abrupt offset between the two trends. We interpret that overpressures are present in the shallow section and it is not hydrostatically pressured as assumed in plotting the black diamonds. Reprinted from Flemings and Lupa (2004) with permission of Elsevier.

5.4 Other Normal Compaction Trend Methods

5.4.1 Other Porosity-Effective Stress Methods

The approach presented in Section 5.2 is one of many approaches that assumes a single relationship between porosity and vertical effective stress. I list two other approaches in Table 5.1: (1) the Geotechnical approach and (2) the Butterfield approach. These approaches differ only in how the compaction behavior is described. For example, the Geotechnical approach assumes that void ratio ($e = \frac{n}{1-n}$) is proportional to the log of effective stress, whereas the Butterfield approach assumes that the specific volume ($v = 1 + e$) is a power law function of the effective stress (Table 5.1). I compare the different approaches on the Eugene Island Pathfinder dataset in the following section.

5.4.2 Velocity-Effective Stress Methods

A common approach is to skip the step of mapping velocity to porosity and instead predict pressure directly from velocity. I present two examples.

Table 5.1 *Different effective stress approaches to predict pore pressure. The examples listed here are only a subsample of the many different approaches that establish a normal compaction trend (left column) and then use this to predict pressure (middle column). Zhang (2011) provides other examples.*

NCT Equation	Pressure Prediction	Name
$n = n_o e^{-\beta\sigma'_v}$	$u = \sigma_v - \frac{1}{\beta}\ln\left(\frac{n_o}{n}\right)$	**Hubbert** Developed: (Hubbert & Rubey, 1959) Applied: (Flemings et al., 2002; Gordon & Flemings, 1998; Hart et al., 1995; Mann & Mackenzie, 1990)
$e_1 = e_o - C_c \log\left(\frac{\sigma'_1}{\sigma'_o}\right)$ $e = \frac{n}{1-n}$	$u = \sigma_v - 10^{\left(\frac{e-e_o}{C_c}\right)}$	**Geotechnical** Developed: (Aplin et al., 1995; Saffer, 2007) Applied: (Saffer, 2007)
$v = v_o\left(\sigma'_v\right)^C$ $v = 1 + e$	$u = \sigma_v - \left(\frac{1+e}{1+e_o}\right)^{\frac{1}{C}}$	**Butterfield** Developed: (Butterfield, 1979) Applied: (Long et al., 2011)
$\sigma'_v = \sigma'_{vh}\left(\frac{V}{V_h}\right)^3$	$u = \sigma_v - \sigma'_{vh}\left(\frac{V}{V_h}\right)^3$ $\log(V_h) = \alpha z + \gamma$	**Eaton – Velocity** Developed: (Eaton, 1975) Applied: (Zhang, 2011)
$V = V_w + A\sigma'^B_v$	$u = \sigma_v - \left(\frac{V-V_w}{A}\right)^{\frac{1}{B}}$	**Bowers** Developed: (Bowers, 1995) Applied: (Zhang, 2011)
$\sigma'_v = \sigma'_{vh}\left(\frac{R}{R_h}\right)^{1.2}$	$u = \sigma_v - \sigma'_{vh}\left(\frac{R}{R_h}\right)^{1.2}$ $\log(R_h) = \alpha z + \gamma$	**Eaton – Resistivity** Developed: (Eaton, 1975) Applied: (Zhang, 2011)

5.4.2.1 Eaton Approach

The Eaton approach (Eaton, 1975) assumes effective stress is a function of velocity,

$$\sigma'_v = \sigma'_{vh}\left(\frac{V}{V_h}\right)^3, \qquad \text{Eq. 5.4}$$

where σ'_{vh} is the hydrostatic effective stress at the depth being predicted and V_h is the velocity at that hydrostatic effective stress. In Eaton's original approach, the log of the velocity was assumed to be linearly proportional to depth under hydrostatic conditions:

$$\log(V_h) = \alpha z + \gamma. \qquad \text{Eq. 5.5}$$

α and γ are derived by regression of velocity versus depth through the hydrostatic section. Bowers (1995) points out that Equation 5.4 implies a velocity-effective stress relationship of the form:

$$V = C\sigma'^{\frac{1}{3}}_v. \qquad \text{Eq. 5.6}$$

Bowers (1995) recommends that V_h be determined directly from Equation 5.6 by cross-plotting the velocity against the effective stress in the hydrostatic section. V_h could also be determined in overpressured sections by calibrating Equation 5.6 with direct pressure measurement (e.g., Section 5.3.2). Finally, perhaps the most common approach to estimate V_h is to calculate it from the Bowers equation (Eq. 5.8) described in the following section.

In Figure 5.10, I have predicted pressure for the Pathfinder well using the Eaton approach (Eq. 5.4) with different 'Eaton' exponents. An exponent of 3 predicts a pressure similar to that predicted with the Hubbert approach (Eq. 5.2, Fig. 5.6). If the exponent is increased to 4, the predicted pressure increases and if the exponent is decreased to 2, the predicted pressure decreases (Fig. 5.10). An exponent greater than 3 amplifies the predicted pressure relative to the normal compaction trend prediction, whereas an exponent less than 3 diminishes the predicted pressure relative to the normal compaction trend. When the Eaton exponent is modified, the velocity versus effective stress relationship is modified in the predicted section relative to the calibrated section. When the exponent is greater than 3, the mudrock is more compacted at a given effective stress in the predicted section than it is in the calibrated section. A common practice is to adjust the Eaton exponent to match the observations. The effect of this is to adjust the velocity versus effective stress relationship in the prediction interval relative to the calibration interval.

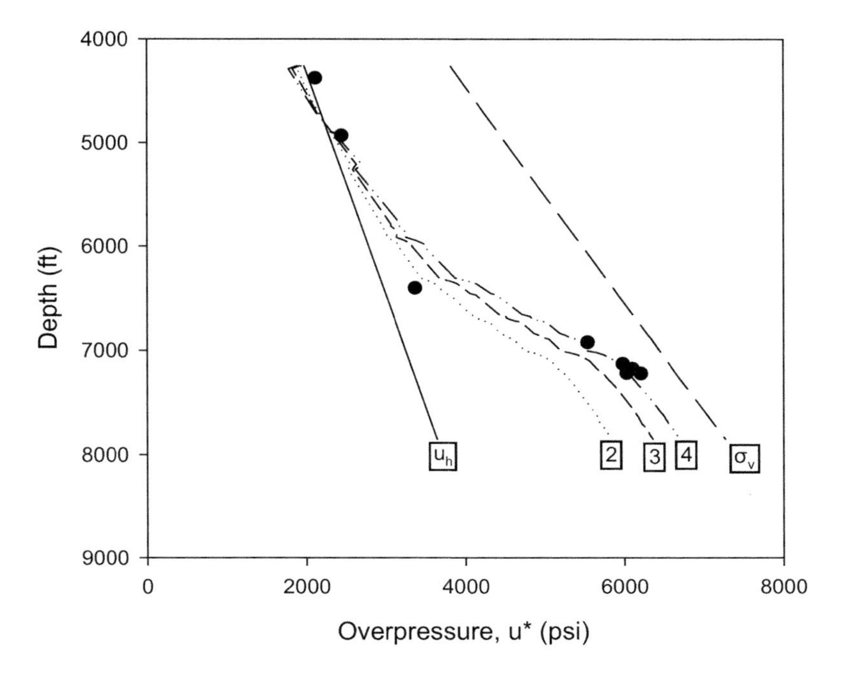

Figure 5.10 Pressure prediction in the Pathfinder well using the Eaton method. Three Eaton exponents are shown. See Section 5.4.2.1 for approach.

5.4.2.2 Bowers Velocity-Effective Stress Approach

Bowers suggested a velocity-effective stress equation of the form:

$$V = V_w + A\sigma_v'^B. \quad \text{Eq. 5.7}$$

Bowers assumed $V_w = 5\,000\,\frac{ft}{s}\left(1\,524\,\frac{m}{s}\right)$. The values for A and B are determined by linear regression through an interval where the velocity and effective stress are known. Equation 5.7 can be rearranged to solve for effective stress:

$$\left(\frac{V - V_\text{W}}{A}\right)^{\frac{1}{B}} = \sigma_v'. \quad \text{Eq. 5.8}$$

This, in turn, can be combined with the effective stress equation (Eq. 2.7) to solve for the pore pressure:

$$u = \sigma_v - \left(\frac{V - V_w}{A}\right)^{\frac{1}{B}}. \quad \text{Eq. 5.9}$$

5.4.3 Resistivity-Based Methods

Eaton's method (Eaton, 1975) is also used to predict pressure from resistivity, where

$$\sigma'_v = \sigma'_{vh}\left(\frac{R}{R_h}\right)^{1.2}, \qquad \text{Eq. 5.10}$$

where R is the measured mudrock resistivity and R_h is the resistivity of that formation if the pore pressure were hydrostatic. As in the case for velocity (Eq. 5.5), the log of the resistivity is assumed to be linearly proportional to depth under hydrostatic conditions:

$$\log(R_h) = \alpha z + \gamma. \qquad \text{Eq. 5.11}$$

The coefficients α and γ are derived by regression of data through knowns where the effective stresses are known. As discussed above, Equation 5.10 implies a resistivity-effective stress relationship of the form:

$$R = C\sigma_v'^{\frac{1}{1.2}}. \qquad \text{Eq. 5.12}$$

As in the case for velocity, R_h can be determined through Equation 5.12 by cross-plotting the resistivity against the known effective stress. One of the challenges with resistivity-based pore pressure prediction is that the resistivity is strongly impacted by changes in fluid salinity and cation exchange capacity. Variation in these properties will impact the ability to establish a normal compaction trend and to predict pressure.

5.5 Comparison of Pressure Prediction Models

I return to the Eugene Island 330 A20-ST2 well to compare results from five pressure prediction approaches described above (Figures 5.11, 5.12, and 5.13). The approach is similar to that presented by Gutierrez et al. (2006) who compared a number of different compaction models to available data. The first step is to determine the normal compaction trend for each approach. I performed a regression of the data between 1 000 ft and 5 300 ft below seafloor in the Shell 331 #1 well where pressures are hydrostatic (Fig. 5.11). Near the seafloor, the porosity changes extremely rapidly with depth and practical experience has suggested it is hard to capture the compaction behavior. For this reason, I exclude the first 1 000 ft of data when I calculate the regressions.

All of the approaches describe compaction adequately (Table 5.2, Fig. 5.11). The regression coefficients for the Hubbert and Eaton methods are 0.96 and 0.95,

Table 5.2 *Normal compaction parameters for five normal compaction trends (Table 5.1). R is the correlation coefficient for linear regression. RMS is the root mean square of the difference between the predicted effective stress and the effective stress predicted by the regression relationship (the horizontal distance between a data point and the dashed line in Figure 5.11).*

Function	Parameter	Value	Parameter	Value	R^2	RMS
Hubbert	β	2.15×10^{-4}	n_o	0.384	0.96	126
Eaton	a	4.019×10^{-5}	γ	3.716	0.95	68
Bowers	A	4.475	B	0.852	0.92	115
Butterfield	v_o	2.6689	C	−0.090	0.92	190
Geotechnical	C_c	-2.976×10^{-1}	e_o	1.329	0.92	174

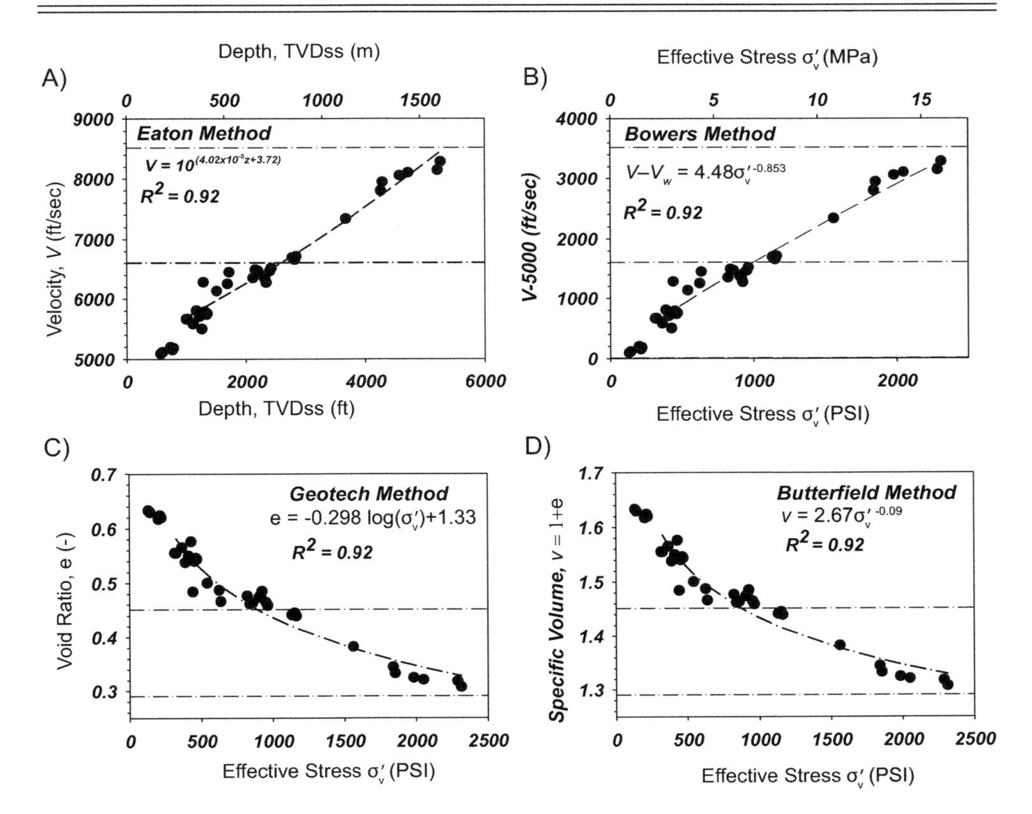

Figure 5.11 Determination of the normal compaction parameters for four common approaches (Table 5.1) for the Eugene Island 330 field.

respectively, and the coefficients for the other approaches are 0.92. A second approach to examine the quality of the normal compaction trend is to predict the pressure or overpressure for each porosity value within the hydrostatic section (Fig. 5.12). In this

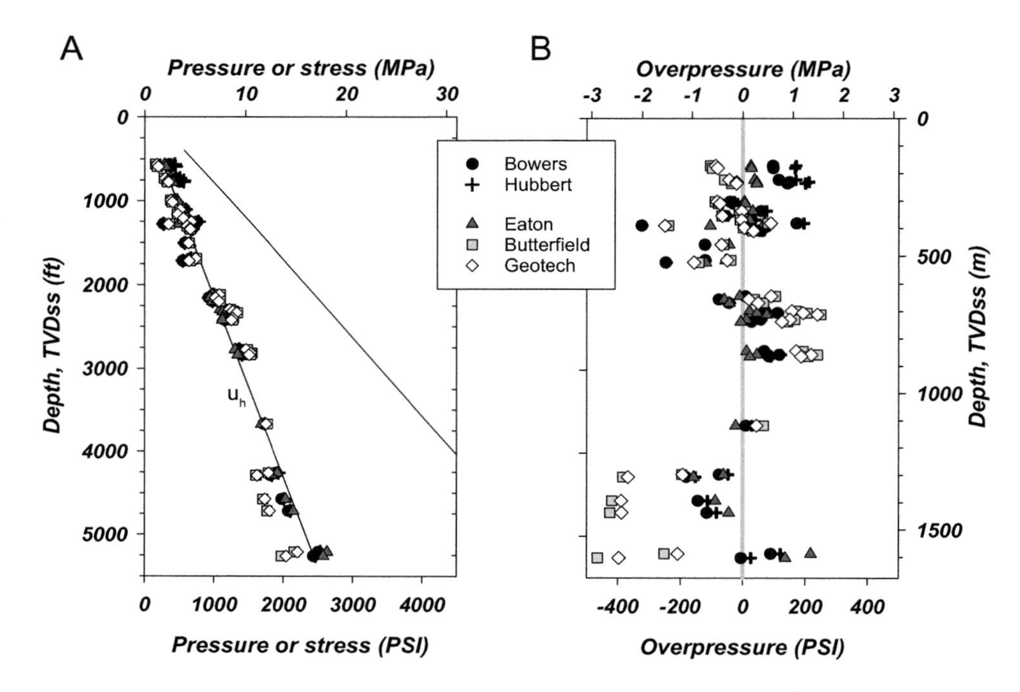

Figure 5.12 Pressure prediction in the 331 #1 well over the hydrostatic section using five different pore pressure prediction models (Table 5.1) with parameters defined in Table 5.2. The normal compaction trend for each approach was defined by regression of the data at the 331 #1 well between 1 000 ft and 5 300 ft. Thus, if the approach were exact, the pore pressure would everywhere be hydrostatic or the overpressure would be zero.

interval, the pore pressure should fall along the hydrostatic line (black line, Fig. 5.12a) and the overpressure should equal zero (grey line, Fig. 5.12b). In this view, two trends are apparent. First, at the base of the section (between 4 000 ft and 5 300 ft), the Butterfield and Geotechnical methods predict pressures lower than the hydrostatic pressure. This is because at higher effective stresses, the Butterfield and Geotechnical regressions overpredict the porosity value for a given effective stress (the regression line is above the values) (Fig. 5.11). Thus, although these methods have reasonable regressions, at higher effective stresses they systematically underpredict the pressure. In contrast, the Eaton method slightly overpredicts the pressure at the deeper locations (Fig. 5.12). The Hubbert and Bowers methods generally follow the hydrostatic pressure over the regressed interval (Fig. 5.12).

I predict pore pressure in the EI-330 A20 ST2 well for each of the normal compaction trends derived for the different compaction functions (Fig. 5.13). The Butterfield and Geotechnical approaches systematically underpredict pressures, and are less than hydrostatic, from the start of the logged interval (~4 300 ft) to

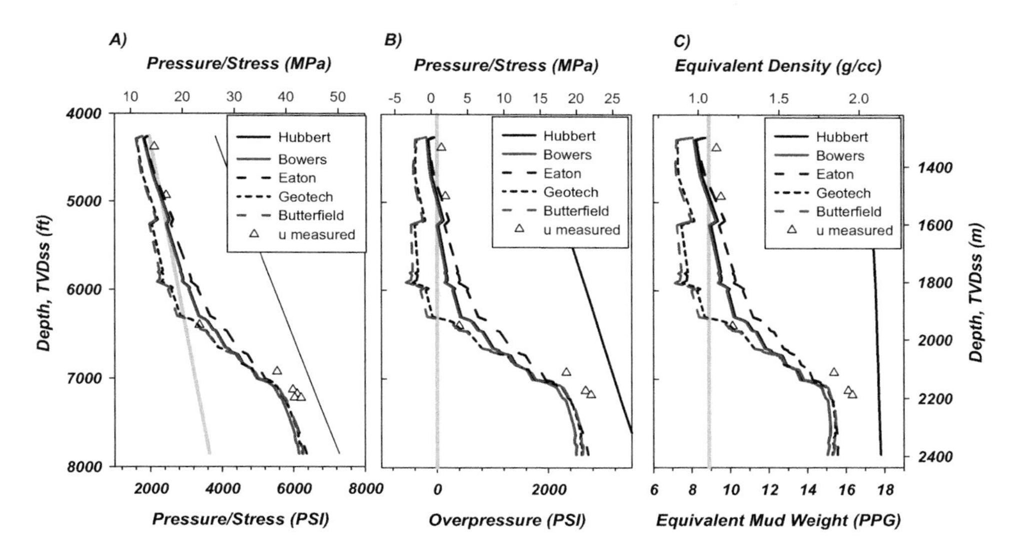

Figure 5.13 Pore pressure prediction at the EI-330 A20 well using five different pore pressure prediction models (Table 5.1) with parameters defined in Table 5.2.

a depth of 6 300 ft before they start to rise (Fig. 5.13). For the first 1 000 ft, this underprediction results from the fact that both of the regressions do not capture the highest stress points (above 2 000 psi, Fig. 5.11 c, d). Between 5 300 ft and 6 300 ft, the porosities actually lie outside the regressed interval (to the left of dash-dot line, Fig. 5.6a). In contrast, through this interval, the Eaton method predicts pressures that are slightly greater than hydrostatic. The Hubbert and Bowers methods more or less track the hydrostatic through this interval. All the prediction methods converge to within 100 psi at a depth of 7 300 ft.

There are a limited number of in situ pressure measurements (triangles, Fig. 5.13). At the shallowest depths, two pore pressure measurements record hydrostatic pressure at ~4 400 ft and ~4 900 ft. The Bowers, Hubbert, and Eaton methods very slightly underpredict these values, while the Butterfield and Geotechnical methods significantly underpredict them. There is an additional pressure measurement at 6 400 ft. The Bowers, Hubbert, and Eaton methods overpredict these values, while the Geotechnical and Butterfield approaches are close to the measured values. Finally, all the pressure prediction techniques underpredict the pressures observed at depth (between 6 900 ft and 7 300 ft). The overpressure and the equivalent mud weight plots (Fig. 5.13b, c) provide a different and expanded view of the pressure field. At a shallow depth, the EMW is underpredicted by almost a pound per gallon and at the base of the well, the EMW plot underpredicts by about a pound per gallon.

This analysis suggests the Hubbert and Bowers approaches most successfully describe the normal compaction trend and most successfully predict overpressure in the EI-330 field. This does not imply that the Hubbert or Bowers approach is the optimal approach in other settings. Some of the approaches that are not as successful in this location may more accurately fit the data in other locations. Furthermore, there may be a different compaction model that more accurately describes the compaction behavior in a particular basin. For example, Hauser et al. (2013) suggest a new compaction approach for pore pressure prediction in Miocene sediments in the Gulf of Mexico.

5.6 Summary

I have demonstrated how to predict pressure from the compaction state of mudrocks. The approach is founded on the concept that compaction state of the mudrock can be used to predict the pore pressure. In this chapter, we assumed that the compaction varies only as a function of vertical effective stress and that there is a single relationship between petrophysical properties (e.g., porosity, resistivity, velocity) and effective stress. There are myriad empirical relationships that describe this compaction behavior. The success of these approaches depends on how well the assumed compaction function describes the compaction behavior.

Ultimately, there is no one optimal pore pressure prediction equation or approach. What is most important is to study the compaction trend (e.g., the relationship of velocity, resistivity, or porosity to effective stress) and rigorously compare proposed compaction models with available data. From this, one can understand where pressure predictions are well constrained by data and where they are not. Finally, all of the approaches presented systematically underpredict the observed pressures, by more or less the same amount, deep in the well (Fig. 5.13). This observation is common, and we discuss different models that account for this behavior in Chapter 6.

6

Pore Pressure Prediction: Unloading, Diagenesis, and Non-Uniaxial Strain

6.1 Introduction

In Chapter 5, I described how to predict pressure from a single normal compaction trend (NCT). Given the porosity, a compaction trend, and the total vertical stress, the pore pressure is determined (e.g., Eq. 5.2). This is a simple way to predict pressure that is useful, fast, and intuitive. Unfortunately, it does not successfully predict pressure in many settings. This is illustrated in Figure 5.4 where the normal compaction trend approach underpredicts the observed pressures.

I now describe approaches that incorporate more complex processes to predict pressure. These approaches can be divided into two topics. One is the case where we continue to assume uniaxial strain but consider more complex material behavior. This includes unloading, diagenesis (e.g., smectite to illite), and creep. The second approach is to relax the assumption of uniaxial strain and include more complex stress states characteristic of different tectonic settings.

6.2 Pore Pressure Prediction in the Presence of Unloading

Unloading occurs when the effective stress decreases. It occurs when the total vertical stress (σ_v) decreases or the pore pressure (u) increases (Eq. 2.7). Unloading can occur by erosion, which decreases total stress, or by an increase in pore pressure, which could be caused by heating and pore fluid expansion, generation of hydrocarbons, or flow into the location. In these cases, even if burial is ongoing (and total stress is increasing), unloading can occur due to the presence of pressure sources during burial (e.g., Flemings et al., 2002).

The impact of unloading on compaction can be viewed through a simple uniaxial experiment as discussed in Chapter 3 (point A, Fig. 3.5). When the rock is loaded, it compacts along the virgin compression line from B to C (Fig. 6.1a). At point C, the material is unloaded (vertical stress is reduced) along a different compaction trend to point D. This effect is also observed when velocity is plotted against effective

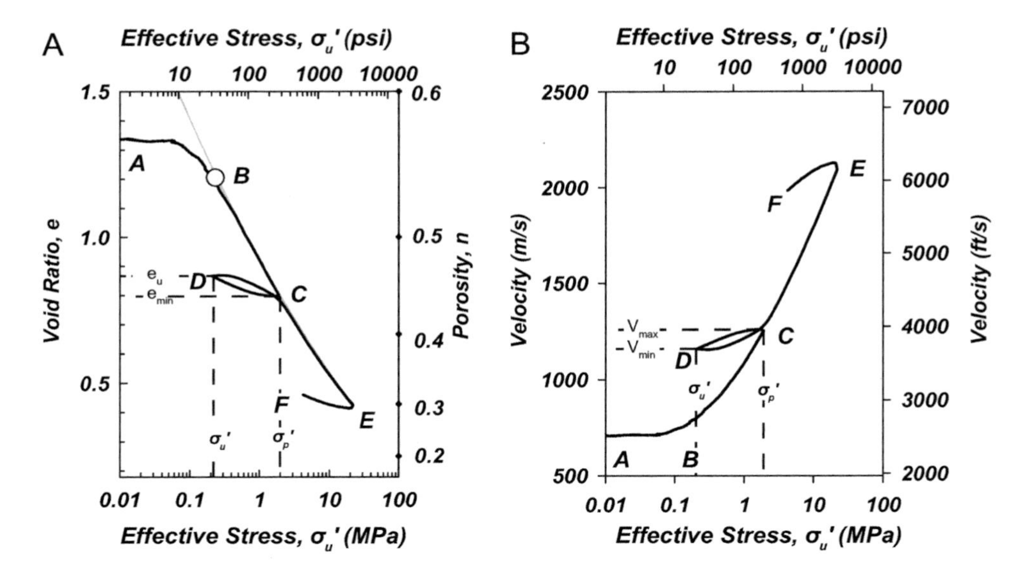

Figure 6.1 (a) Void ratio versus effective stress plot specimen from U1324 C-1 H-1WR at 51.2 mbsf in the Ursa Basin, Gulf of Mexico (see Figure 3.5 for details). (b) Velocity versus effective stress for this experiment. Velocities are calculated from porosity with Equation 5.3 assuming $dt_{ma} = 220\frac{\mu sec}{m}$ and $f = 2.19$.

stress (Fig. 6.1b). During unloading, a different velocity-effective stress path is followed relative to during virgin compression. Below, I present two pore pressure prediction approaches that account for this behavior.

6.2.1 Geotechnical Model

I first incorporate unloading into the Geotechnical model (Table 5.1) to predict pressure. Consider a rock that is loaded along the virgin compression curve to a preconsolidation stress (σ'_p) at point C (Fig. 6.1). The slope of this compression in semi-log space is C_c (Fig. 3.5). It is then unloaded along a slope C_e to an effective stress σ'_u and a void ratio e_u at point D. The void ratio at point C (e_{min}) is described by the virgin compression trend:

$$e_{min} = e_o + C_c log\sigma'_p. \qquad \text{Eq. 6.1}$$

To determine the effective stress after unloading, we project back along the unloading curve to the void ratio (e_u) at point D:

$$\sigma'_u = \sigma'_p 10^{\left(\frac{e_u - e_{min}}{C_e}\right)}. \qquad \text{Eq. 6.2}$$

Equation 6.2 is rearranged to predict the unloaded pore pressure:

$$u = \sigma_v - \sigma'_p 10^{\left(\frac{e_u - e_{min}}{C_e}\right)}. \quad \text{Eq. 6.3}$$

I illustrate this approach with the EI-330 A20 ST2 well (Fig. 6.2). $e_{\min}$ is the void ratio at the preconsolidation stress. I assume that all the mudrock that was unloaded was subjected to this stress. I assume that the maximum velocity (and hence the minimum void ratio) in the well records the preconsolidation stress following the approach of Bowers (1995). This maximum velocity is 8 439 ft/sec at 6 000 ft of depth (Fig. 5.4), which equates to a porosity of 0.23 or a void ratio of 0.30 (Eq. 5.3). Given this void ratio, the preconsolidation stress is then calculated from Equation 6.1 with the result that $\sigma'_p = 2\,940$ psi. I explore the effects of different values of C_e and solve for u (Eq. 6.3) in the Pathfinder well (Fig. 6.2).

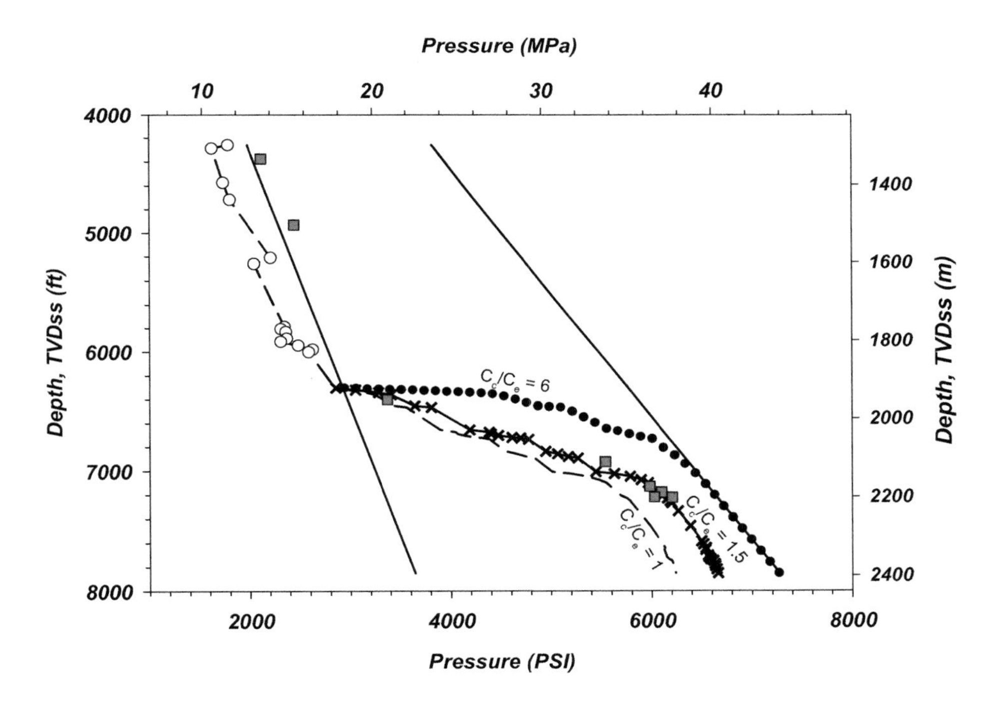

Figure 6.2 Pore pressure prediction for the EI-A20 ST2 well exploring the effect of unloading with the Geotechnical model. The black dashed line illustrates the pressure prediction with no hysteresis ($C_c = C_e$). This is the same pore pressure prediction made with the Geotechnical model in Chapter 5 (Fig. 5.13), using the parameters in Table 5.2. The dotted line illustrates the prediction pressure with $C_c/C_e = 6$, as is observed experimentally (Fig. 6.1). With $C_c/C_e = 1.5$ the pressure prediction (solid line) matches the in situ pressure observations (red squares).

When $C_c/C_e = 1$, the pressure prediction from the Geotechnical example in Chapter 5 is reproduced (Fig. 5.12). When $C_c/C_e = 6$, as is observed in the experimental results (Fig. 3.5), the pressures rapidly converge to the overburden stress and are greater than the observed pressures (red squares, Fig. 6.2). In contrast when $C_c/C_e = 1.5$, the predicted pressures, more or less, match the observed pressures.

The stress path of the mudrocks is illustrated in Figure 6.3. All of the mudrocks beneath 6 000 ft were compressed along the virgin compression trend to point A, which is the present void ratio and effective stress at 6 000 ft ($e_{min} = 0.3$, or $n = 0.23$, $\sigma'_p = 2\,940$) psi (Fig. 6.3). Thereafter, these mudrocks were unloaded to their current state. If we assume $C_c/C_e = 1.5$, then the unload path followed the solid black line (Fig. 6.3). For example, the mudrock at 7 300 ft is presently at point

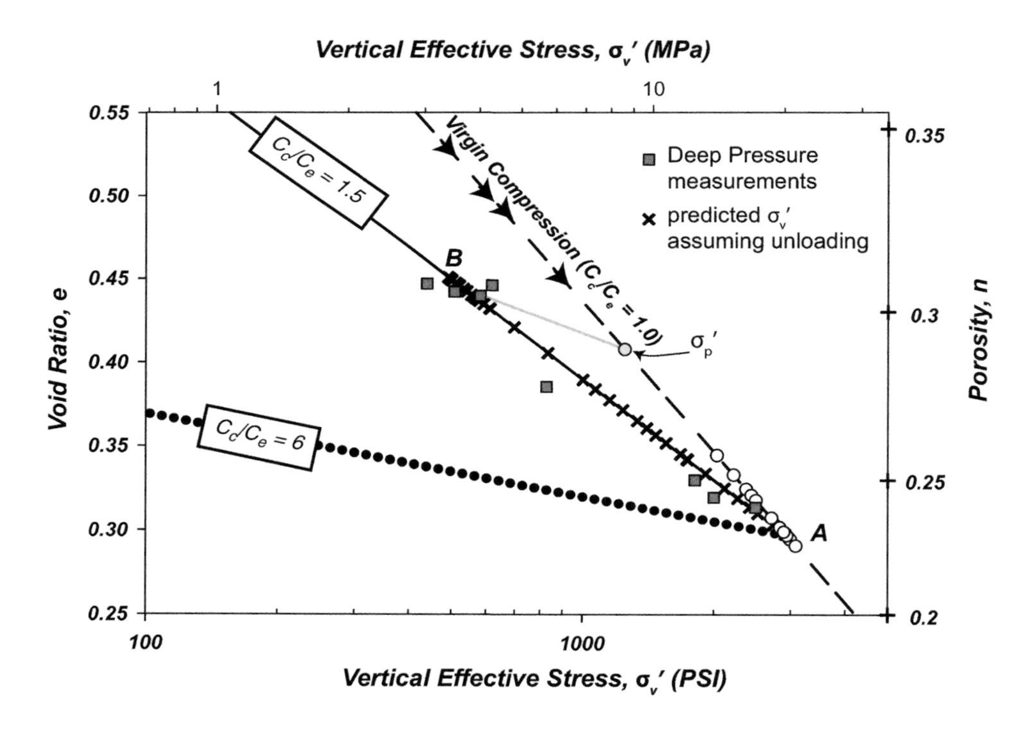

Figure 6.3 Possible stress paths for deep mudrock at EI-A20 ST2 well. The virgin compression trend for the Geotechnical model (parameters in Table 5.2) is shown with the dashed line. The unloading curve for $C_c/C_e = 6$ (dotted line) and 1.5 (solid line) are shown. Actual measured pressures (expressed as effective stresses, $\sigma'_v = \sigma_v - u$) and associated mudrock void ratios are shown with red squares. Predicted void ratios from the well assuming unloading of $C_c/C_e = 1.5$ are shown as X's. Predicted void ratios from above 6 000 ft from the virgin compression trend are shown with open circles. σ'_p is the preconsolidation stress measured on one core from this location (Stump & Flemings, 2001).

B and has unloaded from point A to B (Fig. 6.3). This implies that these rocks were compressed to a void ratio of 0.3 and were then expanded to a void ratio of 0.45. It is not intuitive that this makes sense and in Section 4.3.2.4, I discuss how the present-day effective stress profile may have little to do with the actual stress history that a particular rock undergoes. A second challenge is that to match the observed data, the slope of the virgin compaction curve is only 1.5 times that of the expansion curve. This is much less than is observed experimentally, and it suggests that the assumption that all samples are loaded to the maximum stress in the well may not be appropriate.

6.2.2 Bowers Unloading Model

Bowers (1995) proposed that the velocity after unloading (V_u) is described by:

$$V_u = V_w + A\left(\sigma'_p\left(\frac{\sigma'_v}{\sigma'_p}\right)^{\left(\frac{1}{U}\right)}\right)^B. \qquad \text{Eq. 6.4}$$

A, B, and V_w were previously defined (Eq. 5.7, Table 5.1). σ'_p, is the preconsolidation stress and U describes the slope of the velocity-effective stress unloading curve (Fig. 6.1b). The preconsolidation stress is determined from the velocity log through the original Bowers relationship (Table 5.1):

$$\sigma'_p = \left(\frac{V_{max} - V_w}{A}\right)^{\frac{1}{B}}. \qquad \text{Eq. 6.5}$$

Equations 6.4, 6.5, and 5.2 are combined to solve for pore pressure:

$$u = \sigma_v - \sigma'_p\left[\frac{1}{\sigma'_p}\left(\frac{V - V_w}{A}\right)^{\frac{1}{B}}\right]^U. \qquad \text{Eq. 6.6}$$

I apply Equation 6.6 to predict pressure in the EI-330 A20 well. As described in Section 6.3.1, it is assumed that all the mudrock that was unloaded was subjected to the same preconsolidation stress (σ'_p) and that the maximum velocity in the well records this stress. I explore the effects of different values of U on the predicted pressure in the Pathfinder well (Fig. 6.4). A value of $U = 1$ reproduces the normal compaction trend and no unloading is present. A value of $U = 5$ results in a rapid increase in the pore pressure to the lithostatic pressure. Finally, a value of $U = 1.8$ results in a predicted pressure that is similar to the observed pressures at depth. Bowers (1995) suggests that U is variable but that a value of 3 has been found successful in the Gulf of Mexico.

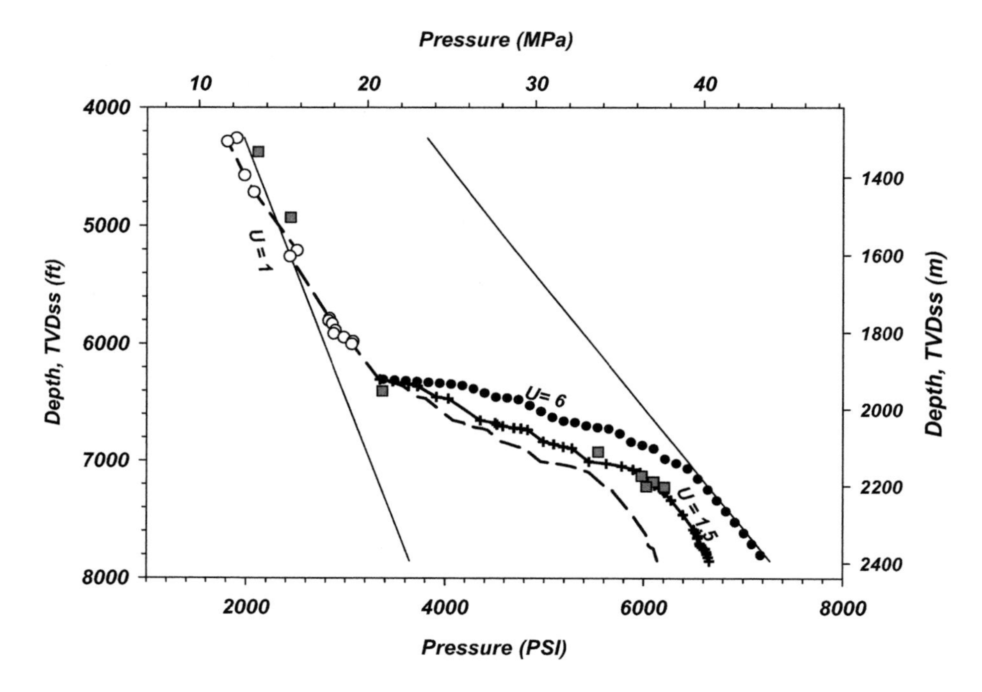

Figure 6.4 Pore pressure prediction for the EI-330 A20 ST2 well exploring the effect of unloading with the Bowers model. The dashed line illustrates the pressure prediction with no unloading ($U = 1$). This is the same pore pressure prediction made with the Bowers model in Chapter 5 (Fig. 5.13), using the parameters in Table 5.2. The solid line illustrates the predicted pressure with $U = 1.5$. The pressure prediction matches the in situ pressure observations. With $U = 6$, the predicted pressure rises to the lithostatic stress and is greater than the measured pressures (dotted line).

The stress paths are illustrated on the velocity versus effective stress plot (Fig. 6.5). In the case of $U = 1.8$, the rocks originally compacted along the normal compaction trend from the seafloor (zero effective stress) to point A. They then unloaded to point B. The velocity decline is quite large during the unload.

6.2.3 Determination of the Preconsolidation Stress

To predict pore pressure with an unloading model that assumes elastoplastic behavior, we must know the preconsolidation stress. In the preceding examples, I assumed that the mudrocks beneath 6 000 ft were preconsolidated to the present stress at 6 000 ft (2 940 psi) where the minimum in porosity or maximum in velocity was encountered. In the Geotechnical model, the mudrock present today at 7 200 ft was compressed to a porosity of 0.23 ($e = 0.3$), and then unloaded to

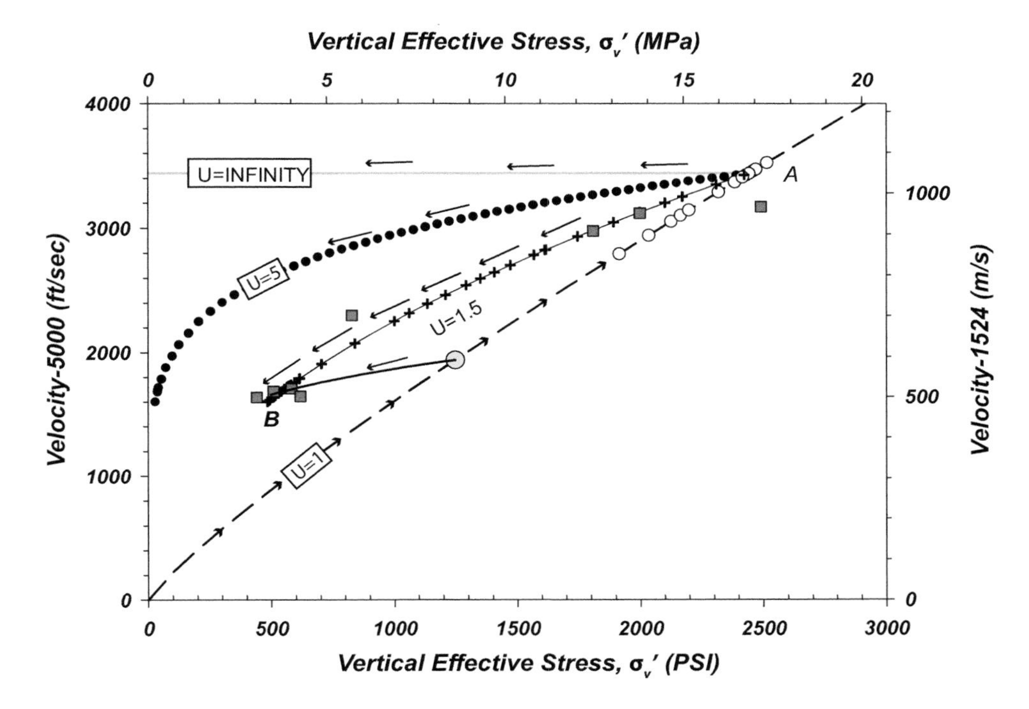

Figure 6.5 Velocity versus effective stress. The black dashed line is the normal compaction trend for the Bowers model (Table 5.1) with the parameters in Table 5.2. In this example all mudrocks that are unloaded are first loaded to point A before they are unloaded. The solid, dotted, and grey lines represent model predictions for values of $U = 1.5$, 5, and infinity, respectively. The red squares mark the actual measured pressures in the A20 ST2 well. σ'_p (yellow circle) is the preconsolidation stress measured on one core from this location (Stump & Flemings, 2001).

its present-day porosity of 0.3 ($e = 0.45$). In the Bowers model, the mudrocks were compressed to a velocity of 8 439 ft/sec and then unloaded to a velocity of 6 636 ft/sec. These are very large differences in both porosity and velocity due to unloading. In fact, as shown in Chapter 4 (e.g., Fig. 4.17), it is unlikely that the deep mudrocks underwent compaction to the maximum effective stress state that is observed in the well today. Thus, the approach of picking the minimum in porosity (or maximum in velocity) as the unloaded effective stress overestimates the amount of unloading.

The EI-330 A20 ST2 well is fairly unique because preconsolidation stresses were derived from uniaxial consolidation experiments on samples taken from deep in the well (Stump & Flemings, 2001). Samples taken from 7 350 ft record a preconsolidation stress of 1 248 psi (yellow circles, Figures 6.3 and 6.5). This value is far less than the original assumption of 2 940 psi. With 1 248 psi as an estimate of the

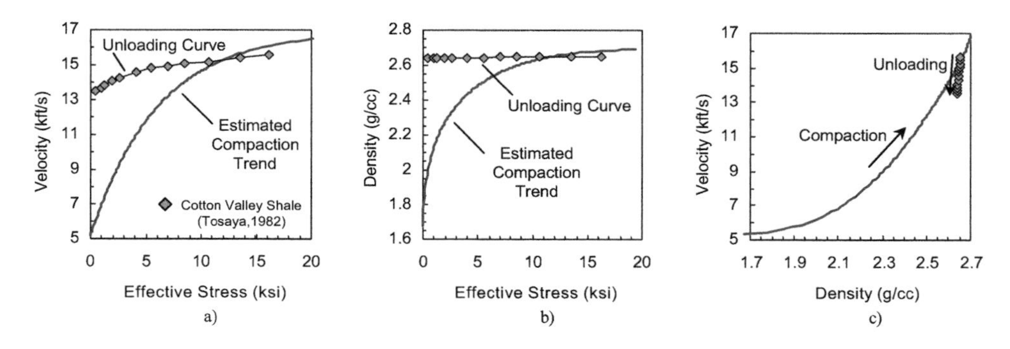

Figure 6.6 Mudrock compaction and unloading behavior. (a) Velocity versus effective stress for virgin compression trend (blue) and unloading (diamonds). (b) Density versus effective stress for virgin compression trend (blue) and for unloading (diamonds). (c) Velocity versus density cross plot. The unloading trend is distinct from the loading trend. Reprinted with permission of Offshore Technology Conference, from Bowers (2001).

preconsolidation stress for these samples, a C_c/C_e ratio of 3 or a U value of 5 successfully matches the observed measured pressures (Figures 6.3 and 6.5).

Unfortunately, it is unusual to have core from which the preconsolidation stress can be determined. Thus, it is challenging to know what value of σ'_p to use for the unloading calculations. Bowers (2001) suggested that transport properties (e.g., sonic velocity, permeability, resistivity) were more sensitive to unloading than bulk properties (e.g., density) and that this behavior could be used to determine the preconsolidation stress. He suggests that for the same drop in effective stress, the velocity drops fractionally more than the density, which does not change significantly (Fig. 6.6a, b). If this is the case, then a velocity-density cross plot can be used to interpret the preconsolidation stress (Fig. 6.6c).

Bowers (2001) proposed that the velocity-density relationship during normal compaction can be described by

$$V = V_o + A(\rho - \rho_0)^B, \qquad \text{Eq. 6.7}$$

where V_o, ρ_0, A, and B are parameters derived from a regression of the normal compaction trend. A and B are not the same as used in the Bowers equation (Table 5.2). Bowers (2001) then proposes that the unloading path is described by

$$\rho - \rho_0 = (\rho_{max} - \rho_0)\left(\frac{\rho_V - \rho_0}{\rho_{max} - \rho_0}\right)^{\mu}, \qquad \text{Eq. 6.8}$$

where ρ is the density, ρ_0 is the reference porosity in Equation 6.7, and ρ_V is the density obtained by substituting the observed velocity into Equation 6.7. ρ_{max} is the density at the preconsolidation stress: this is where the unloading curve (Eq. 6.8) intersects the virgin compaction curve (Eq. 6.7). μ is an unloading coefficient. Bowers suggests using $\mu = \frac{1}{U}$ as a default value. Equation 6.8 can be rearranged to solve for ρ_{max}:

$$\rho_{max} = \rho_0 + \left[\frac{\rho - \rho_0}{(\rho_V - \rho_0)^{\mu}}\right]^{\frac{1}{(1-\mu)}}. \qquad \text{Eq. 6.9}$$

V_{max}, the velocity at the preconsolidation stress, is obtained by substituting ρ_{max} into Equation 6.7.

Bowers (2001) applied this approach to the EI-330 A20 ST2 Pathfinder well (Figures 6.7 and 6.8). The blue line represents the virgin compression trend. The red '+' signs represent the observed velocities and densities. The cyan line records the interpreted density and pressures at the preconsolidation stress. The velocity-density cross plot (Fig. 6.7c) illustrates that material above approximately 6 000 ft lie on the normal compression line whereas material below this (red '+' signs) lie off that line. For example, from point A to B (6 000 ft to 7 000 ft in depth), the velocity density cross plot gets successively further from the normal compression line. However, from B to C, the unloaded points (red) get successively closer to

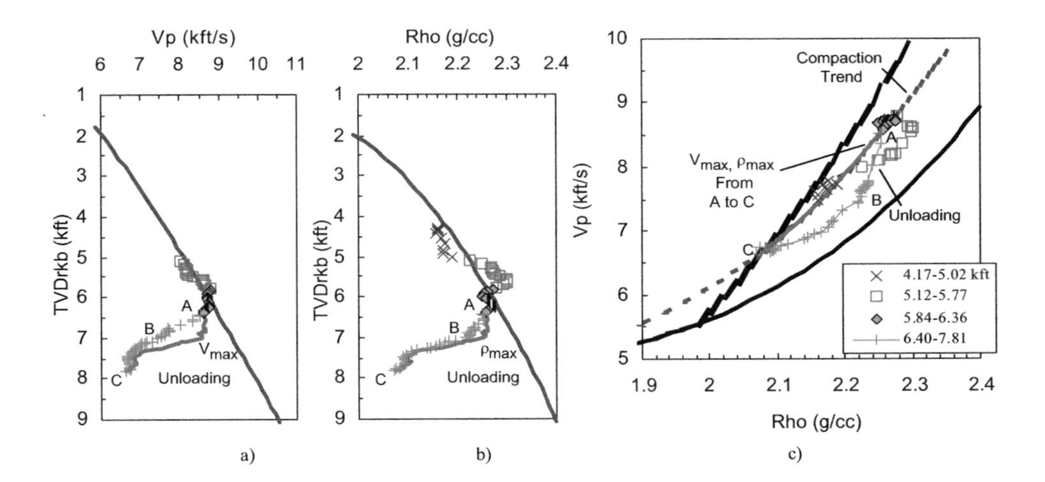

Figure 6.7 (a) Observed velocity versus depth (red crosses), and modeled velocity at preconsolidation stress (V_{max}, blue line) for EI-330 Pathfinder well. (b) Observed density versus depth (red crosses), and modeled density at preconsolidation stress (blue line). (c) Velocity-density cross plot. Normal compaction line is shown with the blue dashed line. Unloaded zone is shown with red '+' signs where data deviate from normal compaction trend. Difference between normal compression trend and unloaded compression trend is greatest at point B. Reprinted with permission of Offshore Technology Conference, from Bowers (2001).

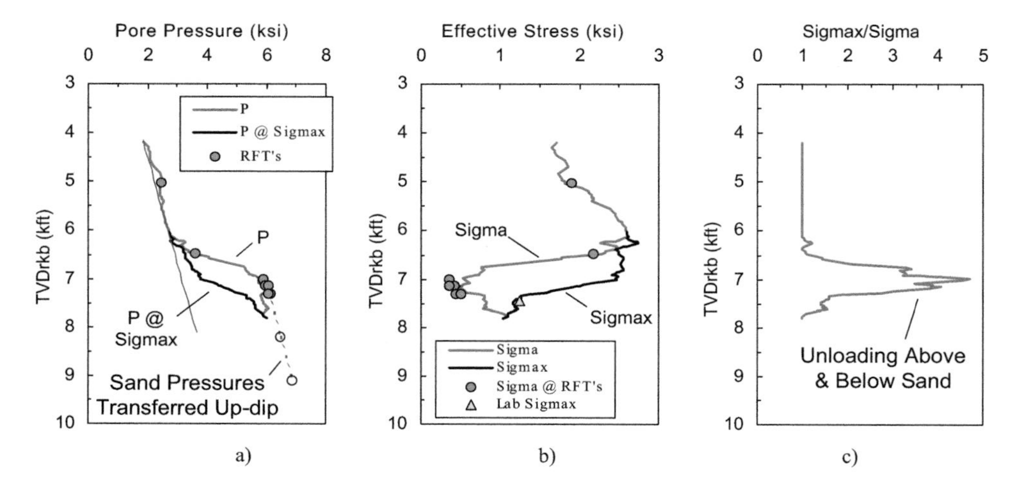

Figure 6.8 (a) Predicted pore pressure in EI-330 Pathfinder well. Black line records pressure predicted from Bowers' traditional approach (Table 5.1, Eq. 5.9). Red line shows pore pressure predicted with unloading accounted for. (b) Effective stress as a function of depth. (c) The ratio of maximum past stress to present effective stress (Sigmax/Sigma). Reprinted with permission of Offshore Technology Conference, from Bowers (2001).

the normal compression trend. This suggests that the degree of unloading varies over the depth of the well.

Figure 6.8 illustrates the final pore pressure prediction. The unloaded pore pressure prediction (red line) matches the observed pressures (green circles) (Fig. 6.8a). The experimentally derived preconsolidation stress lies on the modeled preconsolidation stress (yellow triangle overlies black line, Fig. 6.8b). Bowers (2001) interprets a discrete interval of unloading (Fig. 6.8c). This approach has significant promise, but it remains to be seen whether it can be applied consistently.

6.3 Pore Pressure Prediction with Smectite-Illite Transformation

An increase in pore pressure above what is predicted by the normal compaction trend can also be caused by weakening of the soil framework. In this context, the strength of the spring (the bulk rock) declines and the pore fluid takes on a greater fraction of the load. An example of this is a sensitive clay, where when the natural structure of the soil is destroyed, the material is much weaker (Craig, 2004). A second example of framework weakening, which I describe below, is where the framework becomes weakened due to diagenetic reactions during burial.

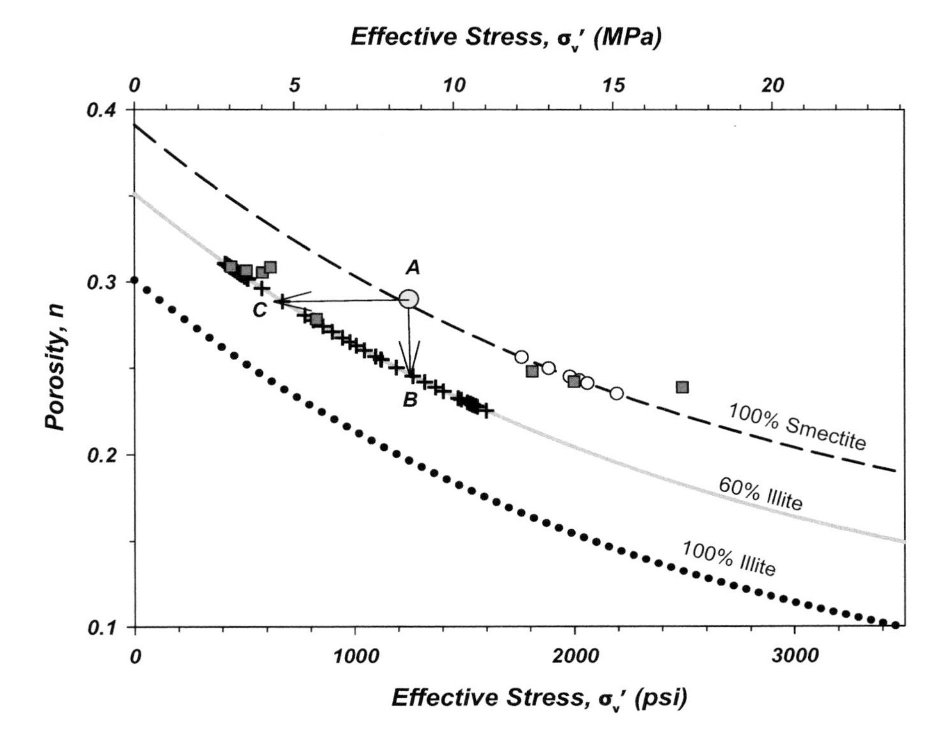

Figure 6.9 Compression curve for smectite-rich and illite-rich mudrocks at the EI-330 Pathfinder well. The dashed curve describes the compaction of the smectite mudrock (Equation 6.10 with $n_m = 0.12$). The dotted line describes the compaction curve for material completely converted to illite (Equation 6.10 with $n_m = 0.03$). The grey line describes material that has retained 40% of the smectite within its layers ($n_m = 0.08$). The red squares represent the in situ pressure measurements, expressed as effective stress, plotted against the porosity in the bounding mudrock. The yellow circle records the preconsolidation stress determined by Stump and Flemings (2001).

6.3.1 Smectite-Illite Diagenesis

The conversion of smectite to illite (illitization) has been incorporated into pore pressure prediction (Lahann, 2002; Lahann & Swarbrick, 2011; Lopez et al., 2004; Wilhelm et al., 1998). While these approaches differ in their details, they all assume that the compaction behavior shifts to allow greater compaction for a given effective stress as illitization proceeds. Lahann (2002) and Lahann and Swarbrick (2011) propose that as smectite is converted to illite, the bound water is expelled, and the normal compression curve is shifted: at the same effective stress the illite is more compressed than the smectite (Fig. 6.9). If the system is drained (fluid is expelled from the system), then the material will move from point A to point B (Fig. 6.9) as the mudrock becomes more illite rich. Alternatively, if no fluid escapes, then the system moves horizontally to the left from A to C (Fig. 6.9). In this case, the increase in pore pressure is equal to the decrease in effective stress. Lahann (2002) terms this load transfer. It is also unloading (a decrease in effective stress), but unloading associated with a weakening of the framework.

Lahann (2002) proposed a modified form of the Hubbert equation to describe the compression curves for smectite- and illite-rich mudrocks as

$$n = n_m + n_o e^{-\beta\sigma_v{}'}, \qquad \text{Eq. 6.10}$$

where n_m is the bound water in the clay structure: $n_m = 0.12$ for smectitic material and $n_m = 0.03$ for illitic material. Equation 6.10 is rearranged to solve for pore pressure:

$$u = \sigma_v - \frac{1}{\beta}\ln\left(\frac{n_o}{n - n_m}\right). \qquad \text{Eq. 6.11}$$

6.3.2 Pore Pressure Prediction at EI-330 A20 ST2 Well

To predict pressure in the A20 ST2 well, we first solve for n_o and β in the shallow section where we assume there has been no conversion to illite ($n_m = 0.12$). n_o and β are derived from a regression of $n - n_m$ versus $n_o e^{-\beta\sigma_v{}'}$ using the shallow data in the 331 well, employing the approach described in Chapter 5. We find $n_o = 0.27$ and $\beta = -3.91 * 10^{-4} psi^{-1}$. These values are different from those derived using the Geotechnical method (Table 5.2) because the equation is different (compare Equation 6.11 to Equation 5.2).

Losh et al. (1999) reported the variation of mixed-layer clay expandability (the ratio of smectite to illite and smectite (S/(S + I)) with depth for the A20 ST2 well (Fig. 6.10). Below 1 500 m (~5 000 ft), the maximum smectite fraction is 0.7, substantially less than the values above 1 500 m. Lahann (2002) interpreted the top of the clay transition at 1 500 m (~5 000 ft) and a conversion to 40% smectite (60% illite) in the well at depth. Compression curves for 100% smectite ($n_m = 0.12$), 60% illite ($n_m = 0.8$), and 100% illite ($n_m = 0.03$) are progressively shifted downwards (Fig. 6.9).

We predict pressures for the three compression curves in Figure 6.9. The 100% smectite curve (dashed line, Fig. 6.11) is similar to the previous pore pressure predictions in Chapter 5 (Fig. 5.6). The shallow hydrostatically pressured zone is captured, but at depth the pressures are underpredicted. The 100% illite curve (dotted line) overpredicts the pressure everywhere. The 60% illite curve successfully captures the deeper pressures, but is overpressured in the shallow section (grey line, Fig. 6.11). Finally, a model where the illite fraction linearly increases with

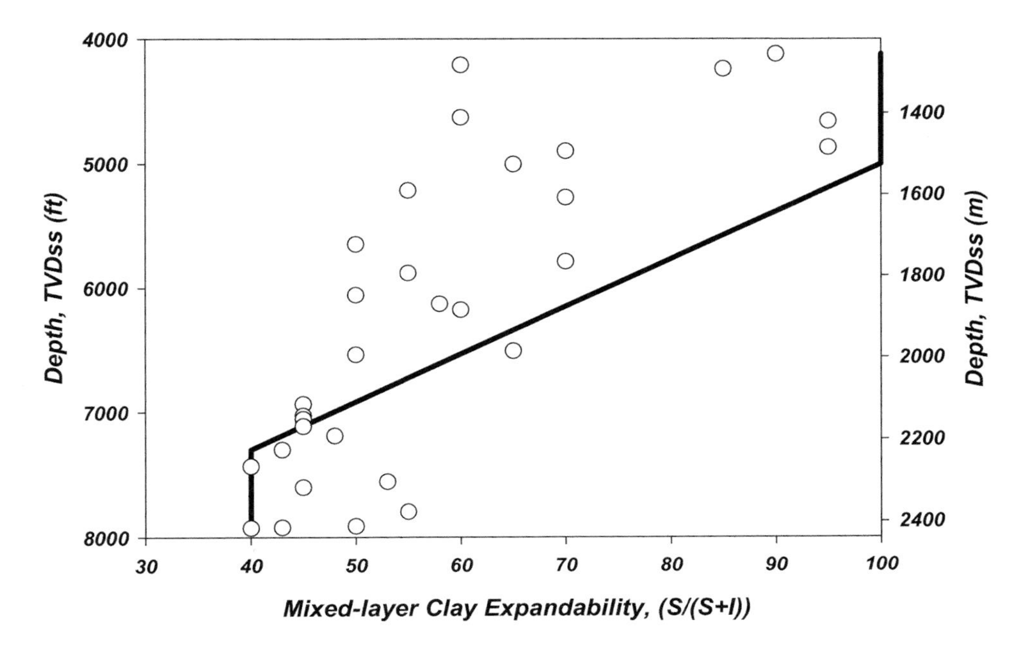

Figure 6.10 Depth variation of mixed-layer clay expandability, smectite/(smectite + illite) (S/(S+I)) for the Pathfinder well. Data from Table 1 of Losh et al. (1999). Note that below 1 500 m, the maximum smectite fraction is 0.7, substantially less than the maximum value above 1 500 m. A top of the clay transition of 1 500 m is interpreted for this well. Black line is the assumed variation in smectite as a function of depth used in the pore pressure model.

depth from a smectite fraction of 100% at 5 300 ft to 40% at 7 300 ft successfully captures the pore pressure behavior although it overestimates the measured pressure at 6 400 ft (black solid line, Fig. 6.11).

We can envision the stress path of the deeper sediments where we have measured pressure through the porosity-effective stress plot (Fig. 6.9). In this view, the framework of the rock weakened as illitization proceeded. As the framework weakened, the pore pressure took on a greater fraction of the overburden stress and the system unloaded from A to C (Fig. 6.9).

This approach was also used to predict the pore pressure at the Macondo well, in the Gulf of Mexico (blue line, Fig. 6.12a and 6.12b) (Pinkston & Flemings, 2019). As in the EI-330 example, the pressures encountered were greater than the predicted pressures when a single compaction trend was used. We interpreted that the increase in temperature with depth drove the transformation from a smectite-rich compaction curve (green line, Fig. 6.12b) to an illite-rich compaction curve (red dashed line, Fig. 6.12). The approach was applied at the Macondo well with success.

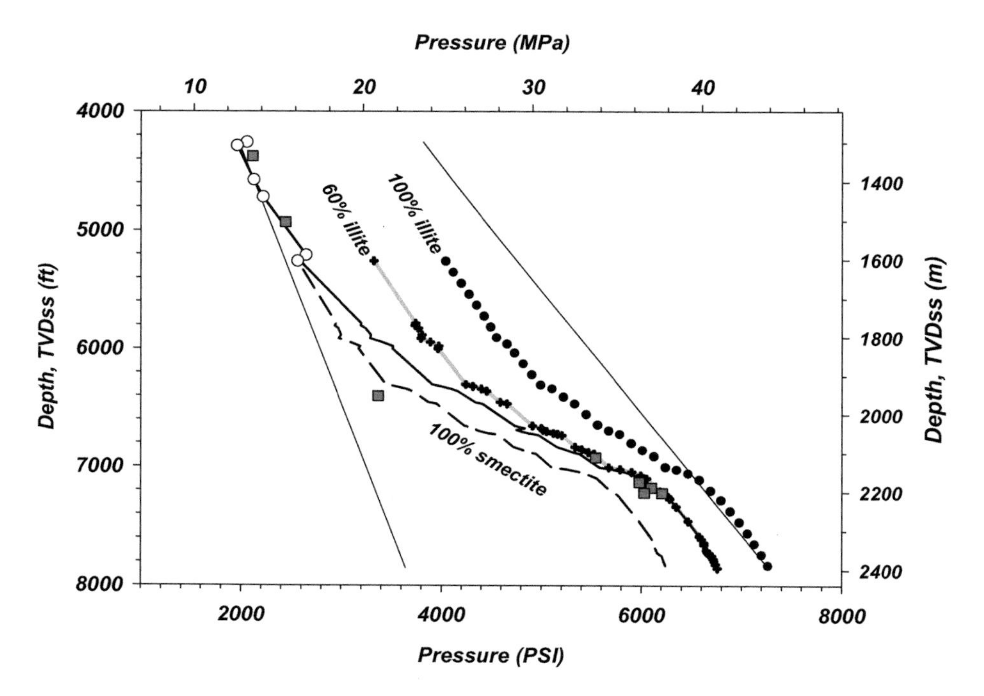

Figure 6.11 Pore pressure prediction for EI-330 A20 ST2 well incorporating the effects of illitization. The dashed line assumes no diagenesis. The dotted line assumes complete conversion to illite. The grey line assumes conversion to 60% illite. The solid black line assumes a linear conversion from no illite to 60% illite from 6 000 to 7 200 ft. This model successfully simulates the observed pressures.

In these examples, I assumed that the compression coefficient, β, does not change during the conversion from smectite to illite. However, Lahann (2002) and Lahann and Swarbrick (2011) suggest that β increases based on empirical calibration. They showed that they needed to increase the value of β in order to reach the measured pore pressures. A fundamental question is what the appropriate compression curves are for smectitic versus illitic material, which is discussed in Chapter 3.

The approach presented could be applied to pressure prediction in a variety of ways. Thermal modeling could be combined with a reaction model to estimate the degree of smectite-illite conversion ahead of the drill bit and predict pressure (Dutta, 1983, 2002a; Lopez et al., 2004; Wilhelm et al., 1998). Alternatively, Katahara (2006) has suggested that density neutron and bulk density logs could be used to discriminate the fraction of bound water in the well, and thus it may also be possible to estimate the degree of smectite conversion during drilling.

The smectite-illite diagenesis model assumes that there is framework collapse as diagenesis proceeds, which shifts the reference porosity (n_o) and the compaction

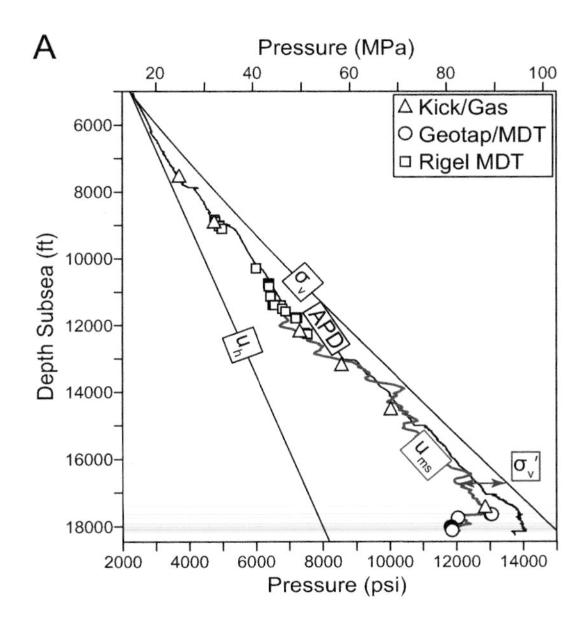

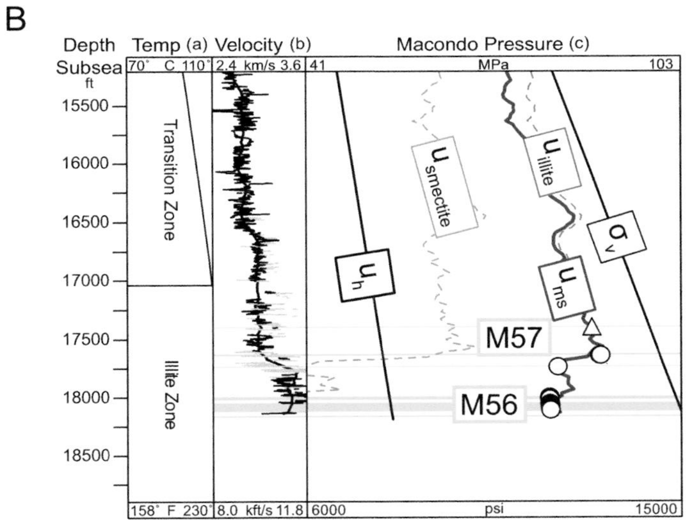

Figure 6.12 Pore pressure prediction for the Macondo well, in the deepwater Gulf of Mexico. (a) Predicted pore pressure based on mudrock velocity shown in blue. (b) Expanded view of the pore pressure prediction over the reservoir interval. Modified from Pinkston and Flemings (2019), licensed under Creative Commons.

behavior (e.g., Fig. 6.9). This reflects a body of work that has observed markedly greater preferred orientation in illitic mudrocks (Day-Stirrat et al., 2008). As pointed out by Lahann and Swarbrick (2011), the model of dissolution and alteration of load-bearing grains and resultant increase in preferred orientation may well apply to other diagenetic systems such as quartz dissolution, kaolinite to illite diagenesis, kaolinite to chlorite diagenesis, or even the maturation of source rocks. Any reaction that results in volume reduction or weakening of the structure of the rock could result in pore pressure increase.

6.4 Pore Pressure Prediction in Different Tectonic Environments

6.4.1 Introduction

I close by describing how to predict pressure under different tectonic stress states where uniaxial strain is not present. In Chapter 3, I described a generalized compaction model, the modified Cam clay model, where compaction depends on average stress and shear stress. Under these conditions, there is a particular compaction curve associated with any stress ratio (η), the ratio of the shear stress to the average stress (Eq. 3.14) as depicted in Figure 3.17. The equation presented was derived for plane strain problems because it is easiest to translate this directly to simple geological stress states. However, the approach could be generalized to complex three-dimensional stress states as long as all three principal stresses are known.

I then simplified the compaction model to the case where one of the principal stresses is vertical (Eq. 3.26), which allows me to explore different Coulomb faulting conditions. Equation 3.26 can be combined with the effective stress equation (Eq. 2.7) to solve for pore pressure:

$$u = \sigma_v - e^{\left(e_o + \Delta e_{s'} + \Delta e_q - e\right)/\lambda}. \qquad \text{Eq. 6.12}$$

The parameters to Equation 6.12 are described in Chapter 3. e_0 is the void ratio at unity (1 MPa) under uniaxial strain, and λ is the slope of the compaction curve on a natural log plot. $\Delta e_{s'}$ (Eq. 3.27) is the compaction that results from the difference in average stress between the tectonic conditions (e.g., thrust or normal-faulting) and the uniaxial condition, and Δe_q is the compaction that results from the difference in shear between the tectonic condition and the uniaxial condition.

6.4.2 Pressure Prediction Examples with the Full Stress Tensor

Flemings and Saffer (2018) used Equation 6.12 to predict pressure in the Nankai Accretionary Prism. At the trench, a portion of the ~1 km thick sediment section on the Philippine Sea Plate is offscraped and has led to the construction of a ~100 km wide and narrowly tapered accretionary prism (Fig. 6.13b). As is typical of a fold and thrust belt, a basal décollement separates offscraped sediments above,

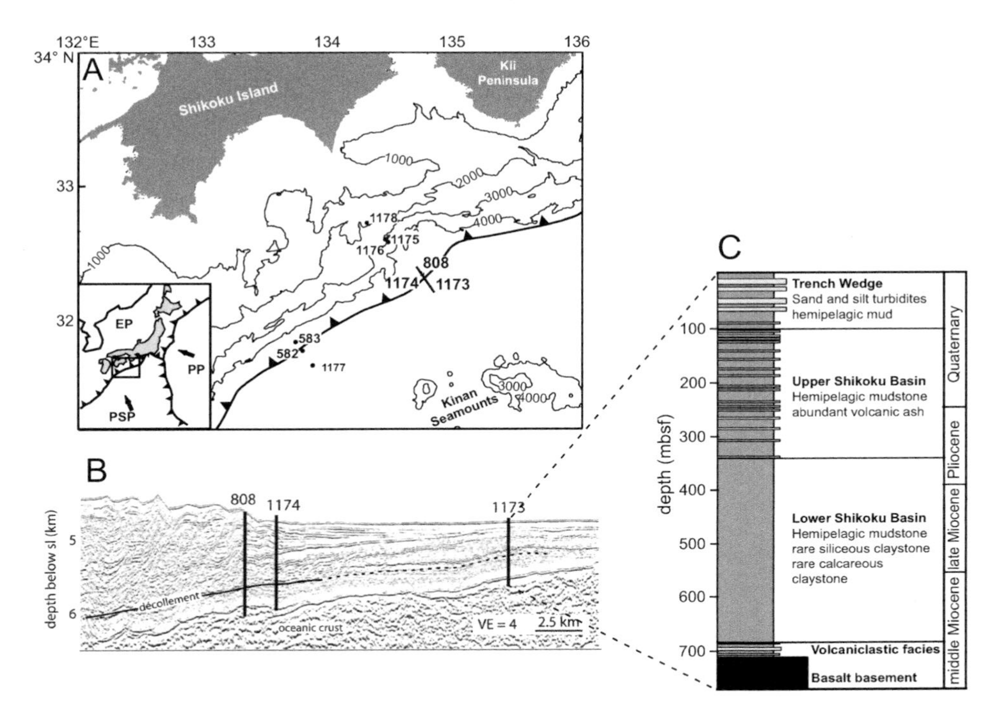

Figure 6.13 (a) Map of the Nankai Trough, with Deep Sea Drilling Project and Ocean Drilling Program drill sites noted; inset shows tectonic setting. PP = Pacific Plate; PSP = Philippine Sea Plate; EP= Eurasian Plate. (b) Seismic depth section (location shown in part a) showing the locations of Sites 1173, 1174, and 808. VE = vertical exaggeration. (c) Stratigraphic section from continuous sampling and core descriptions at Site 1173. Reprinted from Flemings and Saffer (2018) with permission by John Wiley and Sons.

characterized by active folding, thrust faulting, and diffuse lateral shortening, from weakly deformed underthrusting sediments below that exhibit evidence of primarily subvertical shortening.

At Site 808, porosities decrease systematically with depth in the accretionary prism but exhibit a sharp offset to higher values across the décollement, increasing from ~32 to 37% (Fig. 6.14a). The porosity offset has been interpreted to reflect some combination of (1) elevated pore pressure in the underthrust sediments that reduces the effective stress relative to a fully drained vertical burial condition (e.g., Saffer, 2003; Screaton et al., 2002); and/or (2) increased mean effective stress in the accretionary prism driven by tectonic loading (Morgan & Ask, 2004; Tsuji et al., 2008).

Flemings and Saffer (2018) assumed that beneath the décollement, the sediment was deformed uniaxially and that above it, the sediment was undergoing Coulomb failure under compression (Fig. 6.14c). They applied Equation 6.12 for these

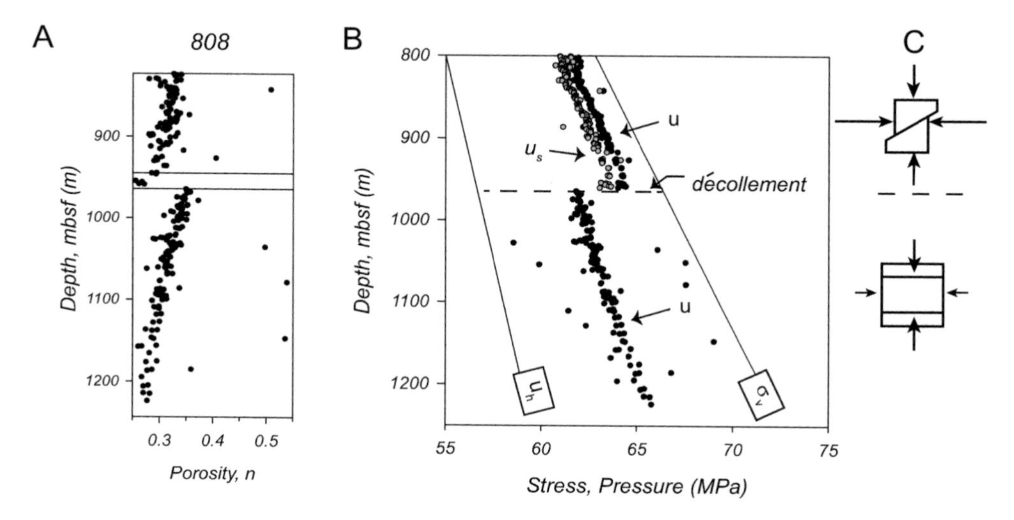

Figure 6.14 (a) Porosity versus depth at Ocean Drilling Program (ODP) Site 808 within the Lower Shikoku Basin (LSB) facies (located in Figure 6.13). (b) Pressure prediction for Site 808. (c) Above the décollement, the system is assumed to be at Coulomb failure under horizontal compression with the result that the horizontal stress is greater than the vertical stress. Below the décollement, the sediments are assumed under vertical uniaxial strain and the horizontal stress is a fraction of the vertical. Reprinted from Flemings and Saffer (2018) with permission by John Wiley and Sons.

conditions and predicted the pore pressure above and below the décollement (Fig. 6.14b). They predicted the pore pressure generated by only the average stress $(u_{s'})$ and the total pressure change (u) (Fig. 6.14b). This distinguishes the relative role of shear-induced pore pressure (Fig. 3.9). They predicted that the pressure in the hanging wall is greater than within the underthrust sediments even though the hanging wall is more compacted than the underthrust sediment (Fig. 6.14a). This counterintuitive result is driven by two effects. First, the average stress (and hence the pressure induced by average stress) is greater in the hanging wall than the footwall because of the much larger horizontal stress in the hanging wall. Second, the shear stress (and hence the shear-induced pressure) is much greater in the hanging wall because it is at failure under compression whereas the footwall is undergoing vertical uniaxial strain.

6.5 Summary

In this chapter, I described methods of pore pressure prediction that go beyond the use of single compaction trend. I first described how to predict pressure when unloading occurs (Section 6.2) and then described how to predict pressure when smectite-rich rocks undergo illitization (Section 6.3). Both techniques address the common observation that in many basins we underpredict the pore pressure in deeper hotter sediments when using a normal trend developed in the younger, cooler, shallow sediments (e.g., Figures 5.4 and 5.13). In these cases, the mudrock at depth is more compacted for a given effective stress than is predicted by the normal compaction trend. In the examples given, the approaches are equally successful in predicting the observed pore pressures. Furthermore, in Chapter 3 (Section 3.2.5 and Figure 3.11) and in Chapter 4 (Section 4.5), I suggested that creep also could drive this process. We have debated which of these processes is dominant for the last 20 years, and this debate will probably continue. This is why the Eaton equation (Equations 5.4 and 5.10) is so popular. The Eaton exponent is commonly changed to empirically correct for this general observation that deeper hotter rocks are more compacted than predicted by the normal compaction trend derived in the shallow section.

There are differences between these two pore pressure mechanisms that may be testable. For example, Bowers (2001) suggests, based on comparison of the density and velocity, that there is a localized zone of unloading and that the deeper section is not unloaded and hence at a lower pressure. In contrast, in the diagenesis model, higher pore pressures can only increase with depth as the smectite-illite reaction proceeds. Important steps forward in this discussion will come from refinement of our ability to interpret unloading (e.g., Fig. 6.8c), a better understanding of the soil behavior that occurs when a smectite-rich material under loading undergoes smectite to illite diagenesis, and a better understanding of creep over geologic time in overpressure. Ultimately, I interpret that either diagenesis (e.g., illite-smectite) or creep may shift the compaction curve so that deeper, older, and hotter rocks are more compacted than their younger, cooler, and more shallow equivalents.

In Chapter 3, I explored the effect of different stress states on pore pressure prediction. The use of a single normal compaction trend implies that only a single stress state is considered and the most commonly assumed one is the stress state that results from uniaxial compaction (Fig. 3.15). However, in many geologic conditions the stress state may vary systematically from these conditions (e.g., Fig. 3.14). Building from Chapter 3, I showed how to extend a single normal compaction trend to predict pressure in multiple tectonic conditions. This approach allows us to make pressure predictions in a range of complex stress regimes such as thrust belts. Furthermore, the approach goes beyond just predicting pore pressure and can directly predict the full stress tensor. This is a considerable advance on previous pore pressure prediction approaches.

7

Pressure and Stress from Seismic Velocity

7.1 Introduction

Seismic velocities are derived from multichannel seismic reflection data. These velocities provide a two- and three-dimensional image of the subsurface but are of lower resolution than the log-based approaches discussed in Chapters 5 and 6. Seismic velocities are often the first information available to predict pressure in frontier basins. I discuss how to invert velocity from seismic data and some of the challenges therein. Once the seismic velocities are derived from the seismic data set, the approaches to predict pressure are identical to the techniques presented in Chapters 5 and 6. I then present two examples of how to predict pressure with the vertical effective stress method. I close with a discussion of how to predict pressure where complex stress states are present with an approach called the full effective stress method. Several recent review papers summarize the approach of pressure prediction from seismic velocity (Chopra & Huffman, 2006; Dutta, 2002b; Sayers et al., 2002a).

7.2 Seismic Velocity

7.2.1 Dix Velocity Analysis

The derivation of velocity from reflection seismic data is easy to visualize through a two-dimensional example. Consider a horizontally stratified earth model (Fig. 7.1). A ship trails a streamer that has 4 receivers (Fig. 7.1). Explosions (generally sourced by compressed air) originate from the source (S) at the rear of the ship. The ray paths for the source event reflect off the seafloor midway between the source and receiver (Fig. 7.1a). The horizontal distance between the source and the receiver is the offset. The time it takes for the seismic wave to travel along the ray path from source to receiver is the two-way travel time. If the ship advances half of the receiver spacing ($\Delta x/2$) between source explosions, then each image point will receive as many ray paths as there are receivers. As the ship advances,

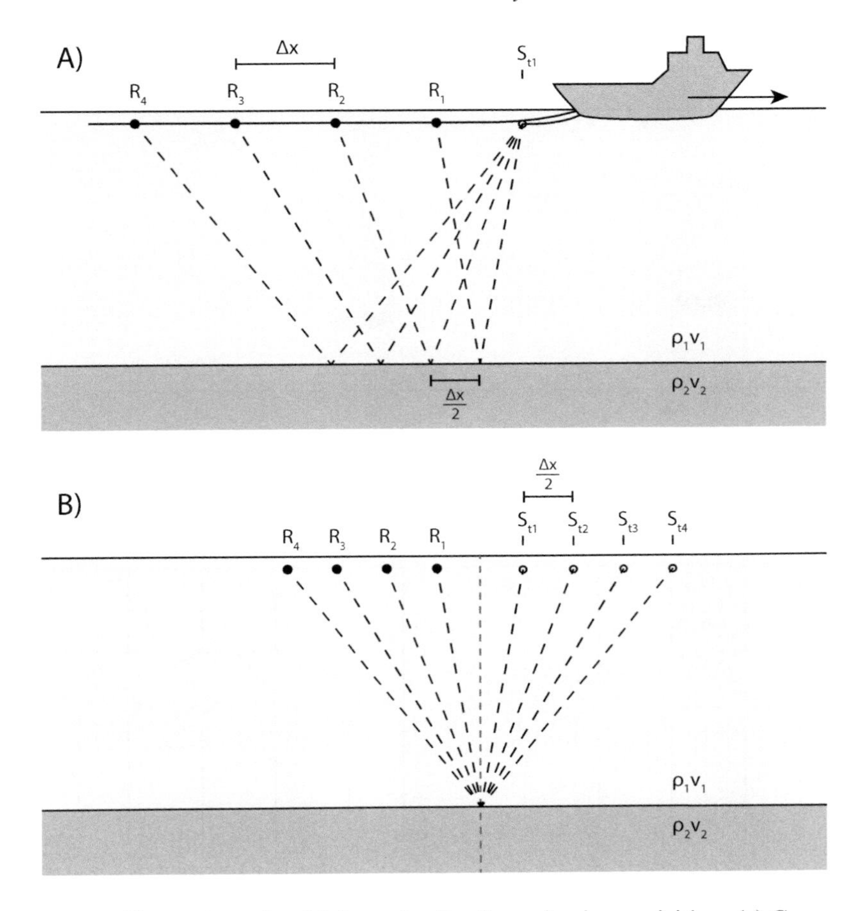

Figure 7.1 Illustration of multichannel reflection seismic acquisition. (a) Common shot gather. The ship moves from left to right setting off a source explosion (S). It trails a streamer with four receivers (R1 ... R4) spaced a distance Δx apart. At a given shot, each receiver records a reflection from the midpoint between the source and receiver. (b) Common depth point gather. If the ship moves forward a distance $\Delta x/2$ between shots, then there will be full-fold acquisition and a single depth point will be imaged by four different ray paths with increasing offset.

successive ray paths that image a single point will have increasing offset (Figures 7.1b and 7.2).

Consider the seismic pulse that is recorded after reflection off the seafloor (Fig. 7.2). The travel time increases according to:

$$t^2(x) = c_o + c_1x^2 + c_2x^4 \dots \quad , \qquad \text{Eq. 7.1}$$

where $t(x)$ is the travel time at offset x. The first two coefficients are,

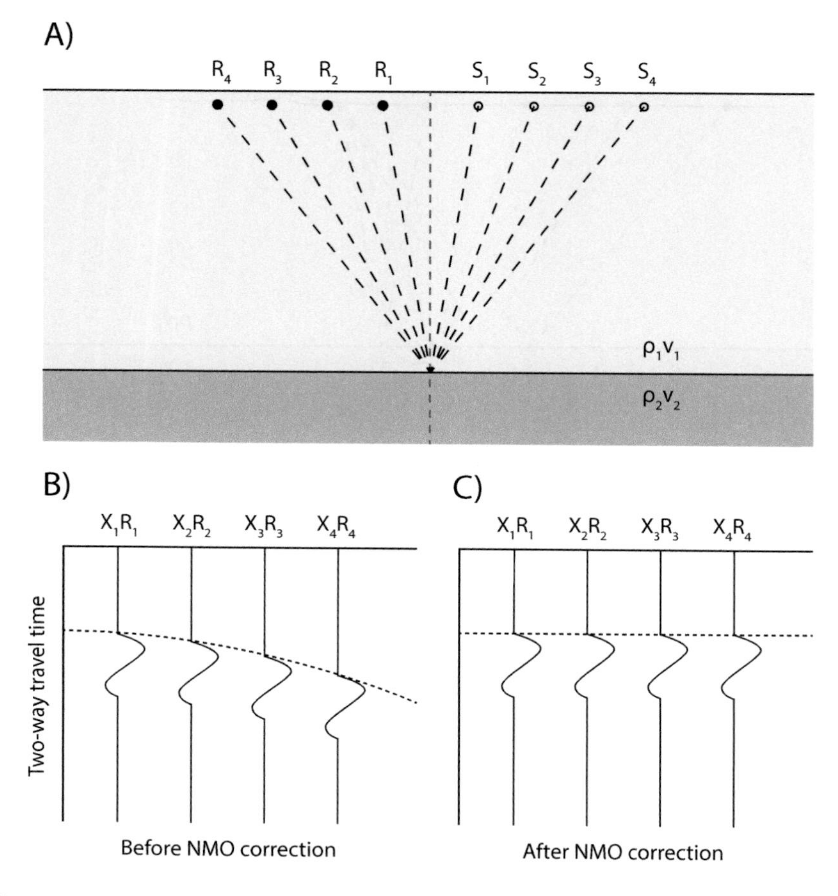

Figure 7.2 (a) Common depth point gather. (b) Signal recorded at each receiver. (c) Flattened gathers after normal moveout correction.

$$c_0 = t^2(0),$$ Eq. 7.2

and

$$c_1 = \frac{1}{V_{rms}^2}.$$ Eq. 7.3

V_{rms} is the root mean square velocity. The remaining terms for Equation 7.1 are complicated and depend on layer thickness and interval velocity (Dutta, 2002b). V_{rms}^2 is a function of the interval velocity for each layer (V_N) and the incident travel time through each layer (Δt_N):

$$V_{rms}^2 = t(0)^{-1} \sum_{i=1}^{i=N} V_i^2 \Delta t_i(0)\,.$$ Eq. 7.4

The total incident travel time $t(0)$ for N layers is the sum of the layer incident times:

$$t(0) = \sum_{i=1}^{N} \Delta t_i(0) . \quad \text{Eq. 7.5}$$

If we consider only the first two terms in Equation 7.1, then it can be restated as:

$$t^2(x) = t^2(0) + \frac{x^2}{V_{rms}^2} . \quad \text{Eq. 7.6}$$

This is termed the small spread approximation because it assumes the offset is small compared to the depth, which allows the additional coefficients in Equation 7.1 to be neglected. This is equivalent to the normal moveout (NMO) equation:

$$t^2(x) = t^2(0) + \frac{x^2}{V_{NMO}^2} . \quad \text{Eq. 7.7}$$

Application of the normal moveout equation is illustrated in Figure 7.2. A hyperbola (Eq. 7.1) is fit to the reflections from different offsets, and from this best-fit hyperbola the incident travel time and the normal moveout velocity is obtained. The normal moveout velocity that is recovered is the best fit hyperbola for reflection events in the data. For a small spread, it is the root mean square velocity. However, V_{NMO} is not necessarily equal to V_{rms}.

Figure 7.3 illustrates the picking of seismic velocities in practice. One can manually fit hyperbolas to observed reflections and from this interpret the interval velocity that results. Alternatively, standard seismic processing packages perform a semblance analysis to determine the NMO velocity that will match an observed event (e.g., Fig. 7.3b). Once the velocities are determined, the events can be flattened (Fig. 7.3c). The picks are made on robust reflection events, and this results in an irregular sample interval.

Interval velocities (V_n) are recovered from the root mean square velocities through Equation 7.4:

$$V_n = \sqrt{\frac{V_{rms,N}^2 t_n - V_{rms,N-1}^2 t_{N-1}}{t_N - t_{N-1}}} . \quad \text{Eq. 7.8}$$

Equation 7.8 is Dix's interval velocity equation, where V_n is the interval velocity for Nth layer, $V_{rms,N}^2$ is the root mean square velocity of the Nth layer, and t_n is the incident travel time to the Nth layer. The process is summarized in Figure 7.4. The normal moveout velocities are recovered and assumed to equal the root mean square velocities. The interval velocities are then calculated with Equation 7.8.

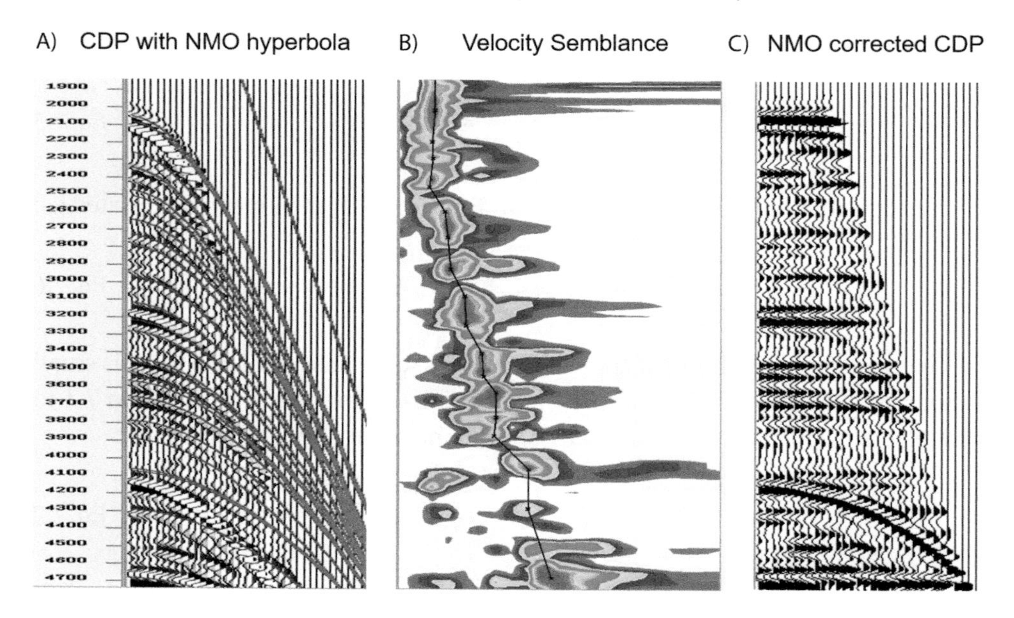

Figure 7.3 Seismic velocity montage. (a) A common depth point gather with normal moveout (NMO) hyperbola overlain. (b) Velocity semblance. (c) NMO-corrected common depth point gather. Example provided by Niven Shumaker via personal communication.

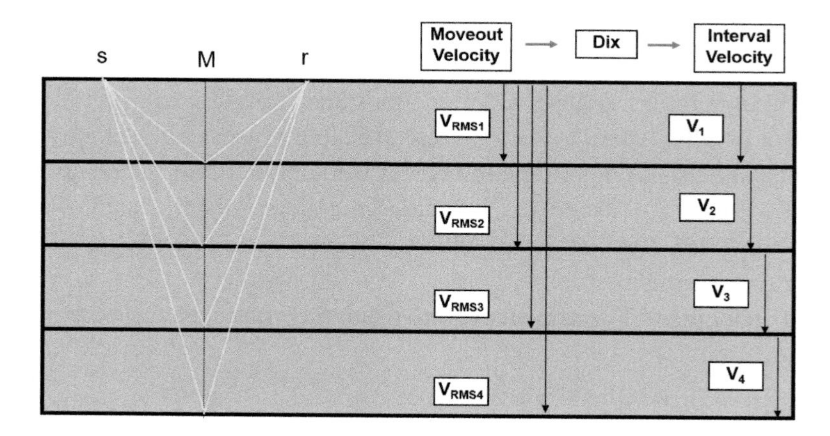

Figure 7.4 Illustration of conversion of moveout velocity to interval velocity through Equation 7.8. Example provided by Niven Shumaker via personal communication.

7.2.2 'Verticalizing' the Data: Calibration and Velocity Anisotropy

The Dix approach assumes flat, homogenous, and isotropic layers. However, a stack of flat isotropic layers with different velocities will behave as if it were anisotropic (Dutta, 2002b; Yilmaz, 1987). Furthermore, mudrocks are not isotropic.

As they compact, clay particles and pores become horizontally aligned (e.g., Fig. 3.3) (Wang, 2002; Wenk et al., 2007) and the horizontal velocity increases relative to the vertical velocity (Nihei et al., 2011; Wang, 2002). This velocity anisotropy is not captured by normal moveout and, as a result, vertical velocities measured by vertical profiling in a well are commonly faster than velocities interpreted through the Dix approach. It is necessary to correct, or 'verticalize,' the seismic velocity to the vertical velocity.

Thomsen (1986) showed that for a vertically transverse isotropic media (VTI), the interval velocities (V_n) could be corrected to the vertical velocity $\left(V_p\right)$:

$$V_n = V_p\sqrt{1+2\delta}. \qquad \text{Eq. 7.9}$$

δ is a combination of elastic parameters. I illustrate an empirical approach to correct seismic velocity to vertical velocity with Eq. 7.9. In the field, a 'check-shot' is commonly performed that measures the vertical velocity by placing a receiver in the well bore and setting off a source at the surface. This represents V_p (dark blue line, Figure 7.5). The interval velocities (V_n) recovered through Eq. 7.8 (magenta line, Fig. 7.5) are faster than the vertical velocities measured with the check-shot, and this difference increases with depth (divergence of blue and magenta lines, Fig. 7.5). Equation 7.9 is used to calculate δ (yellow symbols), and a third order polynomial is fit to this behavior (light blue line). This polynomial is used to correct the interval velocity to the vertical velocity everywhere in the seismic survey for the purpose of effective stress calculation.

Figure 7.5 is one example of calibrating direct measurements of vertical velocity to seismically derived velocity. The example focuses on correcting for velocity anisotropy. However, this calibration step should be generally used to calibrate to correct seismic velocities to vertical velocity (Dutta, 2002b).

7.2.3 Lateral Velocity Variation

Lateral velocity variation can result in incorrect velocity interpretations and hence incorrect pressure predictions when using the Dix approach. I illustrate this with an example from seismic data shot across a salt sheet (Fig. 7.6). At Gather A, the near and the far offset gathers traverse the salt body for their entire ray path. In contrast, the near and the far offsets are both outside of the salt zone at Gather E. Because Gather A traverses the high velocity zone, the incident travel time is less, and the hyperbola is relatively flat, relative to Gather E. Things get more interesting when the gather spans the salt boundary. At Gather B, the near offsets are entirely within the salt. However, the far offsets traverse salt on only half their path. As a result, there is an abrupt downward shift in the normal moveout for the far offsets (Gather

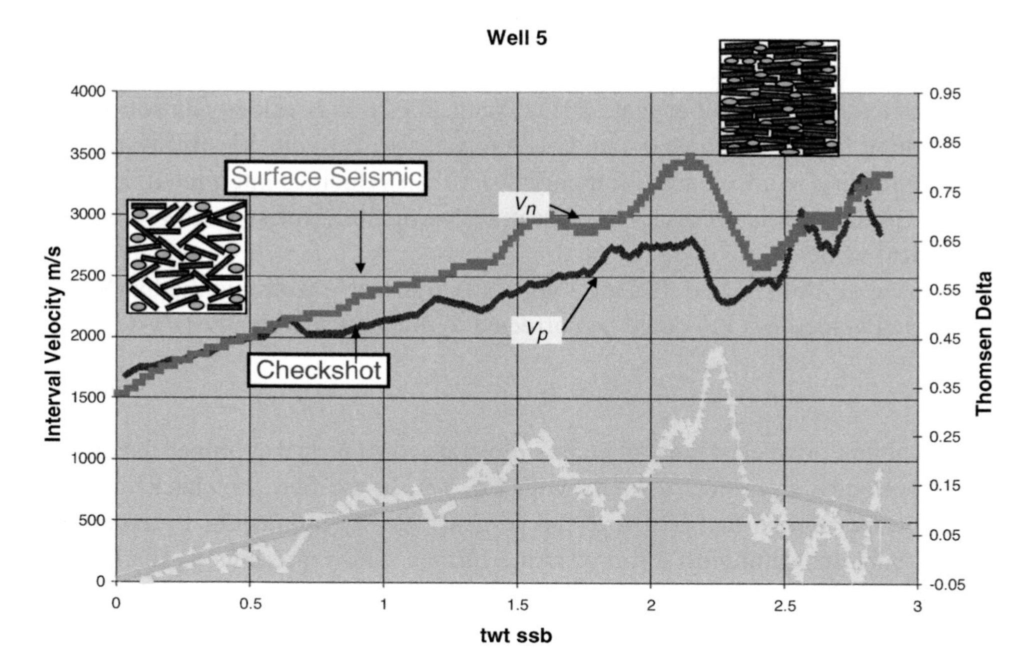

Figure 7.5 An example of how to correct the interval velocity (V_n) to vertical velocity (V_p). The interval velocity (V_n) is generally faster than the vertical velocity (V_p). Equation 7.9 is used to calculate the Thomsen delta (δ) (yellow) that yields this velocity difference. A third order polynomial is fit to the Thomsen delta (light blue line). The insets suggest that the increase in anisotropy from 1 to 2.5 seconds is caused by increasing clay grain alignment during compaction. Republished with permission of the Society of Exploration Geophysicists (SEG), from Shumaker and Vernik (2009).

B, Fig. 7.6b). The opposite occurs at Gather D: the near offsets are entirely outside of salt, but the far offsets traverse the salt over half their journey. As a result, there is an abrupt upward shift in the far offset (Fig. 7.6b, Gather D).

When a single hyperbola is fit to the responses at Gather B and Gather D, the normal moveout velocity that results will be a velocity that is too low (Gather B) and one that is too high (Gather D) (Fig. 7.6c). This will result in a pore pressure that is high where the velocity is low and a pressure that is low were the velocity is high. This velocity variation, and hence the pressure variation that will be predicted from the velocity variation, is entirely the result of the velocity picking process and does not reflect variation in the actual interval velocity (V_p). Chopra and Huffman (2006) emphasizes the need for geologically consistent velocity analysis. Geological heterogeneity must be considered as part of the velocity interpretation. The analysis should carefully consider the geological interpretation first and then interpret the velocity.

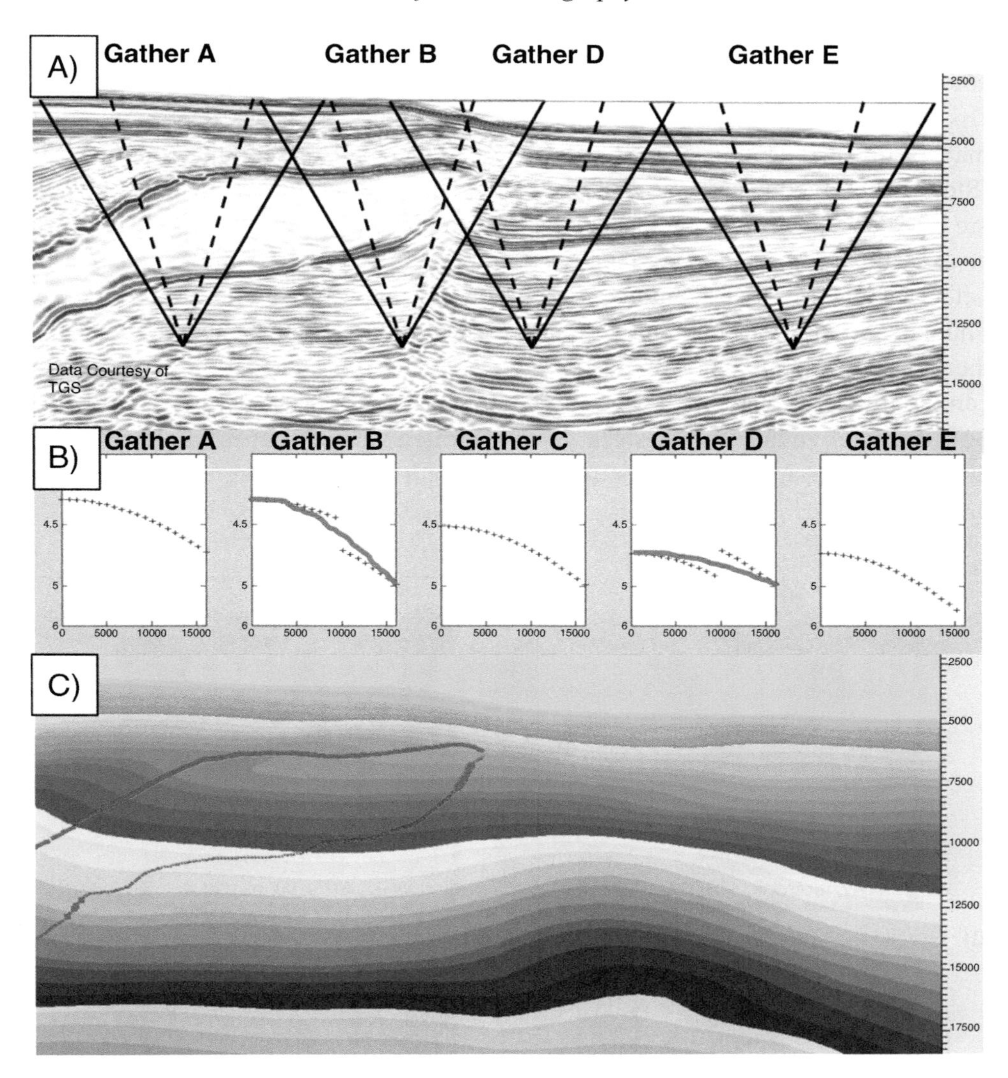

Figure 7.6 Illustration of the implications of lateral variations in velocity on interpreted seismic velocity from the Dix approach. (a) Seismic profile. (b) Common depth point gathers in 5 locations. (c) Interpreted velocity field. Republished with permission of the Society of Exploration Geophysicists (SEG), from Shumaker et al. (2007). Seismic data courtesy of TGS.

7.3 Reflection Tomography

Today, seismic processing and velocity analysis is significantly more sophisticated than the Dix approach described above (see discussion by Chopra and Huffman (2006)). One of the most significant advances is the application of reflection tomography to interpret the velocity. This approach uses prestack depth-migrated common image point (CIP) gathers and applies the same flattening approach as the Dix method. It is assumed that for the correct velocity, poststack depth migration maps a given reflection event to a single depth for all offsets that illuminate it (Stork, 1992; Woodward et al., 1998). 3D tilted transverse isotropic velocity models are often used. This approach accounts for sub-spread changes in velocity. Sayers et al. (2002a) suggest that reflection tomography provides a higher resolution velocity data set that is more appropriate for pressure prediction. However, Shumaker et al. (2007) emphasize that even with these sophisticated approaches, there may be significant error in the predicted velocity and that these errors are commonly due to velocity anisotropy. Shumaker et al. (2014) describe how multiple different vintages of prestack depth-migrated velocity cubes can result from different processing approaches, which can result in very different pore pressure predictions (Fig. 7.7).

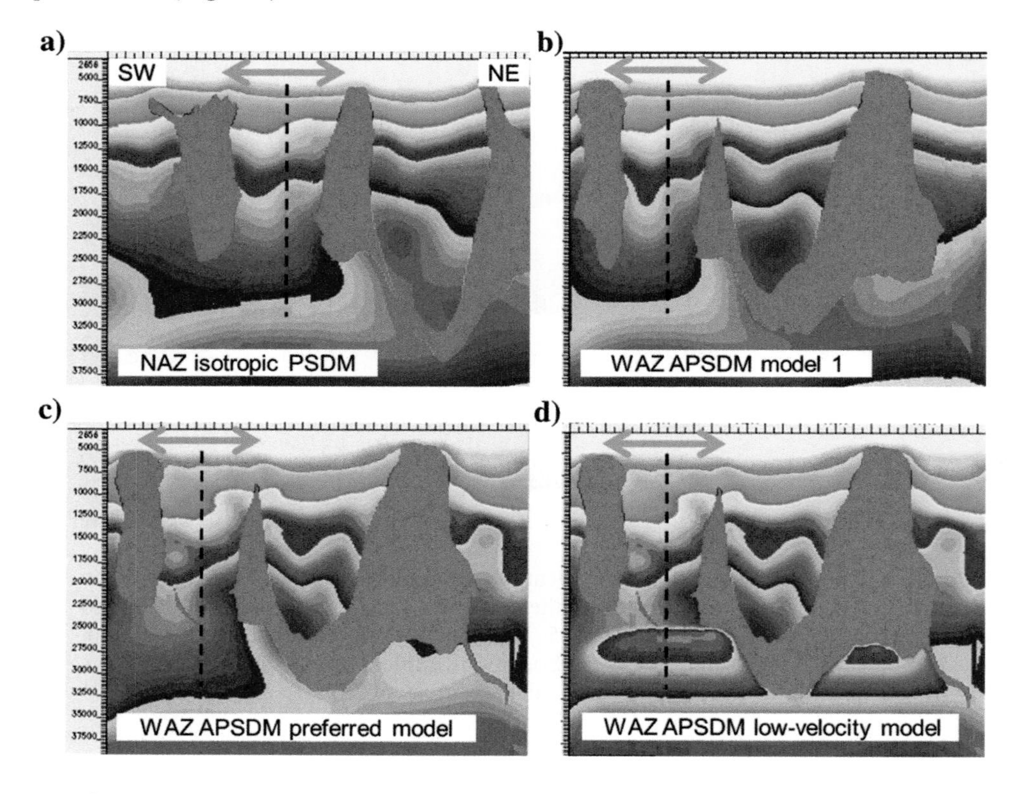

Figure 7.7 Different velocity models presented by Shumaker et al. (2014). Different processing results in significantly different velocity fields, which results in significantly different pore pressure predictions. From Shumaker et al. (2014), licensed under Creative Commons. Republished with permission of the Society of Exploration Geophysicists (SEG) and Niven Shumaker.

7.4 Seismic-Based Pore Pressure Prediction

7.4.1 Vertical Effective Stress Method

7.4.1.1 One-Dimensional Example

I present a one-dimensional case study from offshore West Africa of seismic-based pore pressure prediction that builds from the example in Figure 7.5. The normal moveout velocities (V_{NMO}) are provided as input (blue diamonds, Fig. 7.8a). The interval velocities (V_n, red squares, Fig. 7.8a) are then calculated from the moveout velocities with Equation 7.8. It is not surprising, but it is striking, that the small change in the slope of the moveout velocities with depth results in an abrupt change in the interval velocities. This emphasizes that small differences in the interpretation of the moveout velocities can result in dramatic interval velocity changes.

Next, the interval velocities (V_n) are corrected to the vertical velocities $\left(V_p\right)$ through Equation 7.9 using the result presented in Figure 7.5. The vertical velocities are considerably slower than the interval velocities (compare green triangles to red squares, Fig. 7.8b). After this, the depth is calculated by multiplying the interval velocities by the layer incident travel times and then summing the layer thicknesses.

From here, I revert to the techniques described in Chapters 5 and 6 to predict pressure (Fig. 7.8c, 7.8d). I first approximate the hydrostatic pressure with a pressure gradient of 0.45 psi/ft and the overburden gradient with a pressure gradient of 1.0 psi/ft. I next assume the section is hydrostatically pressured between 500 and 2 500 meters below seafloor (mbsf) and use this interval to constrain the normal compaction trend. I use the Eaton approach (Eq. 5.5) to describe the normal compaction trend (V_h). To constrain V_h, I plot log (V_p) versus depth over the interval from 500 to 2 500 mbsf and use linear regression to constrain α and γ (Eq. 5.5). Given V_h, I apply Eaton's velocity equation with an exponent of 3.0 (Eq. 5.4) and predict pressure versus depth (Fig. 7.8d).

This example is rudimentary, but the result is striking. The seismic velocities image a low velocity zone that is interpreted to record an increase with depth to nearly lithostatic pressures, and below this the pore pressure declines modestly (Fig. 7.8d). In practice, the approach is commonly more sophisticated. As discussed in Chapter 8, it is a gross simplification to assume that the overburden follows a gradient of 1 psi/ft. A more elegant approach would be to determine the overburden by comparing seismic velocity to density data from a nearby well. For example, to estimate the density from seismic velocity, a velocity-density transform is commonly used such as that suggested by Gardner et al. (1974):

$$\rho = \alpha V_P^{\beta}. \qquad \text{Eq. 7.10}$$

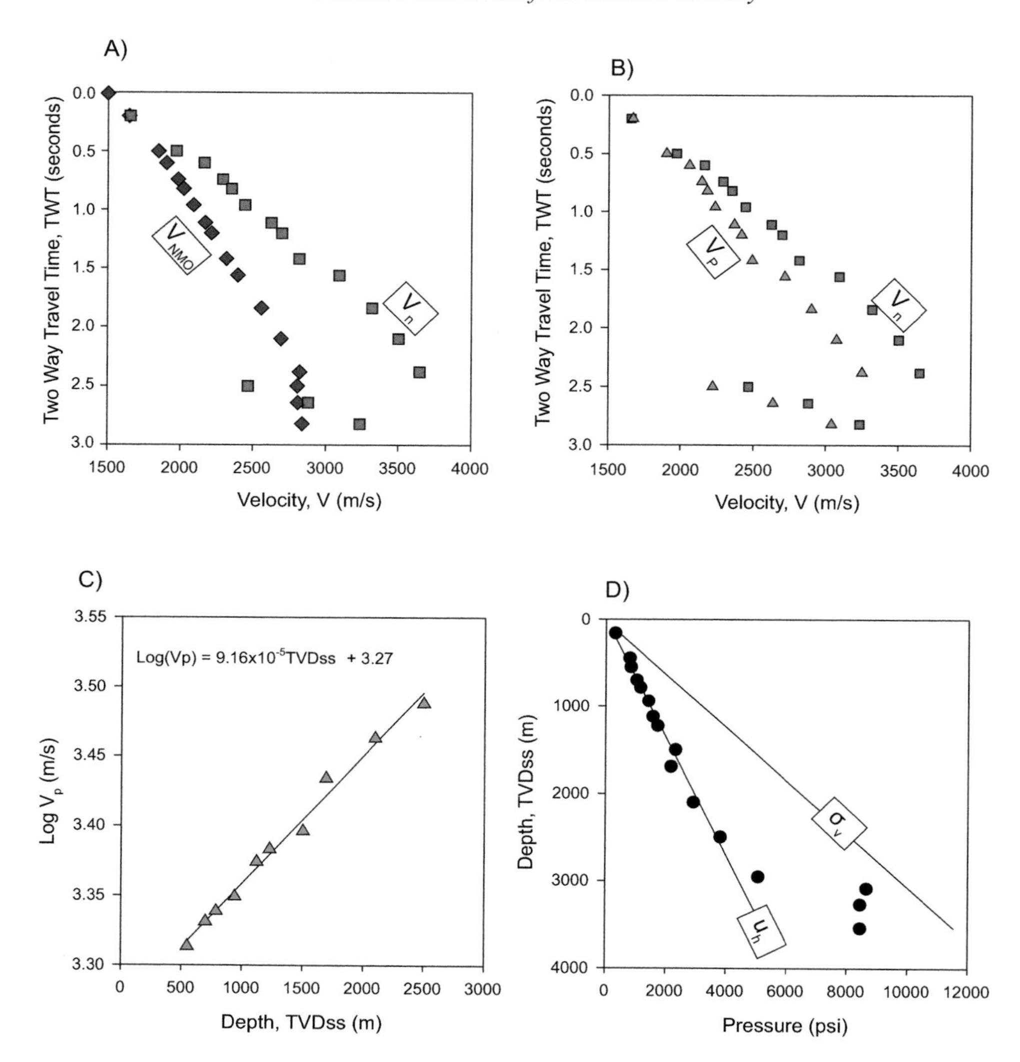

Figure 7.8 An example of pore pressure prediction from seismic data. (a) Normal moveout velocity (V_{nmo}) is shown in blue. The interval velocities (V_n) are calculated from Equation 7.8 (red squares). (b) The corrected vertical velocity (V_p) (green triangles) is calculated from Equation 7.9 using the example in Figure 7.5 (c) The log of V_p versus depth is plotted and a linear correlation is derived over the depth of 500 to 2 500 m to determine the parameters α and γ (Eq. 5.5). (d) Pore pressure is predicted with the Eaton equation (Eq. 5.4). Example based on a case study presented by Shumaker and Vernik (2009).

α and β are empirically constrained by correlating V_p to density data from the well. This relationship is then used to estimate density everywhere there is seismic velocity. These densities are then integrated to estimate the overburden stress throughout the seismic survey. Commonly, where salt is present, a constant density (e.g., 2.2 g/cc) is assigned. Given overburden and seismic velocity, a variety of pore pressure prediction approaches could be applied to predict pressure such as those presented in Chapters 5 and 6.

It is clear from this example that the seismic velocity data are much lower frequency than log velocity data presented in Chapters 5 and 6. Dutta (2002b) suggests that for conventional seismic data, the interval velocities from conventional stacking velocity analysis have frequencies less than 4 Hz, which leads to pressure analysis in layers no thinner than 625 ft (190 m) for typical Gulf Coast sediments. Thus, seismic velocity reveals an extraordinary three-dimensional, but low resolution, view of subsurface pressure.

7.4.1.2 Two-Dimensional Example

I next present a two-dimensional seismic velocity-based pore pressure prediction along a section in the Mad Dog field (Fig. 7.9). The Mad Dog structure is a giant anticline, cored by Jurassic salt, which forms part of the Western Atwater Fold Belt. Reservoirs in this field are in lower Miocene turbidite sands that lie underneath the seaward limit of a major allochthonous salt body, referred to as the Sigsbee Salt (Merrell et al., 2014; Walker et al., 2013).

Seismic imaging of the Mad Dog field is challenging because of imaging issues related to the presence of an irregularly shaped salt allochthon with steep margins and because of the severe seafloor topography at the edge of the Sigsbee Salt (Fig. 7.9). The overall geologic structure of this section and the seismic velocity structure were provided by BP and partners. These were derived from a 3D seismic data volume that had been produced by combining and coprocessing data from multiple seismic surveys, acquired over a 20-year interval. These data volumes included narrow-azimuth and wide-azimuth (WAZ) towed streamer 3D seismic surveys. To account for velocity anisotropy, BP and partners created a 3D tilted transverse isotropic velocity model using seismic tomography, stacking velocities, and well data. The data made available to us consisted of values of interval vertical velocities along the 2D plane of the section, extracted from the 3D velocity model.

To predict pressure, we first determined the overburden throughout the cross-section. We cross plotted the seismic velocity with the density at the GC826-1 well and fitted Equation 7.10 to map seismic velocity to density. A salt density of 2.2 g/cc (Section 8.2.3) was assumed. The bulk density was then integrated over depth across the seismic section to estimate the overburden stress at every location.

We next used the Bowers velocity-effective stress approach (Eq. 5.9) to predict pressure. We built a normal compaction trend using the Bowers velocity model (Eq. 5.7) based on data from the GC826-1 well (Fig. 7.10). This well is overpressured,

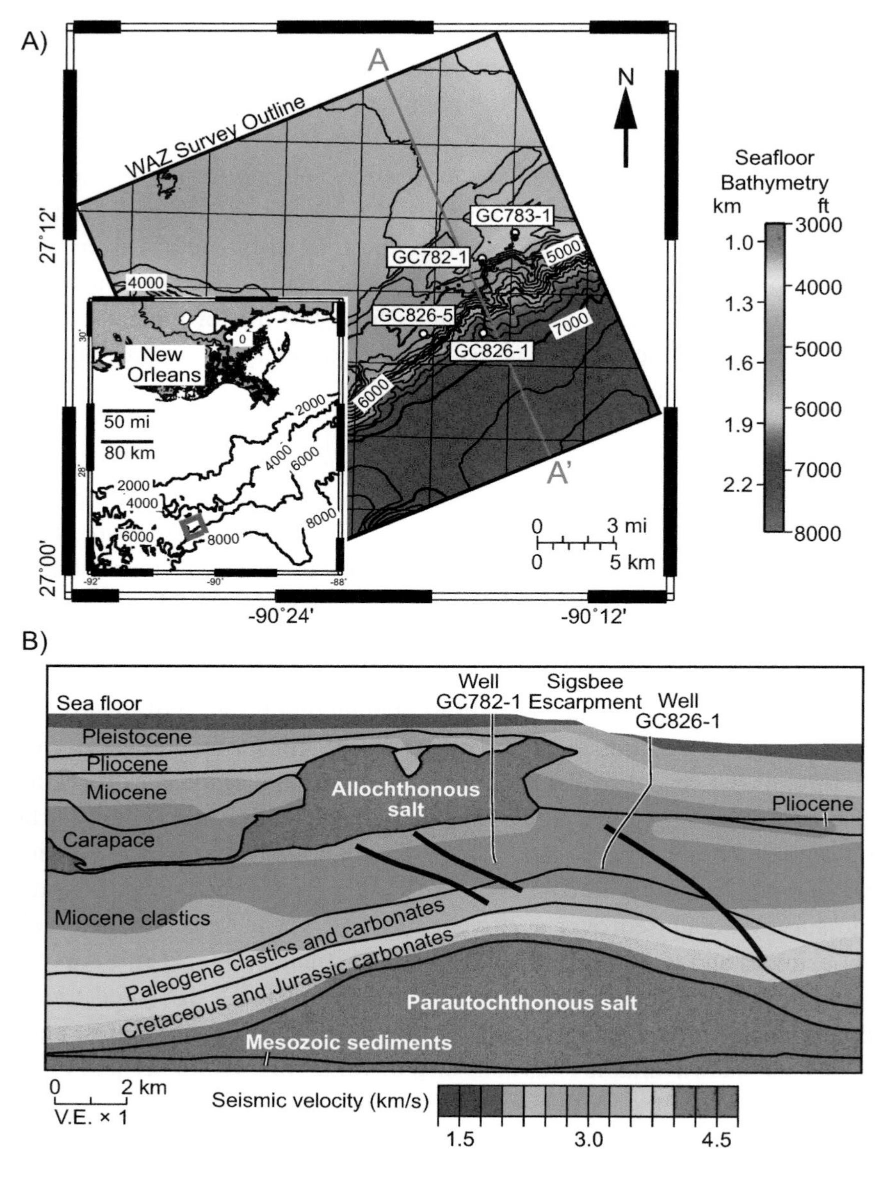

Figure 7.9 Location and velocity cross-section for seismic pressure prediction at Mad Dog field. (a) Footprint of 3D wide azimuth (WAZ) seismic survey in Gulf of Mexico. Location of studied cross-section (red line A-A′). Empty circles represent surface locations of wells, and black-filled circles represent bottom-hole locations. (b) Seismic velocity field and regional stratigraphy. Salt is evident by high velocity values (red). The trajectories of offset wells GC782-1 and GC826-1 are projected onto the cross-section. Part (a) modified after Merrell et al. (2014). Reprinted by permission of the AAPG whose permission is required for further use. AAPG© 2014. Seismic velocity and stratigraphy data are courtesy of BP & Partners. Republished with permission of the Society of Exploration Geophysicists (SEG), from Heidari et al. (2018); permission conveyed through Copyright Clearance Center, Inc.

and we had to build the normal compaction trend within overpressure as described in Section 5.3.2. We first determined the vertical effective stress wherever direct pressure measurements were made by subtracting the measured pore pressure from the overburden (e.g., point P, Fig. 7.10a). We then determined the seismic velocity and the well log velocity at each location where pressure was measured (Fig. 7.10b). The seismic velocity and the log velocities are similar, although the well log data have much higher frequency. Finally, we cross plotted the well log velocity against vertical effective stress and used regression to determine the A and B parameters for the Bowers velocity equation (Eq. 5.7, Fig. 7.10c).

The predicted overpressures are shown in Figure 7.11a. At the largest scale, there is a clear transition from normal pressures in the shallow section (in blue) to overpressures in the deeper section. The transition depth approximately parallels the bathymetry, although it is slightly shallower to the north (left) than to the south (right). More striking, is that there is a large zone of relatively low pressure identified underneath the Mad Dog salt body (green zone under right half of salt body, Fig. 7.11a). This is now recognized to be a low pressure, high effective stress zone where the Mad Dog reservoir lies. This is discussed extensively by Merrell et al. (2014) and also in Section 9.2.4. In Figure 7.12, I compare the predicted overpressure at the GC782–1 well (grey line) with the observed overpressure. The seismic-predicted overpressure captures the high overpressures beneath the salt and the decline in overpressure to the underlying reservoir horizons in the Middle Miocene strata. However, the predicted overpressure within these strata overestimates the observed overpressure in the reservoir. We interpret that either the mudrocks are actually more pressured than the reservoirs (Merrell et al., 2014) or there is insufficient resolution in the seismic velocity to resolve the low pressure interval completely.

In Figure 7.11b, we have plotted the difference between the least principal stress and the pore pressure and expressed the value as an equivalent mud weight. The wellbore pressure is normally kept within this drilling window: inflow (a kick) can occur at wellbore pressures below the pore pressure and lost circulation can occur at wellbore pressures greater than the least principal stress. We determined the least principal stress with Equation 8.8, assuming a stress ratio (K) = 0.8. What is striking in this image is the narrow drilling window (~2 pounds per gallon) present beneath salt (red zone beneath salt body). This narrow drilling window has been identified in multiple wells at Mad Dog and resulted in drilling challenges at the 782–1 well (Merrell et al., 2014). Pressure predictions such as this provide important tools both for exploration and for the design of safe and stable boreholes.

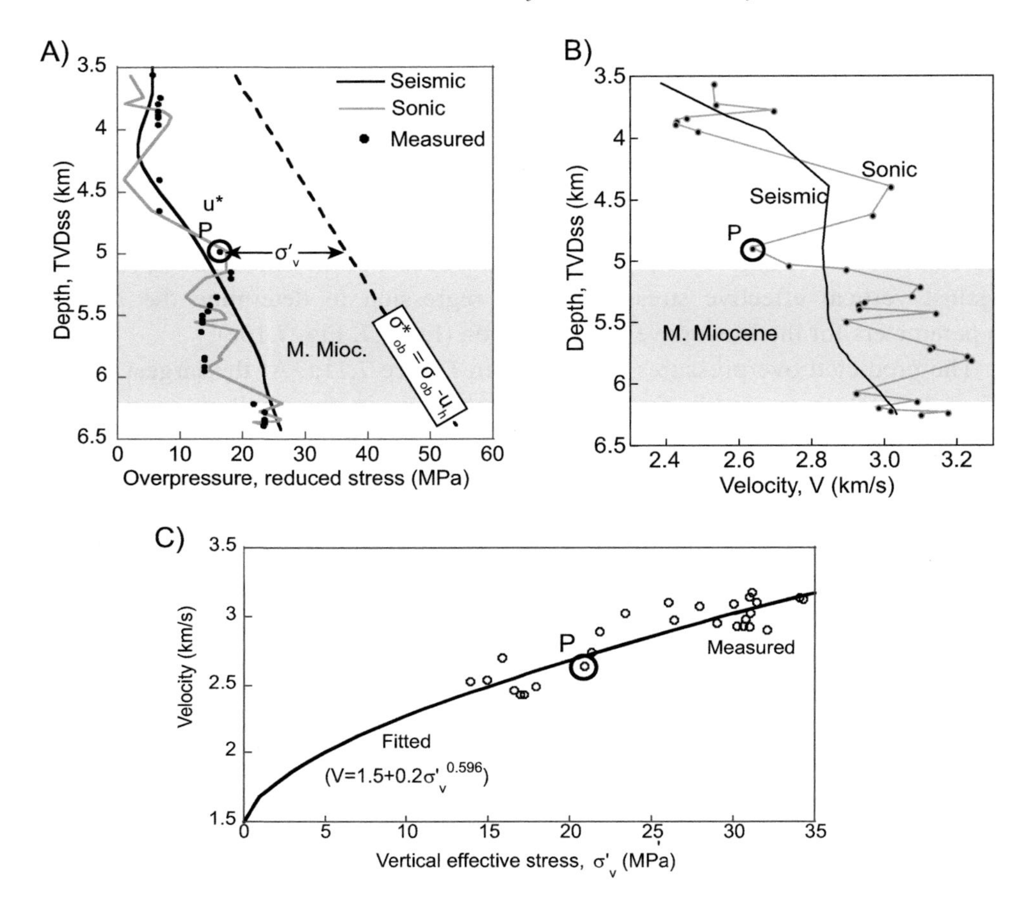

Figure 7.10 Calibration of vertical effective stress-velocity relation (Eq. 5.7) with pressure, stress, and velocity data along well GC826-1 (located in Figure 7.9). (a) Overpressure and reduced overburden stress $\left(\sigma^*_{ob} = \sigma_{ob} - u_h\right)$. (b) Velocity along well. The seismic velocity (black line) is compared to the sonic velocity (grey line) from the well log at locations where there are direct pressure measurements (black dots). (c) Parameters A and B for velocity-vertical effective stress equation (Eq. 5.7) are constrained by a regression of the velocity-effective stress data points. Republished with permission of the Society of Exploration Geophysicists (SEG), modified from Heidari et al. (2018); permission conveyed through Copyright Clearance Center, Inc.

At this point, it should be obvious that the two-dimensional example can easily be extended to three-dimensional pore pressure prediction (e.g., Fig. 7.13). It is now fairly routine to make pore pressure predictions in three dimensions. This provides an exciting view of the subsurface that can be used to optimize well design and envision hydrocarbon entrapment.

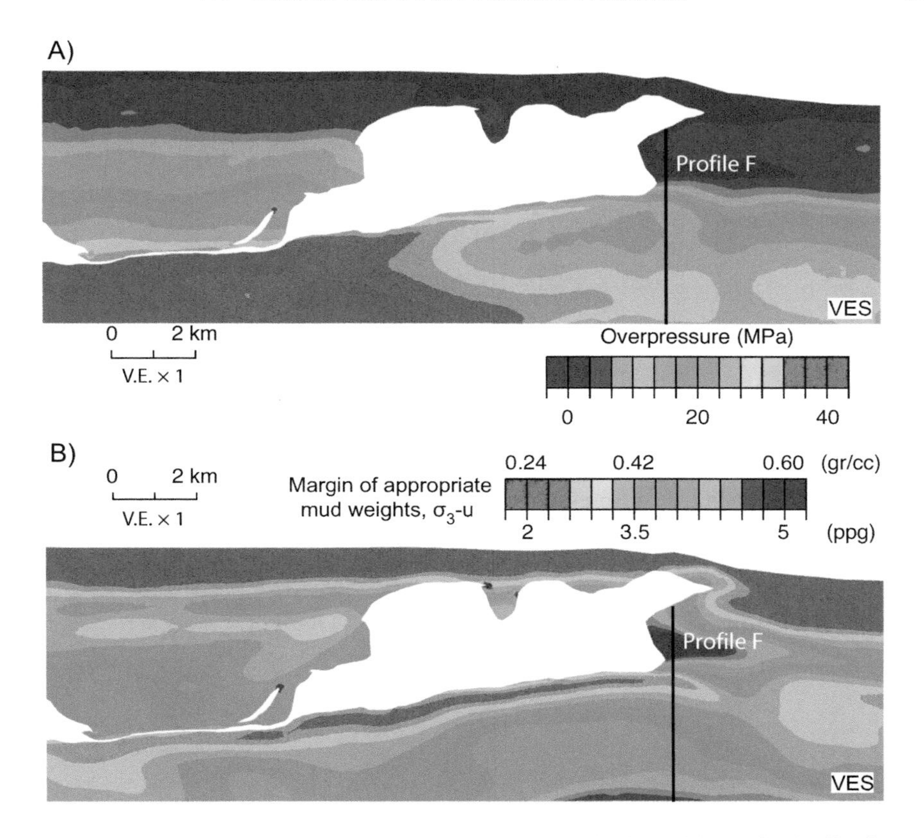

Figure 7.11 Predicted overpressure from seismic velocity with vertical effective stress method. (a) Overpressure field. (b) Drilling margin ($\sigma_3 - u$) in equivalent mud weight. Hotter colors indicate a smaller drilling margin. Colder colors indicate a larger drilling window. Republished with permission of the Society of Exploration Geophysicists (SEG), from Heidari et al. (2018); permission conveyed through Copyright Clearance Center, Inc.

7.4.2 Full Effective Stress Method

I close this chapter with a recent advance in seismic-based pore pressure prediction that we term the full effective stress method. I extend the discussion on how to predict pore pressure in areas with complex stress states that I began in Section 3.4 and applied to a well-based example in Section 6.4. I demonstrate the new approach with the same seismic data set as the vertical effective stress example from the Mad Dog field (Fig. 7.14).

The stress state at Mad Dog is complex (Fig. 7.14). First, modeling suggests the ratio of horizontal to vertical effective stress varies dramatically across the basin, ranging from 0.65 (extensional) to 1.15 (compressional) (Fig. 7.14a). Furthermore,

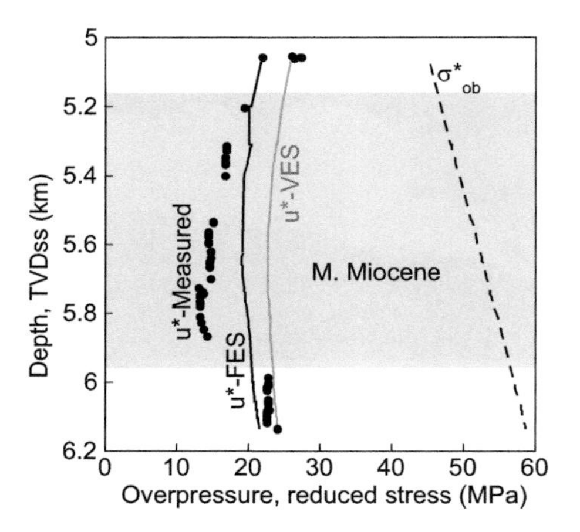

Figure 7.12 Predicted overpressure compared to in situ measurements along well GC782-1 (located in Figure 7.9). Republished with permission of the Society of Exploration Geophysicists (SEG), from Heidari et al. (2018); permission conveyed through Copyright Clearance Center, Inc.

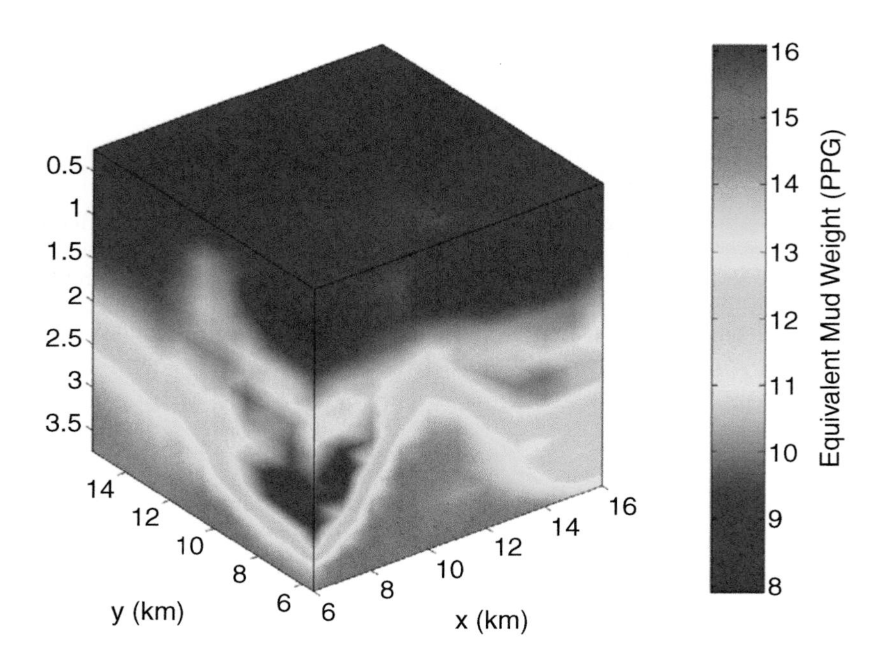

Figure 7.13 Three-dimensional pore-pressure prediction in equivalent mud weight (ppg) for a seismic cube with velocity derived using reflection tomography. Republished with permission of the Society of Exploration Geophysicists (SEG), from Sayers et al. (2002b).

the orientation of the maximum principal stress ranges from vertical to sub-horizontal (Fig. 7.14b). The presence of salt and the significant change in the topography of the seafloor across the section drive these effects. Because of the inability of salt to sustain shear stress, no shear stress is applied to the sediments at the salt-sediment interface and the principal stresses in these sediments are thus normal and parallel to the salt interface (Fig. 7.14b). Because of the inability to sustain shear stress, salt also transfers its entire overburden stress laterally to sediments on its sides. This salt behavior together with the change in the basin bathymetry lead to high lateral stress to the right of the salt and cause the maximum stress to tilt from vertical toward horizontal in this area (Fig. 7.14b). This stress state is far from that present under vertical uniaxial strain, where the maximum principal stress is vertical and the ratio of horizontal to vertical effective stresses is constant for a given material and stress level (Fig. 3.15, Eq. 3.3). To accurately predict pore pressures, this complex stress state should be considered.

Heidari et al. (2018) described the full effective stress approach. It couples a geomechanical model that estimates the total stress tensor with seismic velocities to predict pressure. The effective stress is inverted from velocity, and ultimately the pore pressure is obtained by subtracting effective stress from total stress. In practice, the total stresses and the pore pressure must be iteratively solved until the final solution is obtained.

Heidari et al. (2018) presented their approach in terms of total mean (σ_m) and shear stress (q), where

$$\sigma_m = \frac{\sigma_1 + \sigma_2 + \sigma_3}{3}, \qquad \text{Eq. 7.11}$$

and

$$q = \sqrt{\frac{(\sigma_1 - \sigma_2)^2 + (\sigma_1 - \sigma_3)^2 + (\sigma_2 - \sigma_3)^2}{2}}. \qquad \text{Eq. 7.12}$$

The ratio of shear to mean effective stress (σ'_m) is:

$$\eta = \frac{q}{\sigma'_m}. \qquad \text{Eq. 7.13}$$

A modification of the Bowers equation (Eq. 5.7) is used to describe the relationship between velocity and effective stress:

$$V = V_0 + A_e \sigma_e'^{B_e}. \qquad \text{Eq. 7.14}$$

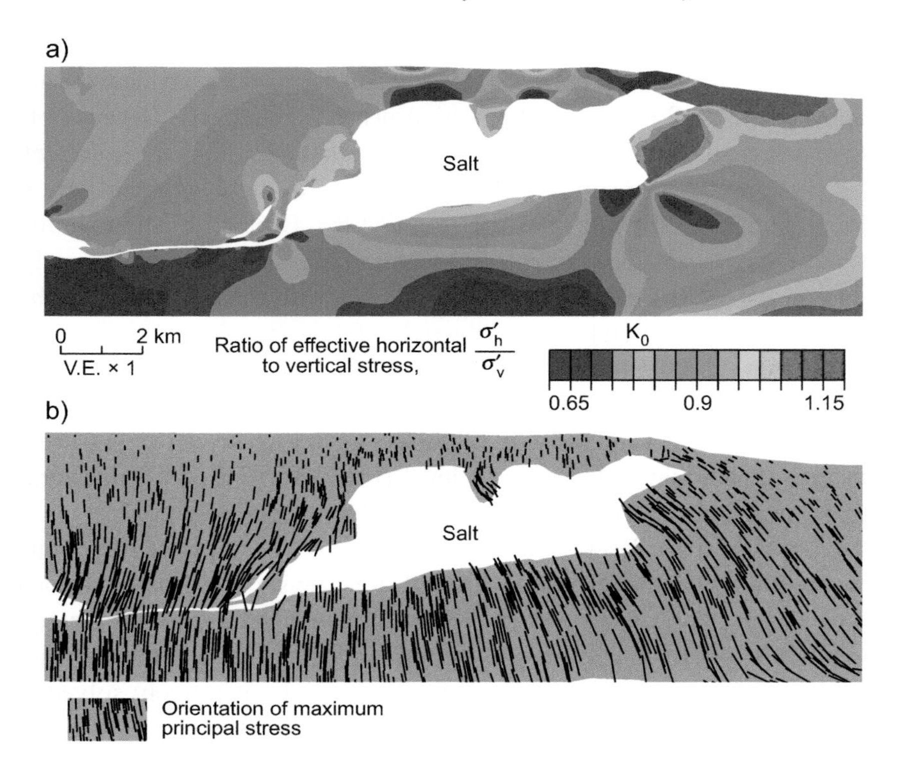

Figure 7.14 Stress field predicted by geomechanical model around salt. (a) Ratio of effective horizontal to vertical effective stress. If deformation were uniaxial, this ratio would be constant and equal to 0.8 for this example. (b) Orientation of maximum principal stress. The minimum principal stress is perpendicular to the maximum stress direction. The magnitude and orientation of the principal stresses differ significantly from those resulting from vertical, uniaxial, strain. Republished with permission of the Society of Exploration Geophysicists (SEG), from Heidari et al. (2018); permission conveyed through Copyright Clearance Center, Inc.

σ'_e is the equivalent stress and is a function of mean stress and shear stress (expressed through η, Eq. 7.13):

$$\sigma'_e = \sigma'_m \left(\frac{(M)^2 + \eta^2}{(M)^2} \right). \qquad \text{Eq. 7.15}$$

M is the value of η at critical state failure (Wood, 1990) and can be calculated from the friction angle (Fig. 3.9, Eq. 3.4).

I first illustrate how to constrain the normal compaction trend (the values A_e and B_e from Eq. 7.14). I use the same pressure and velocity data to calibrate what I used in the two-dimensional example of the vertical effective stress method (Fig. 7.10a,

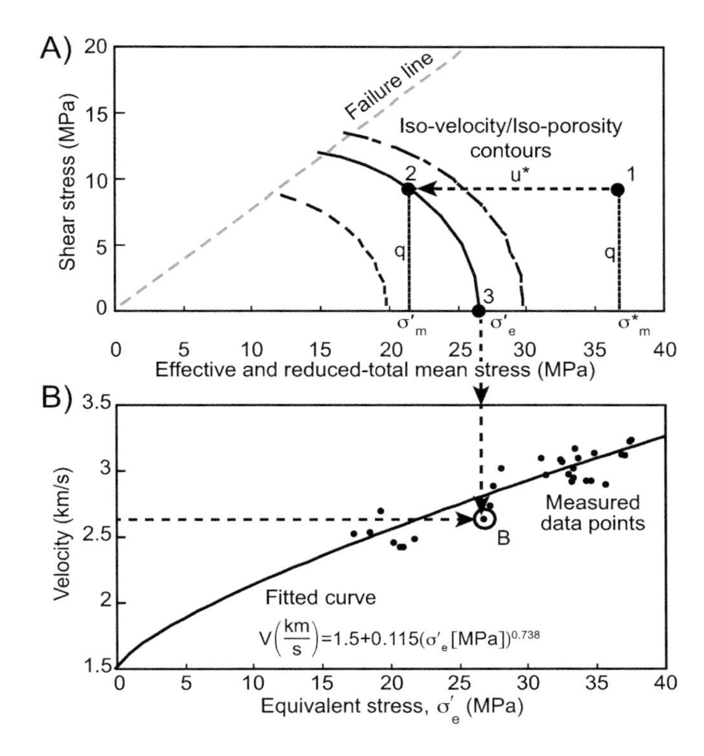

Figure 7.15 Calibration of velocity and equivalent stress relation in the full effective stress model. (a) Determination of equivalent stress at each pressure measurement point. (b) Equivalent stress and velocity at each pressure measurement point are cross plotted and a regression is used to recover parameters of equivalent stress-velocity relation (A_e and B_e, Eq. 7.14). See text for discussion. Republished with permission of the Society of Exploration Geophysicists (SEG), from Heidari et al. (2018); permission conveyed through Copyright Clearance Center, Inc.

b). At each pressure measurement location, I determine the total mean and shear stress from the geomechanical model. Then, I calculate the mean effective stress $\left(\sigma'_m = \sigma_m - u\right)$. Given σ'_m and q (point 2, Fig. 7.15a), I calculate the equivalent stress (σ'_e)(point 3, Fig. 7.15a) by combining Equations 7.15 and 7.13. The equivalent stress and the measured velocity define a data point (point B, Fig. 7.15b) for the equivalent stress-velocity relation. The parameters of this relation (A_e and B_e, Eq. 7.14) are obtained by fitting Equation 7.14 to these data points (Fig. 7.15b).

Once the velocity-effective stress relation is calibrated (Eq. 7.14), it can be used to predict pressure from velocity. The calibrated relation (Eq. 7.14, Fig. 7.16) is used at any point at which seismic velocity is given to calculate the equivalent stress (point 1, Fig. 7.16b). Shear stress, acquired from the geomechanical model, is used in conjunction with the isovelocity contour to calculate the effective mean stress (σ'_m, Eq. 7.15; point 2, Fig. 7.16a). The calculated effective mean stress is subtracted

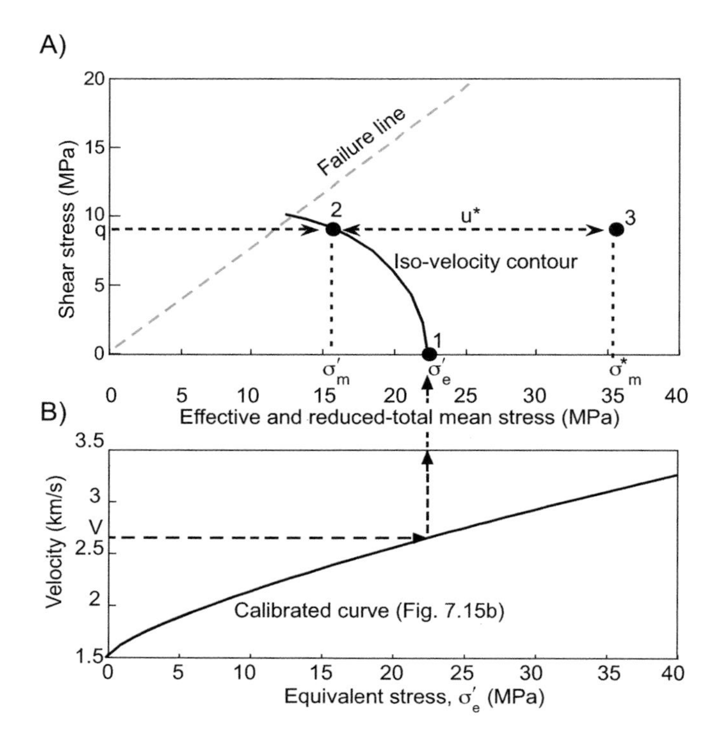

Figure 7.16 Process of pressure prediction in full effective stress method. (a) Calculation of pore pressure from equivalent stress (σ'_e), shear stress (q), and total mean stress (points 1–3). σ^*_m is the reduced-total mean stress: the total mean stress less hydrostatic pressure. σ'_m is the effective mean stress. (b) Calculation of equivalent stress (σ'_e) from seismic. Republished with permission of the Society of Exploration Geophysicists (SEG), from Heidari et al. (2018); permission conveyed through Copyright Clearance Center, Inc.

from the total mean stress, acquired from the geomechanical model (point 3, Fig. 7.16a), to find the pressure (u^*, points 3—2, Fig. 7.16a).

The results from the full effective stress (FES) method (Figures 7.17 and 7.18) are distinctly different from those from the vertical effective stress (VES) method (Figures 7.11 and 7.12). In this example, the FES method predicts lower pressure than the VES method in many locations; this is particularly apparent beneath the salt and in the minibasin to the north (left) of the salt (compare Fig. 7.17a to Fig. 7.11a). Heidari et al. (2018) discussed that the difference in pressures that the two approaches predict at a point indeed derives from the difference in total mean and shear stresses at that point and the stresses at the same burial depth at the calibration well. The FES-VES pressure difference due to the spatial difference in total mean stress and shear stress is shown in Figure 7.18. Shear stress beneath salt and to the north (left) of the salt sheet is less than shear stress at the same burial depths at the

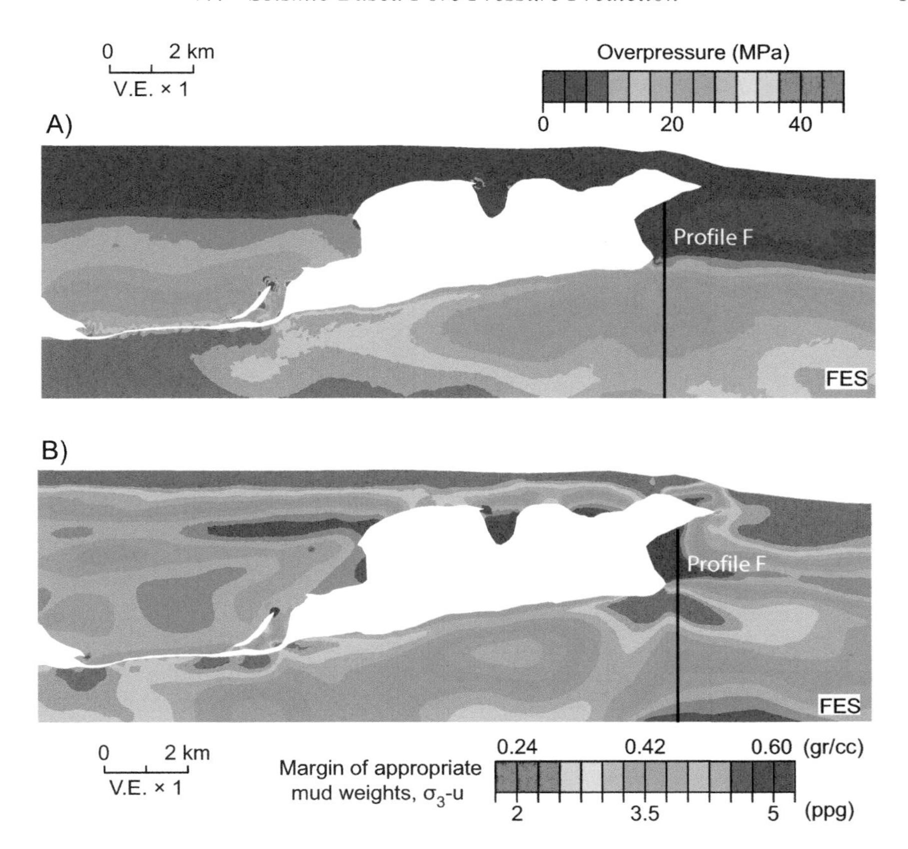

Figure 7.17 Overpressure and drilling margin as predicted by full effective stress method. (a) Predicted overpressure. (b) Drilling margin ($\sigma_3 - u$) expressed in equivalent mud weight. Hotter colors indicate a smaller drilling margin. Republished with permission of the Society of Exploration Geophysicists (SEG), from Heidari et al. (2018); permission conveyed through Copyright Clearance Center, Inc.

calibration well 826–1. This is because shear stress at the calibration well is relatively high due to large lateral stress that salt applies to the area where the calibration well lies.

Figure 7.19 extracts pressure and stresses along profile ‘F’ from in front of the salt. The FES predicted pore pressure is less than the VES predicted pore pressure in this location (Fig. 7.19a). This is because the mean stress (σ_m) is lower in this location than in the calibration well, which is outboard from this location (Fig. 7.19b). The shear stress (q) is not significantly different in the two locations (Fig. 7.19b). In Figure 7.19c both the pore pressure and the least principal stress are illustrated with the VES approach and the FES approach. The most striking point of this plot is that the least

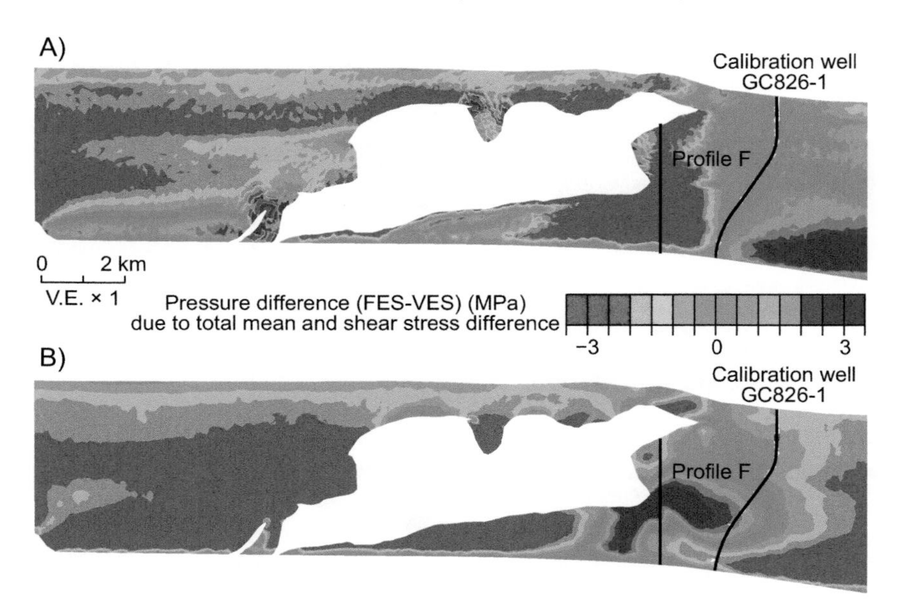

Figure 7.18 Contribution of the total mean and shear stress spatial difference to the pressure difference of the FES and VES models. (a) Contribution of the total mean stress difference. (b) Contribution of the shear stress difference. Republished with permission of the Society of Exploration Geophysicists (SEG), from Heidari et al. (2018); permission conveyed through Copyright Clearance Center, Inc.

principal stress varies independently of the pore pressure in the FES approach, whereas it is a fixed proportion of the pore pressure in the VES approach (Eq. 3.3).

In summary, the full effective stress method for pressure prediction from seismic velocity accounts for non-uniaxial strain and stress states commonly present in complex stress settings. Total stresses are estimated with a geomechanical model, and effective stresses are estimated from velocity, taking into account the effect of the effective mean and shear stress on sediment porosity and velocity. The method identifies stress complexities caused by salt and basin bathymetry. The method can also predict the full stress tensor. Our prediction of minimum stress is in many areas far different, either in orientation or in magnitude, from the one derived with the VES method.

7.5 Summary

I have reviewed methods of pore pressure prediction from seismic velocity. I emphasized that the successful prediction of pore pressure is strongly dependent on inverting the correct velocities, and that common processing techniques can

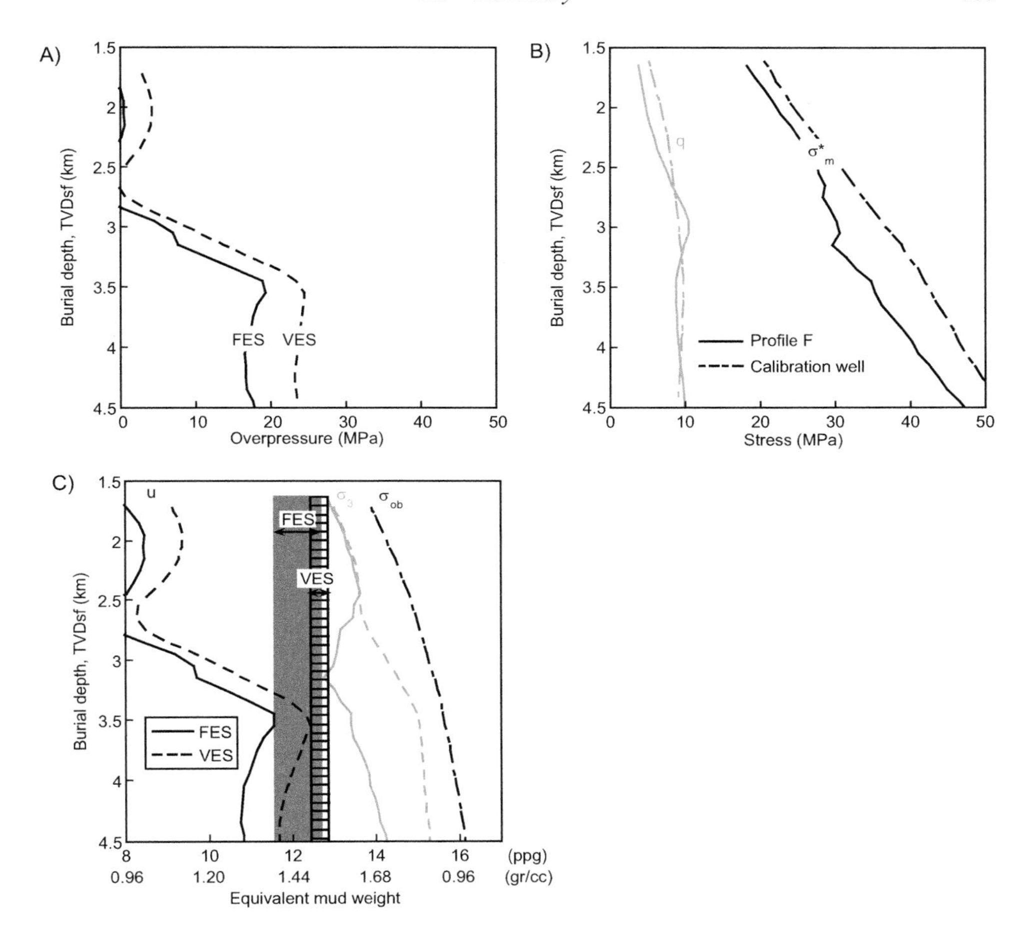

Figure 7.19 Pressure, stresses, and drilling window for vertical profile F (located in Figures 7.11 and 7.17). Depths are total vertical depth below the seafloor (TVDsf). (a) Overpressures predicted by the FES and VES methods. (b) Reduced total mean stress ($\sigma_m^* = \sigma_m - u_h$) and shear stress ($q$) along profile F (solid line) and along calibration well GC826-1 (dashed line). The FES method predicts lower overpressure than the VES method at any burial depth along the profile because the total mean stress is lower at the profile than at the same burial depth at the calibration well. (c) Pore pressures and minimum stresses predicted by the FES (solid lines) and VES methods (dashed lines) in equivalent mud weight. The shaded area represents the drilling window as predicted by the FES method, and the dashed area represents the drilling window as predicted by the VES method. FES method predicts a larger window and lower values for drilling mud weights than the VES method. Republished with permission of the Society of Exploration Geophysicists (SEG), from Heidari et al. (2018); permission conveyed through Copyright Clearance Center, Inc.

result in incorrect velocity predictions that will then map into incorrect pore pressure predictions.

Seismic-based pore pressure prediction provides a fascinating insight into pore pressure and stress in sedimentary basins ahead of the drill bit. The approach yields a three-dimensional view of the subsurface that has much lower resolution than is available through well log data. The results can be used both in the exploration process to estimate trap integrity and migration pathways and for drilling, to design safe and stable boreholes.

I provided two examples to describe vertical effective stress methods that are well established and practiced in industry. I closed with an approach called the full effective stress approach. This approach incorporates geomechanical modeling into velocity-based pore pressure prediction. It has the potential to illuminate the full stress tensor and elucidate how pore pressures can be driven by spatial variation of the stress field. In my view, the method has extraordinary potential.

8

Overburden Stress, Least Principal Stress, and Fracture Initiation Pressure

8.1 Introduction

Total stresses are an important control on pore pressure and fluid flow in sedimentary basins. The magnitude of the total stresses must be known in order to estimate the pore pressure. When pore pressure exceeds the least principal stress, hydraulic fracturing can occur and, if so, fluids are rapidly drained. This can occur naturally or during drilling. I first describe how to estimate the overburden stress, which is commonly assumed to equal the vertical stress (σ_v), and then describe how to estimate and model the least principal stress (σ_3) and the fracture initiation pressure in a borehole. Zoback (2007) provides an extensive overview of these issues.

8.2 Characterization of Overburden Stress

We commonly assume the overburden stress equals the vertical stress in systems where one of the principal stresses is interpreted to be vertical (Figures 3.14 and 3.15). We use overburden stress to understand compaction behavior, and we ultimately subtract the effective stress from the total stress to predict pressure (e.g., Eq. 5.2). We need to know the overburden stress to predict the least principal stress. Furthermore, a range of complicated geomechanical models have been developed that rely on overburden stress as an input. Despite its importance, simplistic estimates of overburden stress are commonly made and this can result in incorrect estimates of pressure and stress.

The overburden stress is calculated by integrating the density profile (Eq. 2.5). I illustrate an example of this approach from the Auger field in the Gulf of Mexico (Fig. 8.1). The bulk density is obtained from the wireline bulk density log. Caliper logs were used to identify zones of borehole washed-out and in these intervals densities were interpolated from data above and below the washed out interval. No logging measurements were made from the seafloor to a depth of 4 900 ft below seafloor (Fig. 8.1a). This is common because in the shallow borehole, the borehole

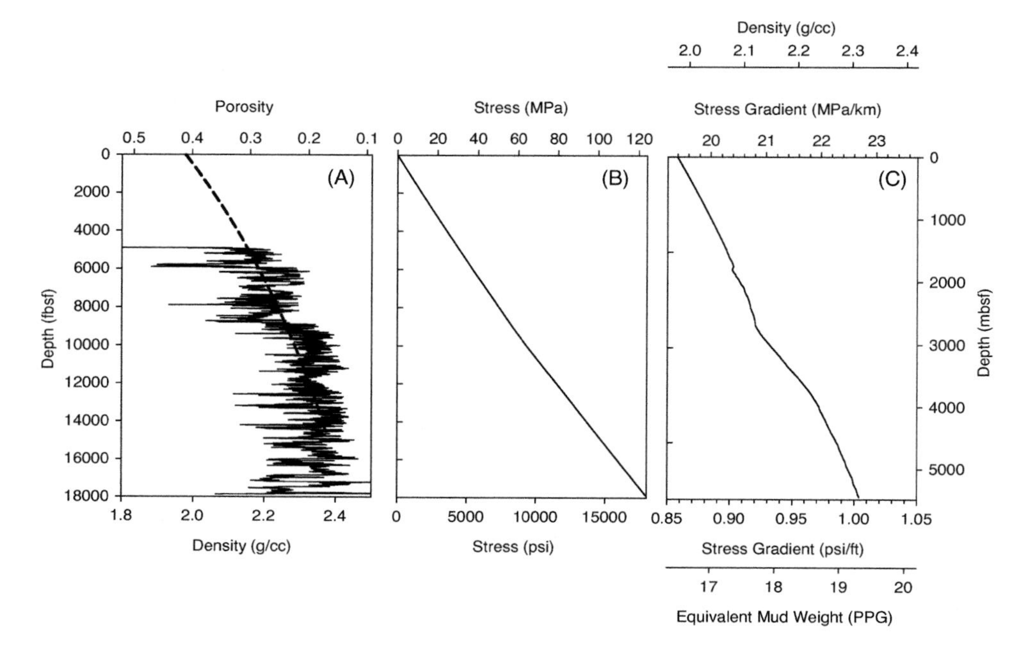

Figure 8.1 (a) Density versus depth below seafloor in the Auger basin from the wireline density log. A matrix density of 2.65 g/cc and a fluid density of 1.15 g/cc was assumed to calculate the porosity. No density measurements were made between the seafloor and 1 500 m below the seafloor. In this interval, density was assumed to increase exponentially with depth with parameters determined by regression of the density data between 1 500 m and 4 500 m (black dashed line). (b) Overburden stress calculated by integrating the bulk density. (c) The average overburden gradient (the vertical total stress divided by the depth). See Reilly (2008) for further details.

diameter is typically large and the logging tool cannot record meaningful values, or because it is considered an unnecessary expense. In this example, Reilly (2008) assumed density was an exponential function of depth in this interval (dashed line, Fig. 8.1a) using the approach of Athy (1930).

The increase in stress with depth appears linear (Fig. 8.1b). However, the slope of the overburden curve flattens slightly with depth, reflecting the increase in bulk density with compaction. The continuous density profile is integrated, multiplied by the gravitational constant, and then divided by depth to yield the average overburden gradient (Fig. 8.1c). In this perspective, the details of the compaction behavior become apparent. Over the depth of the hole, the average overburden gradient ranges from ~0.85 psi/ft to more than 1 psi/ft (Fig. 8.1c). It is common to estimate overburden stress by assuming a constant overburden gradient of 1 psi/ft (Eaton, 1969). This is clearly unreasonable for most of the borehole at Auger.

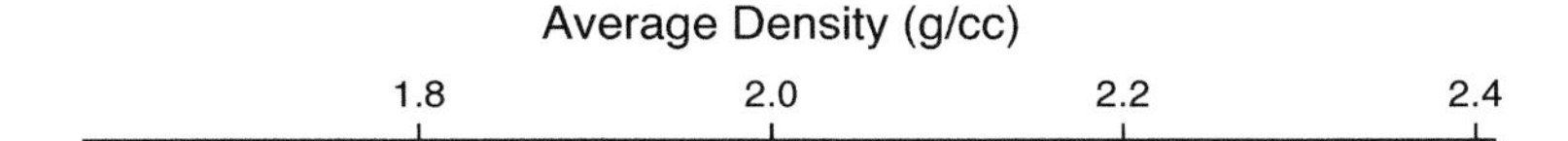

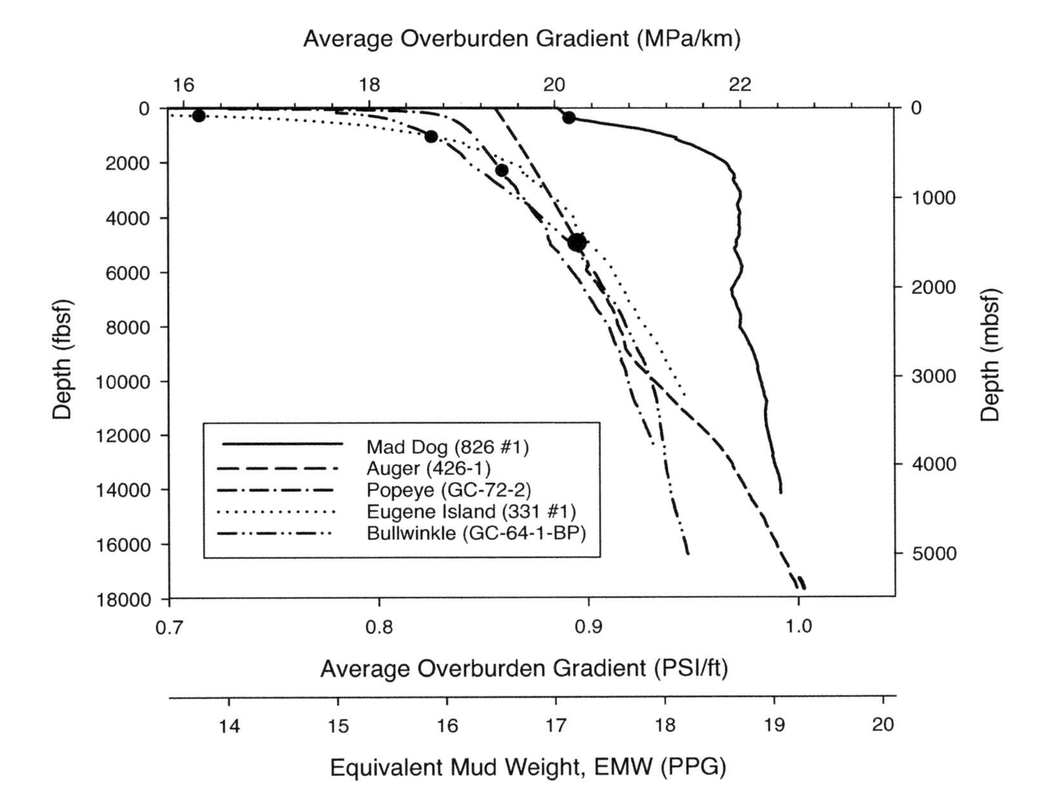

Figure 8.2 Average overburden gradient at five Gulf of Mexico locations. Above the circles, no log data were available and the bulk density was assumed to decline exponentially with depth to merge with the first available data. Bulk density values were integrated to determine the overburden gradient.

Overburden gradients are compared for five Gulf of Mexico locations ranging from the shelf (EI-330 with a water depth of ~100 m) to deepwater (Mad Dog, with a water depth of 6 560 ft) (Fig. 8.2). There are two obvious trends. First, the Mad Dog overburden gradient is larger at all depths than the other examples. The Mad Dog well penetrates older strata (Early Miocene, >16.4 Ma) than the other locations, which are generally younger than Late Pliocene, ~3.6 Ma (Alexander & Flemings, 1994; Flemings & Lupa, 2004; Holman & Robertson, 1994; Rafalowski et al., 1994). A longer duration may have allowed greater compaction at Mad Dog. Second, the EI-330 overburden stress is slightly larger at any given depth than the other younger basins (Bullwinkle, Popeye, and Auger). The EI-330 shelf minibasin (Alexander & Flemings, 1994) has a thick section of normally compacted strata (Chapter 5). In contrast, Auger, Popeye, and Bullwinkle all lie on the continental slope, overpressure begins near the surface, and compaction is limited (Flemings & Lupa, 2004). In summary, the overburden stress is greater in the EI-330 basin because these sediments are more compacted than in the slope basins.

8.2.1 Overburden in the Shallow Section

It is challenging to determine the overburden stress in the very shallow sections where logging data are either not acquired or acquired in a very large diameter borehole. In this zone, the overburden is commonly estimated (e.g., above solid circles, Fig. 8.2).

One of the best approaches to constrain the overburden stress in the shallow section is to use direct measurements made in geotechnical drilling. Recently it has become more common to drill, core, and log the shallow sedimentary section to determine sediment properties for the design of production facilities (Ostermeier et al., 2001). In addition, scientific ocean drilling programs – through the Ocean Drilling Program, the Integrated Ocean Drilling Program, and now the International Ocean Discovery Program – commonly make detailed measurements in the shallow sedimentary section. In the Ursa Basin, in the Gulf of Mexico, density and porosity were directly measured on core samples (MAD, or moisture and density measurements), and bulk density was measured during logging while drilling (LWD) (Fig. 8.3). LWD porosity is generally slightly less than the MAD porosity (Fig. 8.3d). This is due to core expansion (Mikada et al., 2002), or it is possible that oven drying during MAD measurements removed some of the interlayer water, resulting in a higher porosity than would be calculated from the LWD bulk density (Dewhurst et al., 1996). Ursa provides a striking example of the amount of compaction that can occur in mudrocks in the shallow section. Surface porosities of 80% decline to around 40% at a depth of 500 meters. In Figure 8.4, several locations drilled by the ocean drilling programs are compared to illustrate the possible variation in overburden within the shallow section.

8.2.2 Overburden Interpreted from Logs Recorded in Oil versus Water-Based Muds

Logs recorded in wells drilled with water-based mud can record lower bulk densities in logs than are actually present due to the reaction of the water-based muds with the near-borehole mudrock. For example if the borehole fluid is fresher

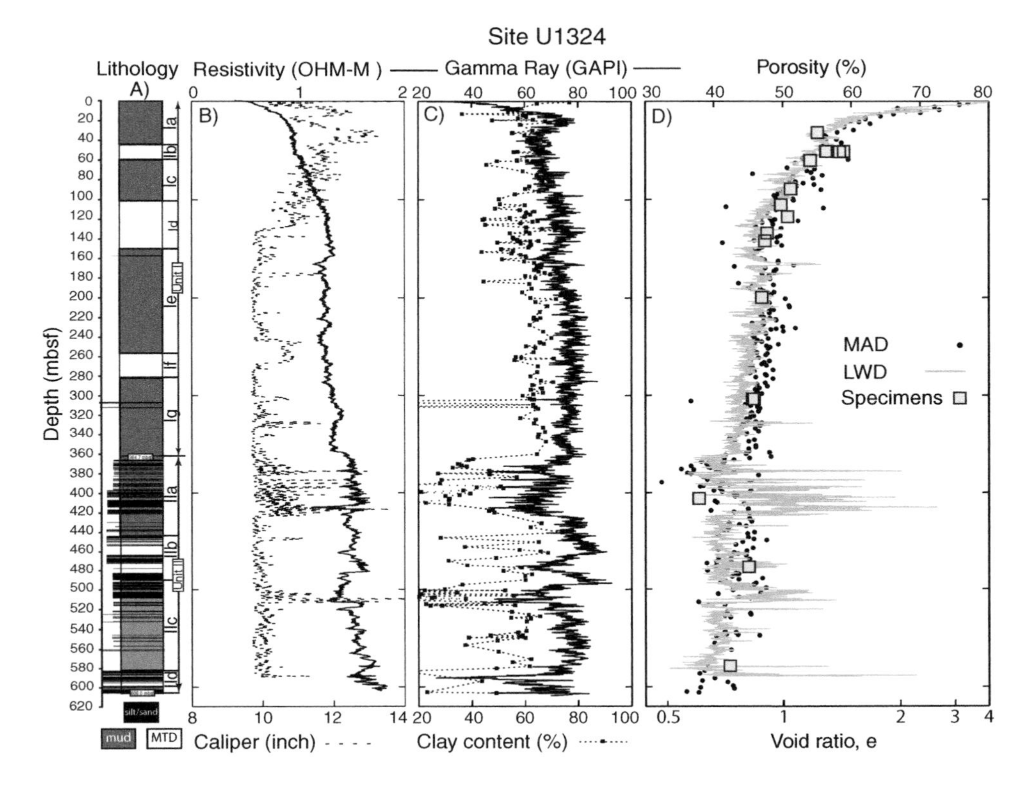

Figure 8.3 Core, log data, and interpretations from IODP Site U1324 in the Ursa Basin, Gulf of Mexico. (a) Lithologic interpretation. (b) Logging while drilling (LWD) deep resistivity and caliper logs. (c) Gamma-ray log data from hole 1324A and clay-sized fraction from core samples. (d) Porosity of tested specimens (squares), shipboard moisture and density (MAD) porosity measurements, and LWD porosity. Reprinted from Long et al. (2011) with permission of Elsevier.

than the formation fluid, smectite-rich clays can expand significantly. This alteration is a function of the porosity, how reactive the clay is (the fraction of swelling clays (Anderson et al., 2010)), the diagenetic state of clays, the fluid used in water-based mudrock drilling (contrast between mud and formation salinity), and the length of the exposure to drilling fluids. LWD logs in wells drilled with water-based mud record higher bulk densities at equivalent depths than subsequent logs run on wireline (Allen et al., 1993); this suggests that the reaction of water with the mudrock takes time. Analysis of repeated density runs over time in wells drilled with water-based mud in the deepwater Gulf of Mexico suggest that the effect of water-based mud can reach 0.03–0.08 g/cm3 (Braunsdorf & Kittridge, 2003). Over the depth of a 20 000 ft well, the overburden difference that results can be as much as 1 000 psi.

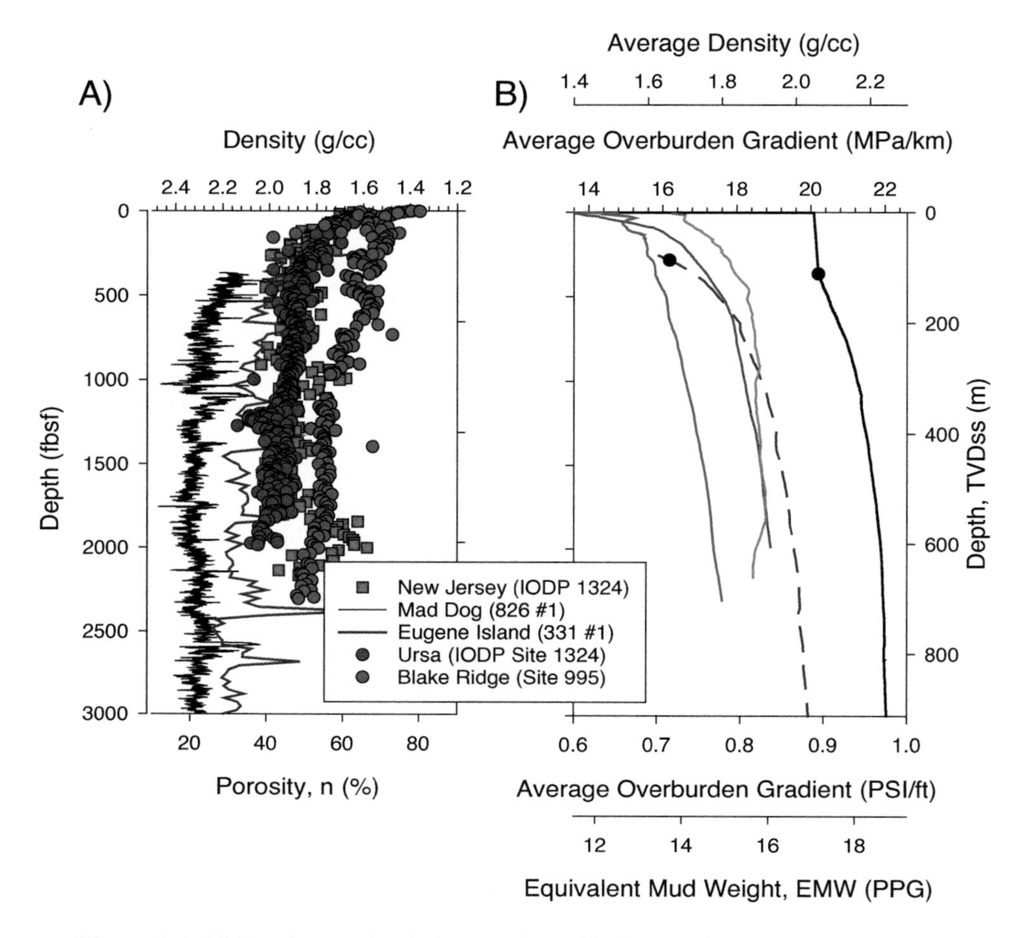

Figure 8.4 (a) Depths are feet below seafloor (fbsf). Density versus depth below seafloor over the first 3 000 ft in several locations in the Gulf of Mexico and the East Coast of the U.S. The symbols record core measurements. The solid lines record the porosity derived from the bulk density logs at Mad Dog (black line) and Eugene Island 330 (Fig. 8.2). (b) Average overburden gradient versus depth.

8.2.3 The Effect of Salt and Bathymetry on Overburden

In the deepwater Gulf of Mexico, it is common to drill through 5 000 to 10 000 ft of salt to reach the reservoir target. This brings a challenge for the estimation of the overburden stress because the bulk density log is calibrated to sandstone and does not record accurate salt densities. Thus, to estimate overburden, the salt density must be assumed.

The density of pure halite is 2.15 g/cc while that of anhydrite is 2.97 g/cc. The Louann salt, the source of salt in the Gulf of Mexico, is composed of 95% halite and approximately 5% anhydrite where mined or cored (Balk, 1949, 1953; Gera, 1972; Hudec et al., 2009; Lerche & Petersen, 1995; Nance et al., 1979). The density of this mixture is 2.2 g/cc, and this is the value we have used in the Gulf of Mexico (Merrell et al., 2014). Others have used smaller salt densities of 2.076 g/cm^3 at Walker Ridge 285 #1, 2.16 g/cm^3 at Keathley Canyon, and 2.165 g/cm^3 for deepwater Gulf of Mexico seismic modeling (Bird et al., 2005; Fredrich et al., 2007; Yarger et al., 2001). Barker and Meeks (2003) suggested that in situ salt in the Gulf of Mexico has a density that ranges from 2.0 to 2.1 g/cm^3, and Dusseault et al. (2004) pointed out that salt density could be elevated by admixing higher density sediment to pure halite.

We illustrate the overburden stress for the 782 #1 well in the Mad Dog field, which penetrated more than 5 000 ft of salt (Figures 8.5 and 8.6). We assume a salt density of 2.2 g/cm^3. The effect of salt on overburden stress is most visible when the overburden gradient is plotted with depth from the seafloor (Fig. 8.7d). In this

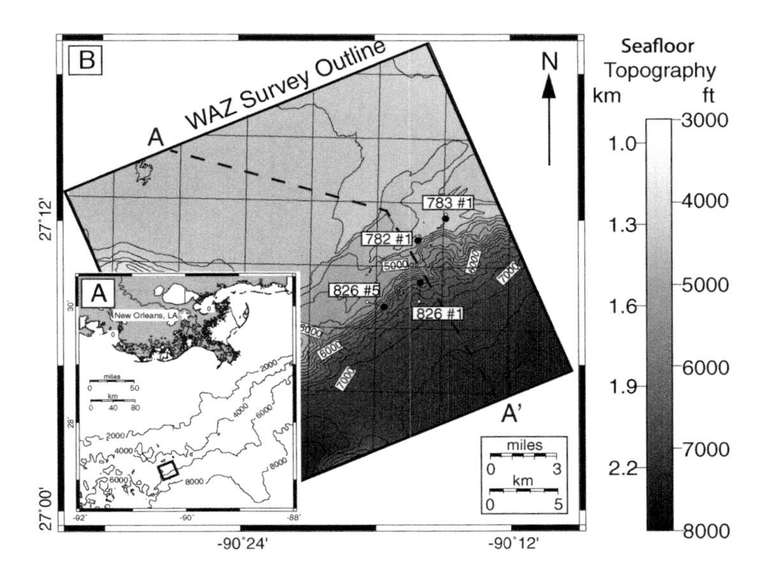

Figure 8.5 Location of the Mad Dog field, Gulf of Mexico. (a) The Mad Dog field (black box) is approximately 190 miles (306 km) southwest of New Orleans. The seafloor bathymetry is illustrated with 2 000 ft (609 m) contours from National Oceanic and Atmospheric Administration (NOAA) bathymetry data. (b) The black outline is the footprint of the three-dimensional seismic survey. The seafloor bathymetry is illustrated with 200 ft (60 m) contours. The water depth decreases between well 826 #1 outboard of the Sigsbee Escarpment and the three wells (826 #6, 783 #1, and 782 #1) inboard of the Sigsbee Escarpment. The empty circles represent the surface well locations, and the black circles represent the bottom-hole locations. From Merrell et al. (2014). Reprinted by permission of the AAPG whose permission is required for further use. AAPG© 2014.

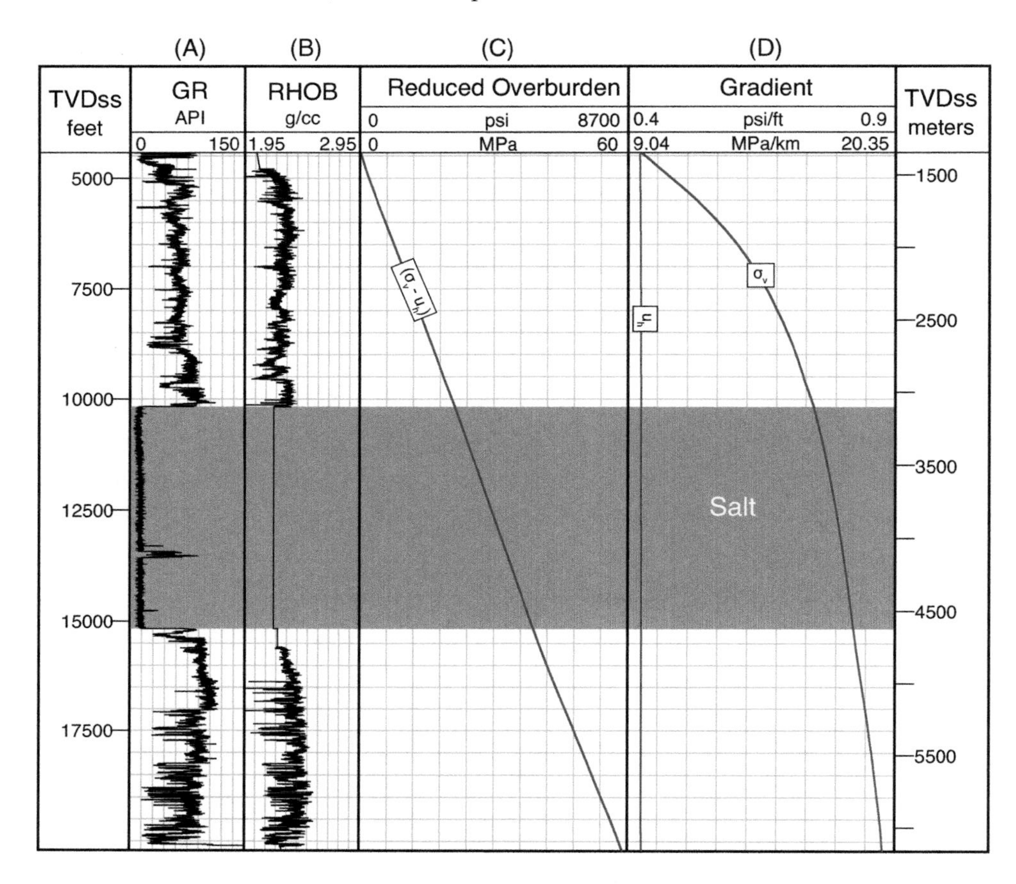

Figure 8.6 Total overburden stress with salt present at Mad Dog from well 782 #1. (a) The gamma-ray log and (b) the bulk density log; a constant bulk density of 2.2 g/cm3 was assumed where salt is present. (c) The total overburden stress (σ_v) is calculated (Eq. 2.6) and plotted as a reduced lithostatic stress ($\sigma_{red} = \sigma_v - u_h$). (d) The overburden gradient (σ_v/TVD_{ss}) is calculated from the sea surface. From Merrell et al. (2014). Reprinted by permission of the AAPG whose permission is required for further use. AAPG© 2014.

perspective, the average overburden gradient decreases with depth in the salt interval. This is because the salt density is less than the overlying sediment density. For the 782 #1 well, the difference is equal to 0.04 psi/ft. When this difference is integrated over the depth of the 782 #1 well, the vertical stress is ~400 psi less than the overburden at an equivalent depth below seafloor in a well where no salt is present (826 #1) (Fig. 8.7e, inset).

Wells drilled inboard of the Sigsbee Escarpment (e.g., 783 #1 and 782 #1) have shallower water depths, and therefore thicker sediment columns above a given depth below sea surface (TVD_{SS}) (Fig. 8.7a, b, c). This results in a higher overburden stress at a given depth below sea surface for these wells relative to those outboard of the Sigsbee escarpment in deeper water (e.g., compare 782 #1 with 826#1, Fig. 8.7b, c). The greatest difference in water depth is 2 137 ft (651.5 meters) between well 826 #1, with a water depth of 6 560 ft (2 000 meters), and well 782 #1, which has a water depth of only 4 423 ft (1 348 meters). The difference in overburden between these two wells is 500 psi (3.4 MPa) at the reservoir interval (M20, Figure 8.7).

Where salt is present at Mad Dog, the water depth is less (Fig. 8.7a). In these locations, the overburden stress is increased because of the thicker overburden, but decreased because of thick salt, relative to a well where no salt is present. Ultimately, because water depths at Mad Dog span such a wide range, the overriding control on the overburden at a given depth below the sea surface is the water depth (Fig. 8.7b, c). In summary, the overburden at a given depth varies significantly spatially because of the extraordinary changes in water depth and salt thickness. As a result, a deviated well at Mad Dog will undergo changes in overburden that result from both changes in overlying salt thickness and water depth and this should be accounted for.

8.2.4 Overburden Stress: Summary

The estimation of overburden stress is conceptually simple (Eq. 2.5). However, capricious application of Equation 2.5 can result in significant error in its estimation. The density profile for any specific location results from the complex interaction of lithology, water depth, and degree of compaction. For these reasons, the most reasonable approach is to develop an overburden model based on nearby data. I also discussed that it is possible to estimate the overburden from seismic velocity in Chapter 7. As well paths become more complex and as we tackle areas with significant lateral variation in overburden, it will be necessary to build two- and three-dimensional models of overburden stress.

8.3 Least Principal Stress and Fracture Initiation Pressure

We want to know the natural, in situ, least principal stress in the Earth's crust. It plays a critical role in deformation and fluid migration. We also want to know the fracture initiation pressure. This is the borehole pressure at which fluid losses occur through fractures and is a critical parameter for the design of safe and stable boreholes. The fracture initiation pressure is impacted both by the far-field least principal stress and by the stress perturbation induced by the presence of the borehole. Different theoretical models describe the borehole pressure at which

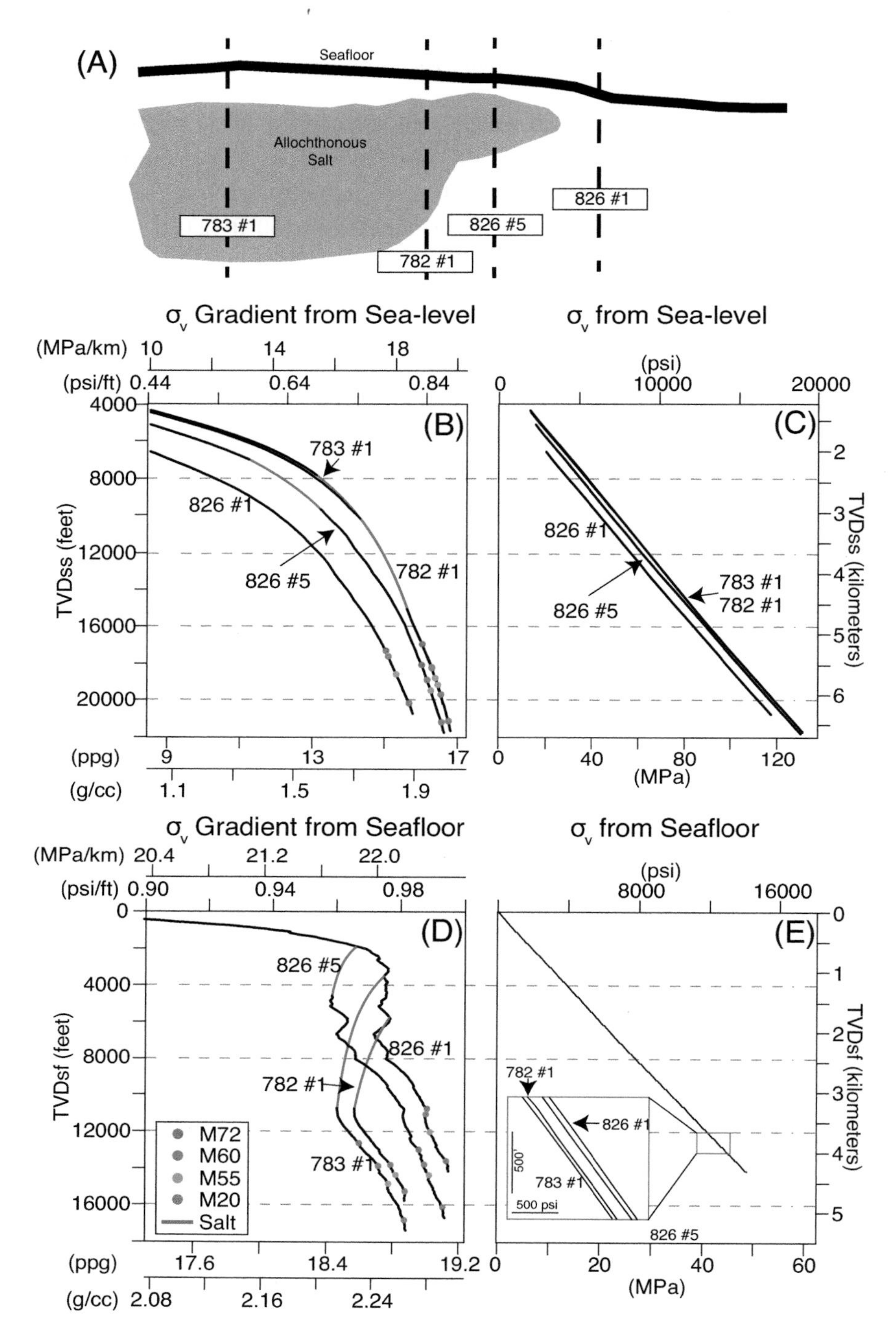

Figure 8.7 Overburdens and overburden gradients from the sea surface and seafloor for four wells at Mad Dog. Where salt is present, a constant density of 2.2 g/cm^3 is assumed. (a) Well locations. (b) Overburden gradient from the sea surface. Well 826 #1 has the smallest gradient because it is in the deepest water. (c) Overburden stress from the sea surface. (d) Overburden gradient from the seafloor. (e) Overburden stress from the seafloor. From Merrell et al. (2014). Reprinted by permission of the AAPG whose permission is required for further use. AAPG© 2014.

fracture initiation occurs. In practice, fracture initiation occurs at borehole pressures that range from the least principal stress to much higher values.

8.3.1 Characterization of Formation Pressure Integrity Tests

The least principal stress (σ_3) and the fracture initiation pressure are most commonly determined from hydraulic fracturing of the formation. The formation pressure integrity test (FPIT) is one class of hydraulic fracturing experiments. This generic name describes pressure tests that are conducted immediately after drilling out a casing or liner shoe (Alberty & McLean, 2014; Kunze & Steiger, 1992; White et al., 2002; Zoback, 2007). The casing is usually set in mudrock. After the casing is set and the cement around it has hardened, 10–20 ft of fresh formation are drilled. Then the test is conducted by pumping drilling fluid into the wellbore at a constant rate between ½ to 1 barrel/min (0.04–0.16 m^3/min). FPITs are used to verify the integrity of the cement at the casing shoe, measure the stress state of the exposed formations for well planning and operations, and determine the maximum pressure that a casing shoe can be safely exposed to (Alberty & McLean, 2014).

Formation integrity tests (FITs), leak-off tests (LOTs), and extended leak-off tests (XLOTs) (Kunze & Steiger, 1992) are examples of formation pressure integrity tests. During an extended leak-off test, fluid is pumped into the wellbore at a constant rate and then pumping is ceased ('shut in,' Fig. 8.8). Borehole pressure rises approximately linearly at first. With continued pumping, there is a break in slope of the pressure versus time curve. The break in slope is the leak-off pressure (LOP). With further pumping, pressure may rise at a lower rate until the peak pressure or formation breakdown pressure (FBP) is reached (Fig. 8.8). With further pumping, pressure drops to the fracture propagation pressure (FPP), a lower but approximately constant value. Pumping is then terminated, and there is an immediate drop in pressure. The first pressure measurement made after shut-in is the instantaneous shut-in pressure (ISIP). The fracture closure pressure (FCP) is interpreted to record the moment the fracture closes. In low flow rate tests (e.g., 1 barrel/min) with a low viscosity fluid (water or thin oil), the LOP, FPP, and ISIP often have approximately the same values (Zoback, 2007). Repeated extended leak-off tests commonly result in declining values of the FBP, ISIP, and the LOP (Fig. 8.9), and it is generally thought that the ISIPs recorded by the later tests record the most accurate estimate of the least principal stress (Zoback, 2007).

Most FPITS are not extended leak-off tests such as the example in Figure 8.8. Rather, the pump-in phase is terminated before the LOP (a formation integrity test) or soon after the LOP is reached (a leak-off test). In a formation integrity test, the borehole pressure is raised to a fixed value based on the planned mud weight to be

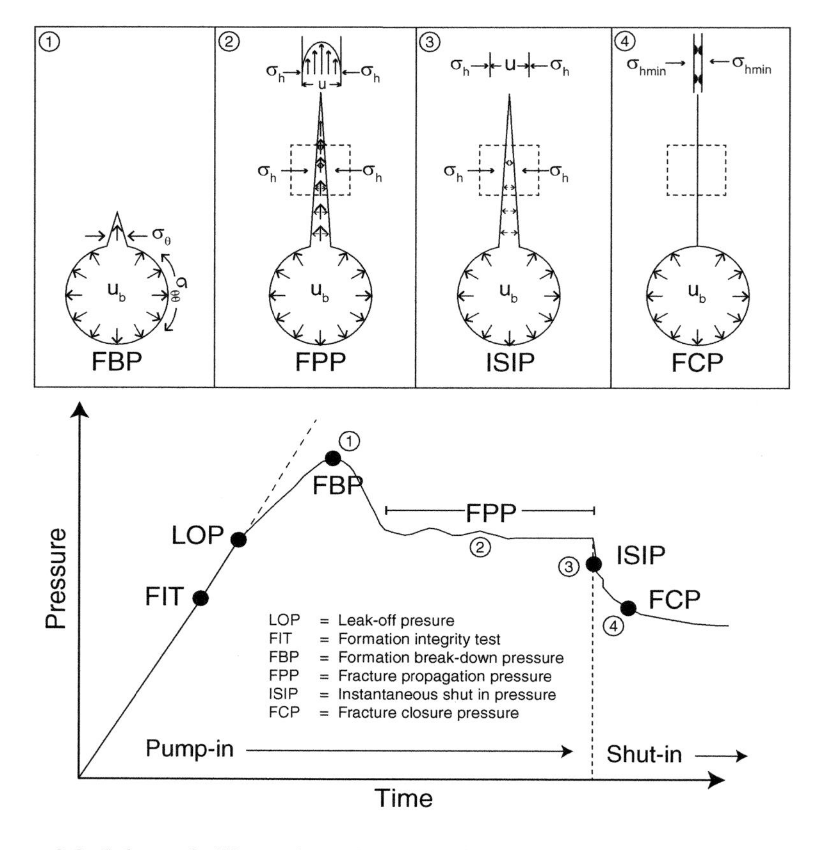

Figure 8.8 Schematic illustration of an extended leak-off test (XLOT). The insets (1 to 4) illustrate the interpretation of the physical processes ongoing in the borehole at specific moments during the leak-off test. Modified from Gaarenstroom et al. (1993). Reprinted with permission of the Geological Society of London.

used to the next casing point. If successful, there is no fluid loss into the formation at that pressure due to fracturing (Fig. 8.8). FITs provide a borehole pressure that cannot be exceeded during drilling but tell little about the stress state. In a leak-off test, only the leak-off point is determined; I discuss the relationship between the LOP and least principal stress below.

8.3.2 The Interpretation of Pressure Integrity Tests

The interpretation of stress from formation pressure integrity tests is based on a model for stress concentration around a borehole in an isotropic linear elastic

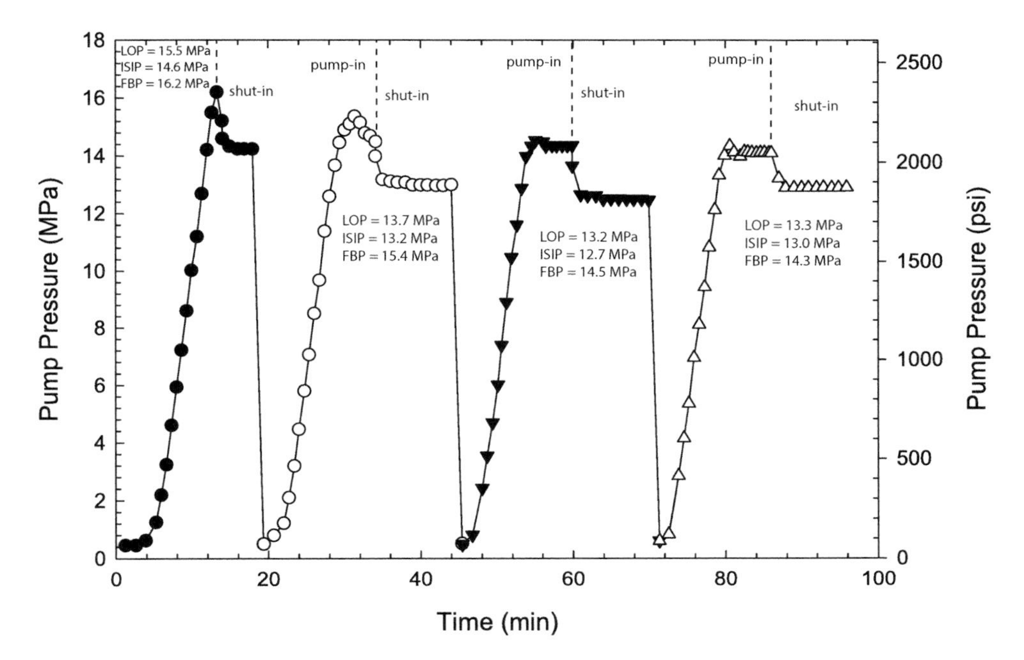

Figure 8.9 Pressure versus time during repeated extended leak-off tests. Fluid is pumped in at a constant rate (pump-in), and then pumping is stopped (shut-in) while borehole pressure is observed. Modified from White et al. (2002). Reprinted with permission of the Geological Society of London.

medium (e.g., Kirsch, 1898), on two-dimensional hydraulic fracture models that rely on propagation of a crack in a linear elastic medium under plane-strain conditions (e.g., England & Green, 1963), and on empirical experience. I review the interpretation of the FBP, FPP, ISIP, FCP, and LOP below.

Formation Breakdown Pressure (FBP): Hubbert and Willis (1957) demonstrated that hydraulic fractures form normal to the least principal stress and they derived the equation for hydraulic fracturing break-out (the formation breakdown pressure) from the Kirsch equations (Kirsch, 1898) (Fig. 8.11).

The formation breakdown pressure is the borehole pressure necessary to make the effective tangential stress equal the tensile strength (T) at the borehole wall:

$$u_b = 3\sigma_h - \sigma_H - u + T, \qquad \text{Eq. 8.1}$$

where u_b is the borehole pressure, u is the pore pressure, and T is the tensile strength: this is the pressure at which a tensile fracture forms (Fig. 8.8, #1). Raleigh et al. (1976) suggested that with successive reopening of an existing hydraulic fracture, the tensile strength (T) can be ignored. In highly fractured or

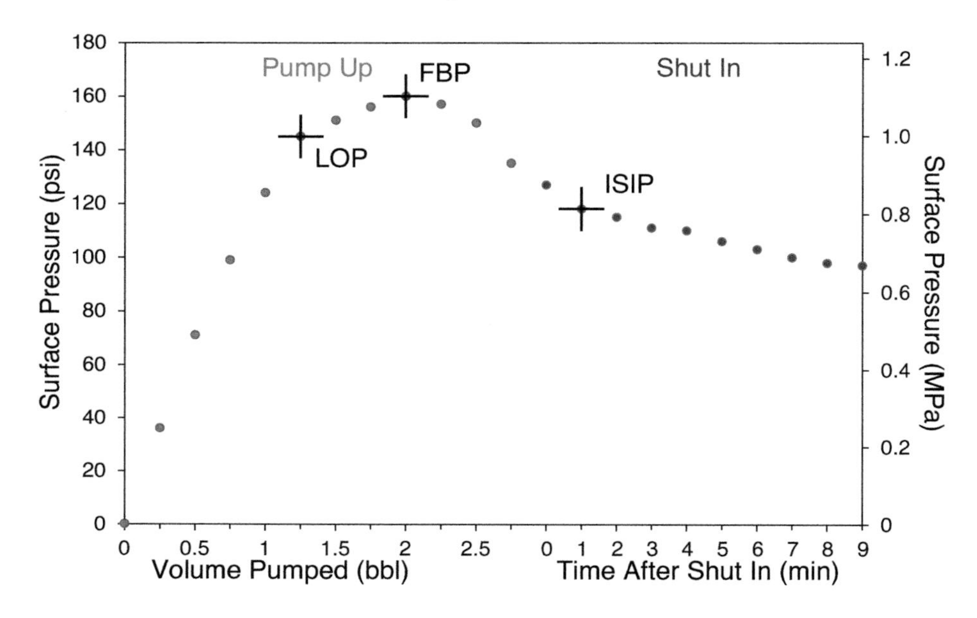

Figure 8.10 The extended leak-off test at the 22" shoe at the Macondo well. Test #8 of 8 tests. The pump-in stopped before a stable FPP was displayed. Reprinted with permission from Pinkston (2017).

poorly indurated rocks typical of an oil field, it is generally assumed that tensile strength is negligible and Equation 8.2 is valid even without multiple reopenings:

$$u_b = 3\sigma_h - \sigma_H - u. \quad \text{Eq. 8.2}$$

For the case where the two horizontal stresses are equal, the breakdown pressure is twice the least principal stress less pore pressure. The breakdown pressure declines as the contrast in horizontal stresses increases; when $\sigma_H = 2\sigma_h$, the breakdown pressure equals the least principal stress less the pore pressure.

Su and Onaisi (2019) suggest that fluids may also be lost due to the initiation of shear failure at the borehole wall:

$$u_b = 3\sigma_h - \sigma_H - u - (\sigma_v - UCS - u)K. \quad \text{Eq. 8.3}$$

UCS is the unconfined strength and *K* is the ratio of the least principal effective stress to the maximum principal effective stress at Coulomb failure (Eq. 3.8). Equation 8.3 predicts lower value for the breakdown pressure than Equation 8.2.

Fracture Propagation Pressure (FPP): The fracture propagation pressure is the borehole pressure necessary to propagate the fracture away from the well (Fig. 8.8, #2). It has to keep the fracture open and drive the fluid flow along the fracture. It

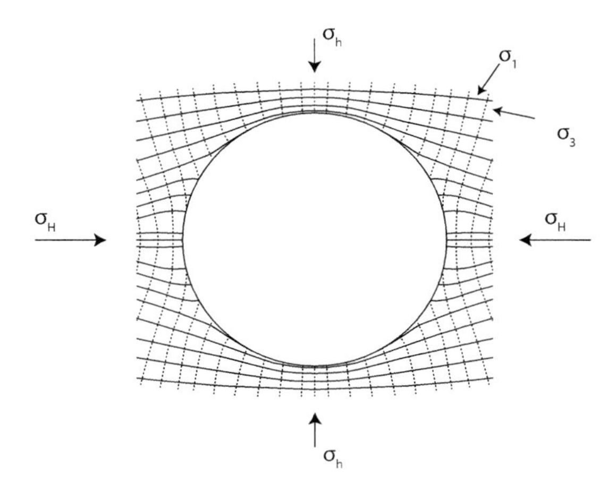

Figure 8.11 Principal stress trajectories around a vertical cylindrical opening based on the Kirsch equations (Kirsch, 1898). The wellbore is a free surface, and thus the principal stress trajectories deviate parallel and perpendicular to it. This causes convergence of maximum stress (σ_H) trajectories, where stresses are most compressive, and divergence of minimum stress (σ_h) trajectories, where stresses are least compressive. This approach assumes that two of the principal stresses are oriented in the horizontal plane, and that the material is isotropic and linear elastic. Modified from Zoback (2007). Reprinted with permission by Cambridge University Press.

thus exceeds the least principal stress and is a function of the injection rate, the fluid viscosity, and the elastic properties of the formation. With low viscosity fluids and low injection rates, the FPP converges on the least principal stress.

Instantaneous Shut-In Pressure (ISIP): The instantaneous shut-in pressure is measured immediately after the pumps are shut in (Fig. 8.8, #3). In this case, the pressure associated with the fluid flow during pumping is eliminated; this is the pressure necessary to keep the fracture wide open.

Fracture Closure Pressure (FCP): The FCP records the moment where the fracture walls start to touch (Fig. 8.8, #4). It is thus thought to record the least principal stress. It is determined by plotting pressure as a function of the square root of time and detecting a change in the linearity of the pressure decay (Economides et al., 1989). Zoback (2007) suggests that the FCP is the more appropriate estimate of least principal stress in cases where a viscous fluid is used during fracturing; otherwise, this method can underestimate the least principal stress.

Leak-Off Pressure (LOP): The leak-off pressure is perhaps the least understood characteristic of a pressure integrity test (Fig. 8.8). Yet, it is one of the most important values because the pump-in phase is commonly terminated soon after the LOP is reached and thus no ISIP is recorded. It is interpreted that the break in slope at the LOP (Fig. 8.8) records an increase in effective borehole volume caused by migration of fluid into fractures that propagate away from the wellbore. This could record the breakdown pressure (Eq. 8.1) in cases where there is a large contrast in the horizontal stresses, because under these conditions there is no stress barrier caused by local borehole stresses (Alberty & McLean, 2004). However, breakdown pressures significantly above the far-field least principal stress are commonly encountered after the leak-off point is reached (e.g., Fig. 8.10).

Empirical experience suggests that many leak-off values occur at or near the least principal stress (Alberty & McLean, 2004; Zoback, 2007). Alberty and McLean (2004) suggest that this occurs when borehole fluid seeps into preexisting conductive fractures at the wellbore. When the borehole pressure exceeds the tangential stress at the borehole wall, fluids start to propagate into these fractures. This occurs when the borehole pressure equals the far-field least principal stress for the case where the two horizontal stresses are equal. This interpretation is supported by observations that with repeated pressure integrity tests, the breakdown pressure declines, presumably because conductive fractures are emplaced in the early tests (e.g., Fig. 8.9). It is also possible that shear failure (Eq. 8.3) may be the source of the fluid loss at the leak-off pressure.

Fracture Initiation Pressure: At the end of the day, the fracture initiation pressure is the borehole pressure that results in loss of fluids through fractures in the borehole wall. The breakdown pressure (tensile failure), the shear failure pressure, and the interpretation that leak-off results at the far-field least principal stress, are all attempts to estimate the fracture initiation pressure.

8.3.3 The Interpretation of Lost Circulation Events

Lost circulation occurs when borehole fluid is lost into the formation during drilling and is commonly caused by opening existent borehole fractures or by forming hydraulic fractures. The breakdown, propagation, or closure pressure during lost circulation events cannot be distinguished. Thus, the pressure at the onset of fluid losses is the fracture initiation pressure. It probably exceeds the propagation pressure and could exceed the breakdown pressure. The pressure at which the well is stable before or after the lost circulation event is taken as a lower bound for the fracture initiation pressure. For example, a kick (influx of fluid into the borehole) occurred when drilling reached 2 700 m (Fig. 8.12). The vertical dashed line on the left of the equivalent mud density plot is the borehole mud weight when the kick occurred and represents a lower bound for the fracture initiation pressure. In response, the mud weight was raised as is recorded by the vertical dashed line on

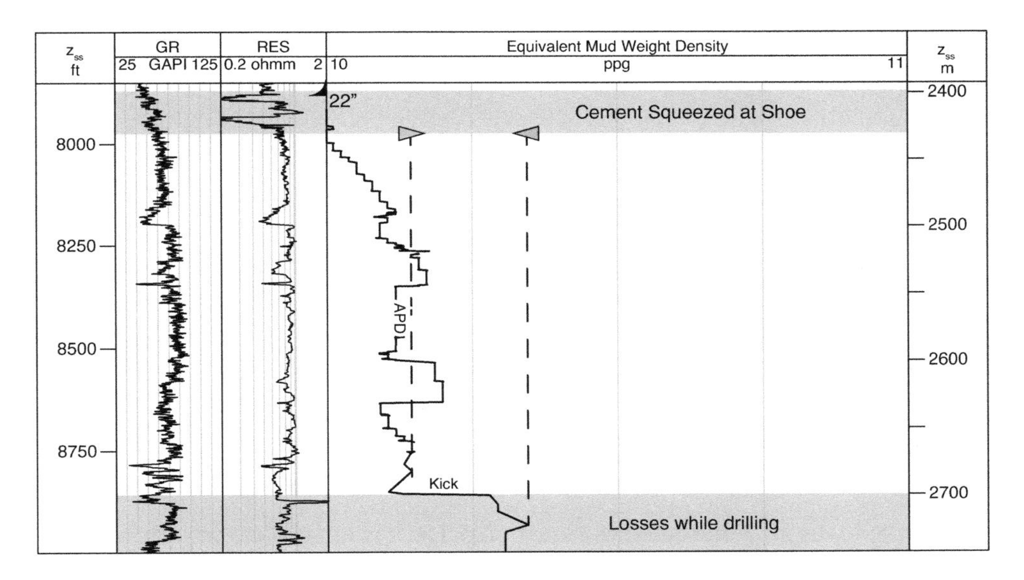

Figure 8.12 Equivalent mud weight versus depth of a kick and mud loss event in the Macondo well. Triangles denote the interpreted loss location. The range of possible fracture initiation pressures is bounded below by the kick mud weight (left, blue triangle) and above by the mud weight at which losses occurred (orange triangle). Dashed black lines record the equivalent mud weight in the borehole during the kick and the loss events. The upper grey box highlights the 22-inch shoe depths reinforced by cement squeeze operations. The lower grey box illustrates the interval that was drilled while losses were occurring. From Pinkston (2017). Reprinted with permission by Will Pinkston.

the right hand side (Fig. 8.12). When this occurred, lost circulation resulted. The right dashed line marks an upper bound to the fracture initiation pressure.

It is challenging to know the exact depth of the lost circulation. For a given ratio of horizontal to vertical effective stress, and if pore pressure does not vary significantly (see ensuing section), this may occur at the shallowest location, the casing shoe, where the borehole pressure is closest to the least principal stress. However, if some part of the wellbore contains natural fractures while other parts do not, fracture initiation may occur where the fractures are present, even if they are located well below the shoe. Furthermore, Growcock et al. (2009) suggest that if one is drilling ahead, and there are changes in torque, or there is a drilling break, the lost circulation occurs at the base of the hole. Finally, sandstones, siltstones, and marls commonly have a lower least principal stress than bounding mudrocks (Alberty & McLean, 2001; Daines, 1982) and thus lost circulation could occur at deeper locations where sands are present. At Macondo locations where the depths of losses were tightly constrained, it was found that losses commonly occurred within siltstones or sandstones well below the casing shoe (Pinkston, 2017).

8.3.4 Least Principal Stress and Fracture Initiation Pressure Models

8.3.4.1 Least Principal Stress

The least principal stress is commonly modeled as a function of pore pressure and overburden stress. The most common approach is to assume that the ratio of the least principal effective stress to maximum principal effective stress is constant, or varies in some systematic fashion:

$$K = \frac{\sigma_3'}{\sigma_1'} = \frac{\sigma_3 - u}{\sigma_1 - u}. \qquad \text{Eq. 8.4}$$

Equation 8.4 was also presented as Equation 3.15. It is arrived at through, at least, three physical models: (1) elastic behavior under uniaxial strain, (2) geotechnical observations under uniaxial strain, and (3) Andersonian faulting (Fig. 3.14).

The poroelastic stress-strain equations, when solved for uniaxial strain conditions, show that

$$K = \frac{\sigma_3'}{\sigma_1'} = \frac{v}{1 - v}, \qquad \text{Eq. 8.5}$$

where v is Poisson's ratio, e.g., Eaton (1969). For $v = 0.25$ (a common value for Poisson's ratio), $K = 0.33$.

In geotechnical engineering, the ratio of horizontal to vertical stress during uniaxial strain is described by Lambe and Whitman (1979):

$$K_0 = \frac{\sigma_3'}{\sigma_1'}. \qquad \text{Eq. 8.6}$$

Most experimental work has focused on materials for which the present stress state acting on the material is the maximum stress state ever applied to the material. Such materials are said to be in a state of normal compression. In this case, K_0 is denoted as K_{0NC}. Semiempirical correlations for estimating K_{0NC} have been proposed (Jaky, 1944; Mesri & Hayat, 1993). All of these models correlated K_{0NC} to friction angle (ϕ') (e.g., Eq. 3.30), where both K_{0NC} and ϕ' are assumed to remain essentially constant with stress level for a particular material.

Figure 3.8 illustrates K_{0NC} for three typical lithologies at different stress levels: smectite-rich Gulf of Mexico mudrock, illite-rich mudrock, and siltstone. Smectite-rich mudrocks have high initial K_{0NC} values that increase rapidly with stress (pink triangles, Fig. 3.8). In contrast, illite-rich mudrocks (RBBC) have slightly lower initial values of K_{0NC} that increase only modestly with effective stress (blue circles, Fig. 3.8). Finally, silt-rich rocks (more granular material) have lower values of K_{0NC} that are approximately constant with increasing effective stress (orange squares, Fig. 3.8). In summary, K_{0NC} varies as a function of lithology (e.g., mudrock versus siltstone) and mineralogy (e.g., smectite-rich versus illite rich).

In the Andersonian faulting model, the Earth's crust is assumed to be at Coulomb failure. The Earth's crust is assumed to contain widely distributed faults, fractures, and planar discontinuities at many different scales and orientations. The difference between the maximum and minimum principal stresses is limited by the frictional strength of these planar discontinuities (Zoback, 2007). The ratio of least to maximum principal effective stresses in this condition is:

$$K = \frac{\sigma_3'}{\sigma_1'} = \frac{\sigma_3 - u}{\sigma_1 - u} = \frac{1 - sin\phi'}{1 + sin\phi'}, \qquad \text{Eq. 8.7}$$

where ϕ' is the friction angle of the favorably oriented discontinuities. Equation 8.7 was also presented as Equation 3.8. A common friction angle (ϕ') is 30° ($\mu = 0.6$) and for this value, the ratio K is 0.33. This approach has been broadly applied to understand the state of stress in the Earth's crust (Zoback, 2007) and has long been applied in geotechnical engineering to describe the state of stress in conditions of Coulomb failure due to lateral compression (passive failure) or lateral extension (active failure) (Lambe & Whitman, 1979)

Figure 3.10 illustrates ϕ' for three typical lithologies at different stress levels: smectite-rich Gulf of Mexico mudrock, illite-rich mudrock, and siltstone. The smectite-rich Gulf of Mexico mudrock (RGoM-EI) has ϕ' values of approximately 27° ($K = 0.37$, Eq. 8.7) at low stress that decrease rapidly with increasing stress to ~10° ($K = 0.7$, Eq. 8.7) (pink triangles, Fig. 3.10). In contrast, illite-rich mudrocks (RBBC) have initial friction angles of approximately 30° that decrease only modestly with effective stress (blue circles, Fig. 3.10). Finally, silt-rich rocks (more granular material) have friction angles that are similar but slightly greater than the illite-rich mudrock. Casey et al. (2016) shows that the variation in friction angle correlates closely with variation in K_{0NC} and that this variation can be described by functions similar to the Jaky equation (Eq. 3.30). In summary, the stress ratio (K) at Coulomb failure varies systematically as a function of lithology (e.g., mudrock versus siltstone) and mineralogy (e.g., smectite-rich versus illite rich). In a similar fashion to the behavior described for K_{0NC} (Eq. 8.6), the stress ratio is lower and relatively constant for illitic mudrock and more granular material. However, smectite-rich mudrocks have higher stress initial stress ratios, and these values increase dramatically with effective stress.

In practice, the stress ratio (K) is constrained empirically by determining least principal stress, pore pressure, and overburden stress at specific locations. Empirical relations are then derived to determine K as a function of depth or stress. Historically, these data are an aggregation of leak-off pressures that are assumed to reflect the least principal stress (e.g., Eaton, 1969). Matthews and Kelly (1967) assumed a constant overburden gradient (1.0 psi/ft) and derived a function for K that increased with depth. Eaton (1969) expanded on this by including a depth-dependent overburden gradient to determine overburden stress. Breckels and van Eekelen (1982) show that a range of relationships for K (Christman, 1973; Eaton, 1969; Matthews & Kelly, 1967; Pennebaker, 1968) show remarkable similarity. Breckels and van Eekelen (1982) emphasized that these relationships vary by basin and developed relationships for the U.S. Gulf Coast, Venezuela, Brunei, the Netherlands, and the North Sea. They also demonstrated that the stress ratio for sandstones is less than that for mudrocks.

Eaton's stress ratio increases from 0.3 at low effective stress to a value of 1 at high effective stresses (Fig. 8.13a). This stress level dependence results in a least principal stress that rises from ~1/3 of the way between the pore pressure and the overburden stress to ultimately converge on the overburden stress at great depths (grey dashed line, Fig. 8.13b, c). This long-observed relationship suggests that stress state becomes more isotropic with depth. I compare Eaton's model with the three uniaxial strain experiments performed on resedimented material (Fig. 3.8). Eaton's model is most similar to the smectite-rich RGoM-EI mudrock (Fig. 8.13). It is also striking how much lower the least principal stress of the RPC siltstone is relative to Eaton's prediction (Fig. 8.13c). This behavior is commonly observed and is one reason hydraulic fractures can be contained within reservoirs sealed by mudrocks.

In summary, both experiments (e.g., Figures 3.7 and 3.8) and observations suggest there is both a stress dependence and a lithologic dependence of least principal stress under either uniaxial strain (Eq. 8.6) or Coulomb failure (Eq. 8.7). This results in significant stress anisotropy in lithologic successions in sedimentary basins (Fig. 8.14). More sand-rich material has a much lower horizontal stress than its bounding mudrocks. This reflects that the more granular (sand-rich) material is stronger and can maintain a larger differential stress (as reflected by lower values of K_{0NC} and higher friction angles for more granular material). Furthermore, at higher effective stresses (deeper), the stress ratio for any individual lithology increases and the behavior is much more dramatic for the mudrock than for the sandstone. This results from the strength dependence of the mudrock as described above.

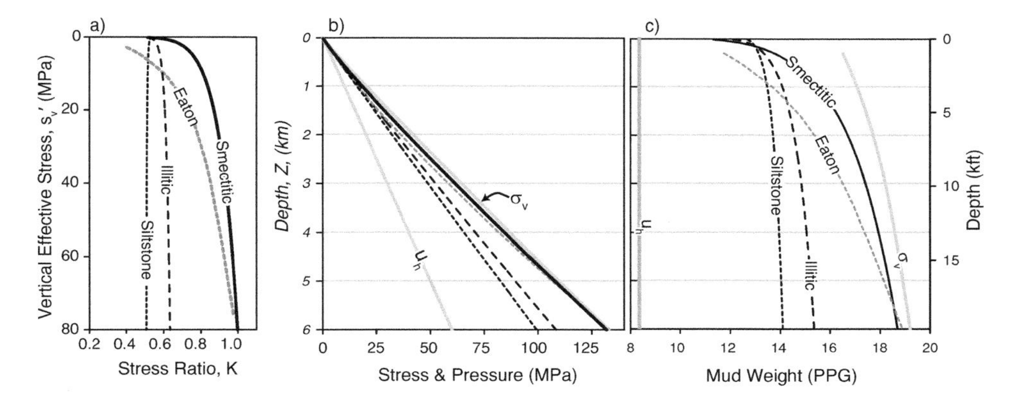

Figure 8.13 Comparison of Eaton (1969) fracture initiation model to least principal stresses during uniaxial consolidation of smectite-rich mudrock, illite-rich mudrock, and siltstone. (a) The variation in stress ratio (K) with vertical effective stress (left) or depth under hydrostatic pressure (far right). (b) The variation in least principal stress ($\sigma_3 = \sigma_h$) with depth. (c) The variation in least principal stress expressed as equivalent mud weight. The "Siltstone," "Illitic," and "Smectitic" are the RPC, RBBC, and RGoM-EI results presented in Figure 3.8. The overburden stress presented in B and C is from the overburden function of Eaton (1969).

Zoback and Kohli (2019) have proposed that an additional process drives stress anisotropy. They suggest that tectonic stresses are driven by Equation 8.7 but that these stresses are modified by creep or relaxation. The creep occurs faster in more clay-rich rocks with the result that mudrocks have a higher stress ratio than sands and that deeper, hotter shales have undergone more creep and thus have a higher stress ratio. While creep may play a role, the observed stress anisotropy can be explained by the dependence of the friction angle (ϕ') and stress ratio (K_0) on stress and composition alone.

8.3.4.2 Modeling the Fracture Initiation Pressure

The borehole pressure at which fluids are lost through tensile (Equations 8.1 and 8.2) or shear fractures (Eq. 8.3) are both proposed to describe the fracture initiation pressure. I compare these pressures to the far-field least principal stress for a vertical borehole in smectite-rich mudrock that has been normally consolidated under uniaxial strain and hydrostatic pore pressure (Fig. 8.15). I assume the two horizontal stresses are equal, and that the UCS and the tensile strength are zero. The borehole pressure necessary to drive a tensile fracture (Kirsch, Fig. 8.15b, c) is much higher than the overburden stress, and the borehole pressure for shear failure (Shear, Fig. 8.15b, c) is modestly greater than the overburden. Of course, it is unlikely that such high borehole pressures would be achieved because as the

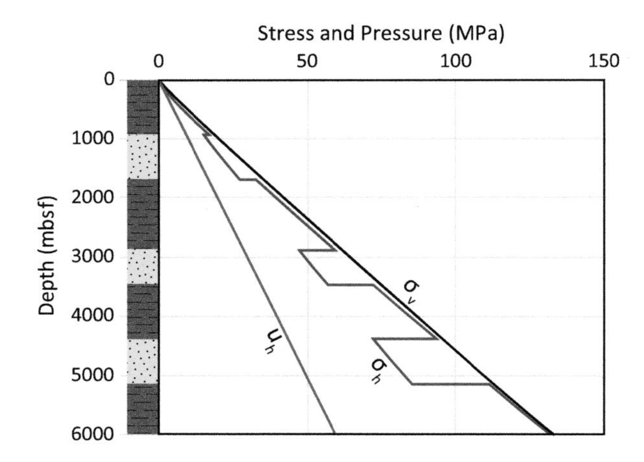

Figure 8.14 Stress anisotropy in interbedded sandstones (yellow) and mudrocks (grey). The least principal stress in the sandstones is less than the least principal stress in the bounding mudrocks. This anisotropy results from the evolution of lateral stress during burial of different lithologies (e.g., Figures 3.7 and 3.8). For this example, the least principal stress in the sandstone is modeled with a stress ratio (K) of approximately 0.5. In contrast, the mudrock stress ratio increases from approximately 0.8 in the shallow section to almost 1.0 at depth.

borehole pressure exceeds the overburden stress, ultimately horizontal fractures (sills) would be generated. However, it is important to note that these very high fracture initiation pressures result from the assumption that the two horizontal stresses are equal, and that lower fracture initiation pressures are possible if the horizontal stresses are different (Su & Onaisi, 2019).

8.3.5 Example: Least Principal Stress at the Macondo Well

I characterize the fracture initiation pressures and the least principal stress in the Macondo well, Gulf of Mexico, from formation pressure integrity tests (FPITs) and lost circulation events (Pinkston, 2017). Because the casing is set in mudrock, the FPITs are recording the mudrock stress state. In contrast, the lost circulation events may record the stress state in any lithology. Based on the ISIP values and the estimated pore pressures, the stress ratio (K) ranges from approximately 0.5 at shallow depths to values close to or above 1.0 for the 13–5/8 inches and 11–7/8 inches casing shoe (green circles, Fig. 8.16). At these deeper locations, the fracture initiation pressure is close to or greater than the overburden stress. These high stress ratios could result from (1) local stress perturbations above the far-field stress; (2) near wellbore stress concentrations that are not eliminated even though

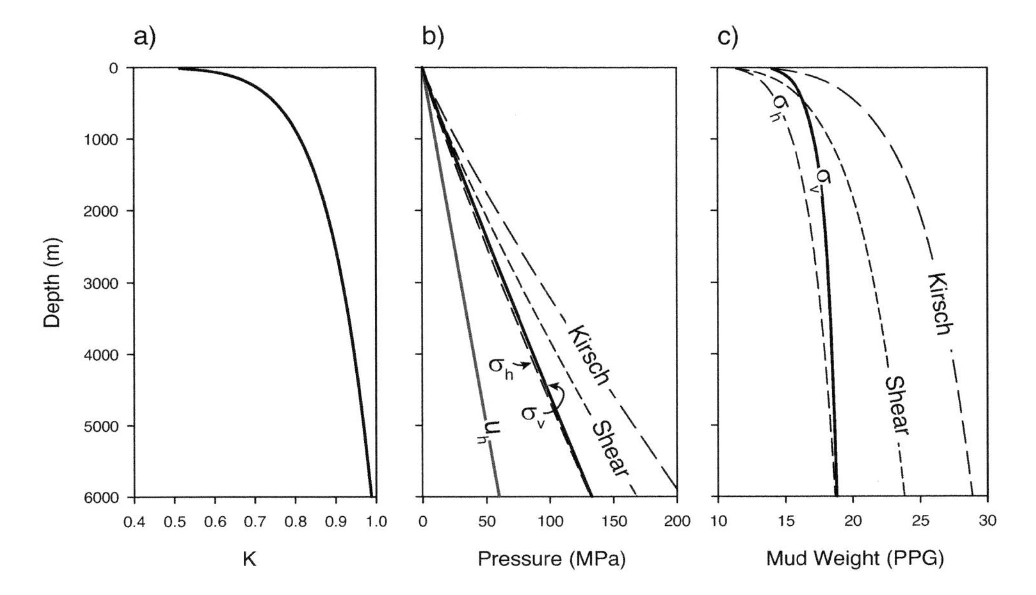

Figure 8.15 Least principal stress and fracture initiation pressures associated with tensile-fracture and shear failure along a vertical borehole in smectitic mudrocks consolidated under uniaxial strain and hydrostatic pore pressure. (a) The variation in stress ratio (*K*) with depth. (b) The variation with depth of least principal stress (σ_3) and fracture pressures for tensile fracture (Kirsch) (Eq. 8.2) and shear failure (Shear) (Eq. 8.3). (c) The variations in (b) expressed as equivalent mud weight.

the ISIPs are generally interpreted to measure the far-field stress; or (3) an underestimation of the overburden stress (the pressures are very close to the overburden and small errors in the overburden will result in large changes in *K*). The mean of the five ISIP-derived effective stress ratios is 0.78 (Fig. 8.16, green dashed line).

The stress ratio (*K*) recorded by the lost circulation events (bounded by the blue and orange triangles, Fig. 8.16) are lower than those recorded by the formation pressure integrity tests (FPITs) (bounded by the red square and green circles, Fig. 8.16). Beneath 15 000 ft, the lost circulation events can be directly correlated to siltstones or sandstones (Pinkston, 2017). It is more difficult to interpret the location of the shallow lost circulation events, but they may also occur in sandstones or siltstones. The stress ratio is commonly lower in sandstones or siltstones than in mudrocks (Alberty & McLean, 2001; Daines, 1982). Thus, it is not surprising that the stress ratio is lower in the lost circulation events in sandstone than it is in the formation pressure integrity tests in mudrock. Furthermore, as described above, this is also observed in our experimental tests where the stress ratio in the silt-rich Presumpscot Clay (RPC) ($K = \sim 0.5$) (short dashed line, Fig. 8.13) is much lower than that present in the smectite-rich Eugene Island mudrock (RGoM-EI) (solid line, Fig. 8.13). I interpret that the lost circulation

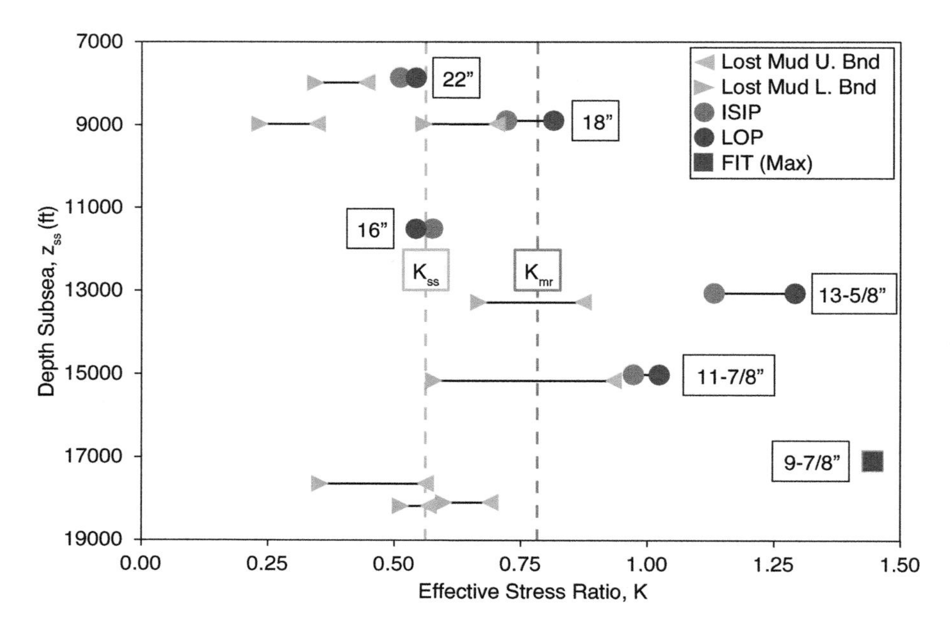

Figure 8.16 Stress ratio (*K*) for the Macondo well estimated from formation pressure integrity tests (FPITS) and lost circulation events. For each FPIT, the effective stress ratio calculated from the Instantaneous Shut-In Pressure (ISIP, green circle) and the Leak-off Pressure (LOP, red circle) are connected with the black line and labeled with the corresponding casing diameter. The vertical green dashed line records the average effective stress ratio for the five ISIP interpretations (green circles). For each mud loss event, two effective stress ratios are calculated: the lower bound (blue triangle) is connected to the upper bound (orange triangle) with a black line. The vertical orange dashed line records the average effective stress ratio for the lower and upper bounds from the lost circulation events. From Pinkston (2017). Reprinted with permission by Will Pinkston.

events record fracturing of sandstones or siltstones where the effective stress ratio is lower than in the mudrock. The average value of the upper and lower bounds for *K* is 0.56 for the lost circulation events, and I use this to describe the stress ratio in sandstone (orange dashed line, Fig. 8.16).

The pore pressure for the Macondo well was presented in Figure 6.12. Given this pore pressure, Equation 8.4 is rearranged to generate a continuous curve for least principal stress:

$$\sigma_h = K(\sigma_v - u) + u. \quad \text{Eq. 8.8}$$

With $K = 0.78$, the mudrock least principal (green line) stress lies 78% of the distance between the pore pressure (blue line) and the overburden stress (black line) (Fig. 8.17). The stress model does not extend above 11 000 ft because there was no

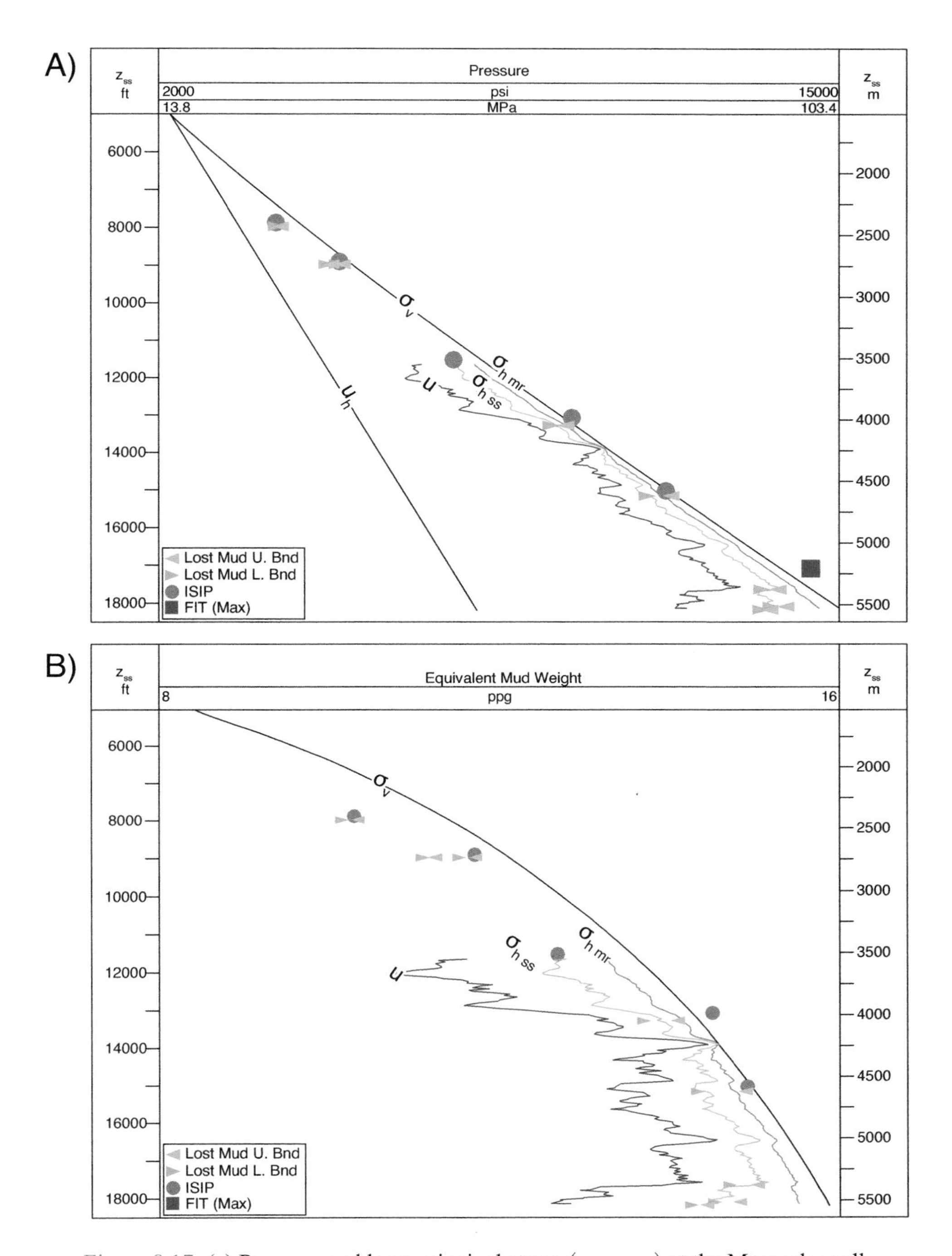

Figure 8.17 (a) Pressure and least principal stress ($\sigma_3 = \sigma_h$) at the Macondo well. (b) Pressures and stresses converted to equivalent mud weight at the Macondo well. The least principal stress within the mudrock (σ_{hmr}, green line) is modeled with $K = 0.78$. The least principal stress within the sandstone (σ_{hss}, orange line), is modeled with $K = 0.56$. Stress interpretations are shown with symbols (triangles, squares, circles). Triangles denote the location and fracture gradient range for each lost circulation event: the lower bound is in blue and the upper bound is in orange (see Fig. 8.16). Green circles denote the downhole ISIP pressures. Modified from Pinkston (2017). Reprinted with permission by Will Pinkston.

sonic log to predict pressure in this interval. With $K = 0.55$, the sandstone least principal stress (yellow line) lies approximately halfway between the pore pressure and the overburden stress. This line more closely matches the lost circulation events.

The least principal stress decreases sharply with the pore pressure regression that occurs below 17 500 ft (Fig. 8.18). A regression in least principal stress for a constant stress ratio (K) is expected from Equation 8.4. However, the decrease in the observed least principal stress is greater than predicted by a stress ratio (K) of 0.8. This is because the stress ratio is lower in the sandstone reservoir than in the bounding mudrock. Pinkston and Flemings (2019) explore the implications of this regression in stress and pore pressure for drilling and completing the Macondo well.

At this point, it should be obvious that while there are models to describe fracture initiation pressure and least principal stress, their actual estimation is fraught with ambiguity, relies on sparse data, and suffers from the fact that there is tremendous heterogeneity in lithology and stress. A common strategy is to assume different stress states in different lithologies (e.g., a different value of K for sandstones and shales as shown in Figure 8.18) and place upper and lower bounds for fracture initiation pressure for each lithology (e.g., by applying both Equations 8.2 and 8.3 as upper and lower bounds).

8.3.6 Summary: Least Principal Stress and Fracture Initiation Pressure

The least principal stress and formation breakdown pressure are interpreted from formation pressure integrity tests (Fig. 8.8). By far the most effective way to interpret least principal stress is through an extended leak-off test, performed multiple times, at low flow rate. In these cases, it has been shown that the instantaneous shut-in pressure converges to the least principal stress (e.g., Fig. 8.9). Unfortunately, these tests are seldom run and often least principal stress and the fracture initiation tests must be inferred from leak-off pressures only.

The fracture initiation pressure is the pressure at which fluids are lost into the borehole through fractures, and it can vary significantly. For a vertical well in a poroelastic medium, an upper bound for this pressure is the pressure at which tensile fractures form (Eq. 8.1). An alternate possibility is that fracture initiation begins when Coulomb failure occurs at the borehole wall (Eq. 8.3). Finally, if significant fractures are already present in the borehole, then the fracture initiation could occur at the least principal stress.

The relationship between least principal stress, overburden stress, and pore pressure is commonly described with the effective stress ratio (K), the ratio of the least principal effective stress to the maximum principal effective stress (Eq. 8.4).

Both experiments and field observations suggest that K varies as a function of lithology, mineralogy, and stress level. Sandstones and siltstones have a lower stress ratio (K) than mudrocks at the same pore pressure and vertical stress. Furthermore, smectite-rich mudrocks have a greater stress ratio than illite-rich mudrocks. Finally, most lithologies demonstrate stress dependency, whereby the stress ratio (K) varies as a function of the vertical effective stress. This behavior is particularly striking in smectite-rich mudrocks where the stress ratio ranges from approximately 0.5 to 1.0 (isotropic conditions) with increasing vertical effective stress.

Ultimately, the prediction of least principal stress and fracture initiation pressure is largely based on empirical approaches that integrate formation pressure integrity tests, lost circulation events, and any other information available (Fig. 8.16). These point measurements are coupled with the stress ratio model (Eq. 8.4) to describe the variation in least principal stress as a function of overburden stress and pore pressure (Fig. 8.17).

Numerous important issues are beyond the scope of this book. I have considered only the poroelastic deformation of rock surrounding a vertically oriented borehole where one principal stress is oriented vertically (Equations 8.1 and 8.2) and under what conditions the stress state reaches Coulomb failure at the borehole wall (Eq. 8.3). The impact of more complex stress states and deviated wellbores is explored by Zoback (2007). Temperature may also impact fracture initiation pressure (Maury & Idelovici, 1995; Pepin et al., 2004). Finally, there are a range of practices used to control losses that are beyond the scope of this book. A particular example is the use of mud additives to increase the fracture initiation pressure (Alberty & McLean, 2001, 2004).

8.4 Summary

To predict pressure, we need to know the overburden stress and to predict trap integrity, or to know the fracture initiation pressure in a wellbore, we need to know the least principal stress. I have described how to characterize these stresses, and I have summarized simple models that describe their interdependency. While these parameters have been studied for decades, advances in technology now allow them to be more accurately measured. Furthermore, experimental analysis of the compaction of rocks over a large range of experimental stresses is providing insight into empirical observations of stress state in the subsurface. Ultimately, a better understanding of the material behavior of rocks as they are buried will illuminate the state and evolution of stresses in basins.

9

Trap Integrity

9.1 Introduction

I describe three mechanisms by which traps fail (Fig. 9.1), and I describe how to predict the maximum column of immiscible fluids (e.g., oil, gas, CO_2) that can be trapped. I first describe capillary sealing (Fig. 9.1a) and then explore two types of mechanical seal: hydraulic fracturing (Fig. 9.1b) and shear failure (Fig. 9.1c).

9.2 Capillary (Membrane) Seal

9.2.1 Overview

Capillary pressure is the difference in pressure between immiscible fluid phases within porous media. In Chapter 2, capillary pressure was defined and its role in reservoir pressure was described. Here, I focus on how hydrocarbons and other non-wetting fluids are trapped within reservoirs by capillary forces.

Hubbert (1953) and Hubbert and Rubey (1959) presented one of the earliest discussions of the entrapment of hydrocarbons beneath a capillary, or membrane, seal. Berg (1975) and Schowalter (1979) present classic reviews of the subject. Smith (1966) applied the concept to how faults seal. Watts (1987) further describes capillary sealing and extends the discussion to three phase systems (e.g., gas, oil, and water).

In a system that is sealed by capillary forces, hydrocarbons are trapped beneath a stratigraphic seal that contains much smaller pore throats than the reservoir beneath it (Fig. 9.2a, b). The hydrocarbons fill the reservoir until the difference between the hydrocarbon pressure and the water pressure (the capillary pressure (u_c)) at the crest of the reservoir equals the migration pressure $\left(u_{cmig}\right)$ within the top seal (Chapter 2). At this point, a connected filament of non-wetting fluid traverses the sample and the non-wetting phase fluid begins to migrate through the rock.

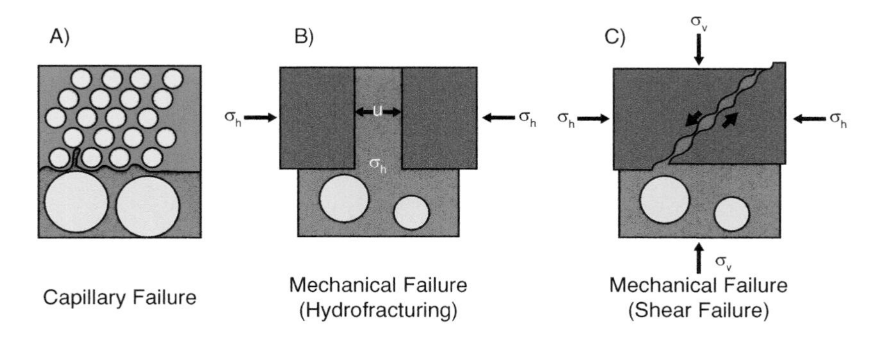

Figure 9.1 Three mechanisms of trap failure. (a) Capillary sealing, (b) hydraulic fracturing, and (c) shear failure.

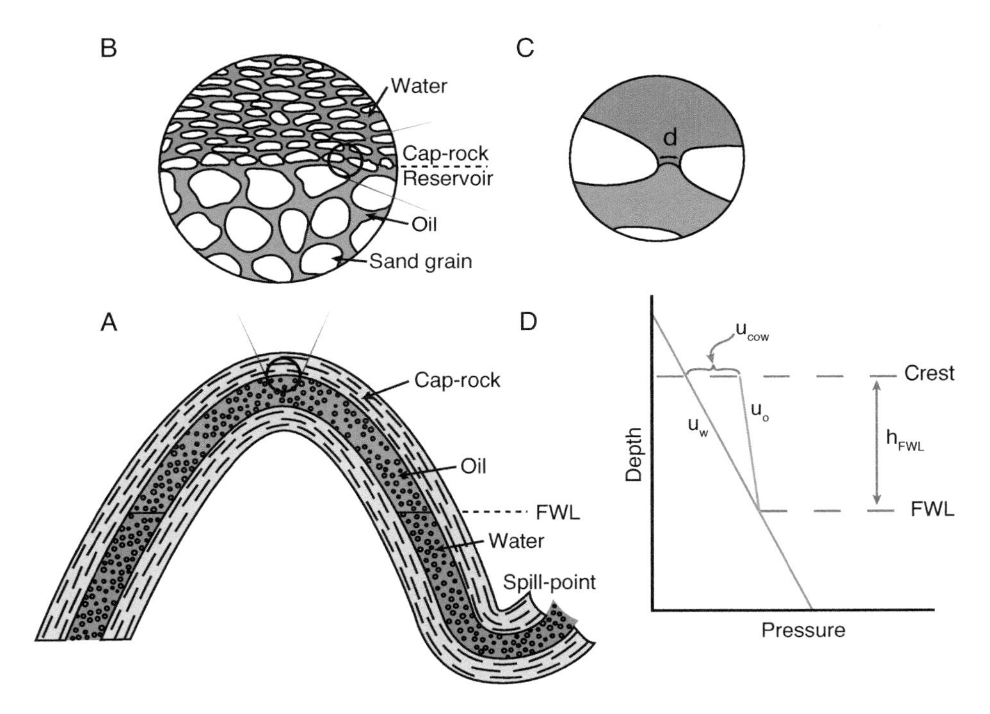

Figure 9.2 Illustration of capillary sealing. (a) A fine-grained seal caps the more coarse-grained reservoir and traps hydrocarbons to the free water level (FWL). (b) and (c) Expanded views of the top seal-reservoir interface. (d) Pressure versus depth profile. Modified from Watts (1987) with permission of Elsevier.

The maximum column height that can be trapped by capillary sealing is:

$$h_{FWL} = \frac{u_{cmig}}{\Delta \rho g}. \qquad \text{Eq. 9.1}$$

The height of the hydrocarbon column (h_{FWL}), measured from the free water level (the depth where the capillary pressure is zero, see discussion of free water level in Chapter 2), is equal to the migration pressure through the seal (u_{cmig}) divided by the difference in the static pressure gradient of each fluid phase $(\Delta \rho g)$. Equation 9.1 illustrates that the height of the trapped column is a function only of the migration pressure and the fluid densities and is independent of the stress state.

This process is illustrated with the aquarium example presented in Chapter 2 (Fig. 2.14). A reservoir (composed of gravel) and its bounding seal (composed of sand) form an anticlinal structure (Fig. 9.3a). The system is originally saturated with water, and then air is injected slowly from below. Air fills the structure to a critical height (h_{FWL}) of 11.9 cm. At that point, the migration pressure is overcome and the gas percolates into the overlying seal. The pressures are shown at the migration pressure (Fig. 9.3b), and the capillary pressure at the crest is 1 127 Pa. The pore throat radius (r) given this capillary pressure (Eq. 2.20) is 0.12 mm. This is on the order of 20% of the grain diameter (0.5 mm) of the seal, which is in the range expected for the ratio of pore throat radius to grain diameter for a well sorted sand. In practice, we observed that this process is cyclical; the reservoir fills to the migration pressure, and then the gas is released through

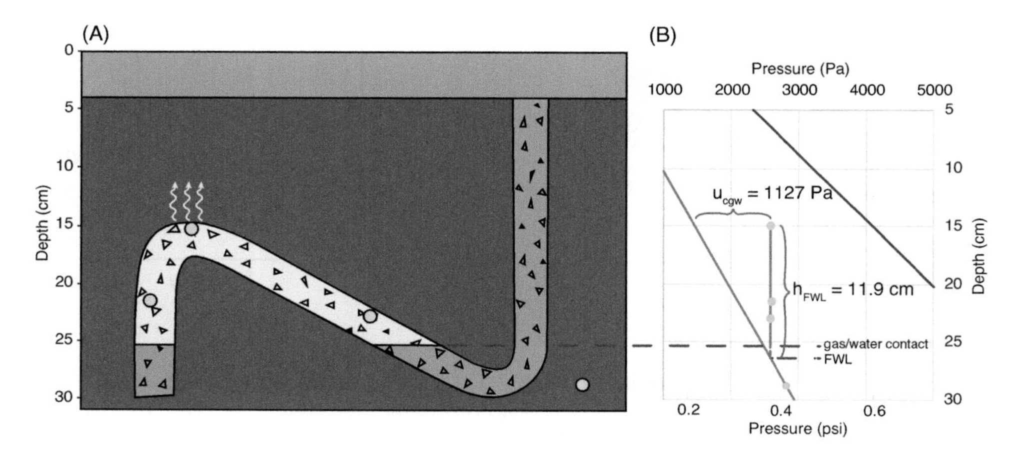

Figure 9.3 Illustration of capillary sealing. (a) A fine-grained seal (composed of 0.5 mm sand) overlies a coarse-grained reservoir (composed of 3–8 mm gravel). Air is trapped within the structure (see also Fig. 2.14). Pressure is measured at each open circle. (b) Pressure versus depth. The capillary pressure at the crest of the structure is 1 127 Pa. At this pressure, the air exceeds the migration pressure and it migrates upward through the top seal.

the seal whereupon the column builds again. Fauria and Rempel (2011) describe these processes with a similar example.

9.2.2 Example

I estimate the potential column height (Eq. 9.1) from mercury injection capillary measurements on resedimented mudrock from the Nankai Trough (Fig. 9.4). I use Thomeer's method (Chapter 2) to estimate the extrapolated displacement pressure (u_{de}) and find it equal to 2 289 psi (15.78 MPa). I assume that this is the migration pressure (u_{cmig}). I apply Equation 2.22 to convert this mercury-air pressure to an oil-brine and a gas-brine migration pressure at in situ conditions. I assume $\gamma_{ow} = 30\frac{dyne}{cm}, \gamma_{gw} = 50\frac{dyne}{cm}, \gamma_{(hg_{air})} = 480\frac{dyne}{cm}, \theta_{ow} = 30°, \theta_{gw} = 0°, \theta_{(air-hg)} = 140°, \rho_o = 718\frac{kg}{m^3}, \rho_w = 1072\frac{kg}{m^3}, \rho_g = 300 kg/m^3$ (e.g., Tables 2.1 and 2.4).

With these parameters, the migration pressure for an oil-water system is 1.11 MPa (162 psi) and for a gas-water system is 2.14 MPa (311 psi) (Eq. 2.22). The gas-water migration pressure is higher than the oil-water migration pressure because the sum of interfacial tension and the cosine of the wetting angle is greater for a gas-water system than for an oil-water system (Eq. 2.22). This is a general result and largely reflects the greater interfacial tension for the gas-water system relative to the oil-water system.

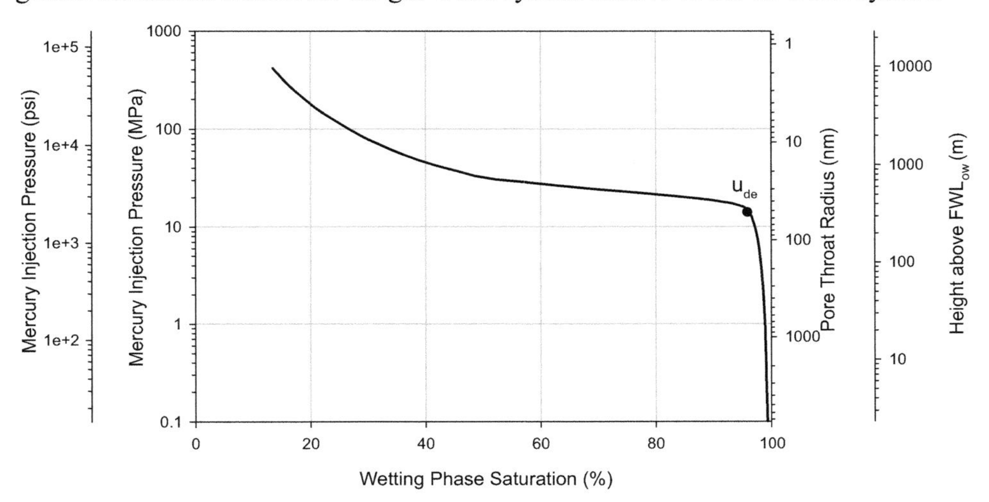

Figure 9.4 Mercury-air capillary injection pressure curve for the Nankai mudstone. The extrapolated displacement pressure (u_{de}) is equal to 2 189 psi and assumed to equal the migration pressure. This is interpreted to be the point at which an interconnected phase of non-wetting fluid can pass through the sample. This sample has 56% clay fraction by mass and was compressed to a vertical effective stress equal to 21 MPa. Fluid parameters are from Table 2.1 and Table 2.4. Modified from Reece (2013). Reprinted with permission by Julia S. Reece.

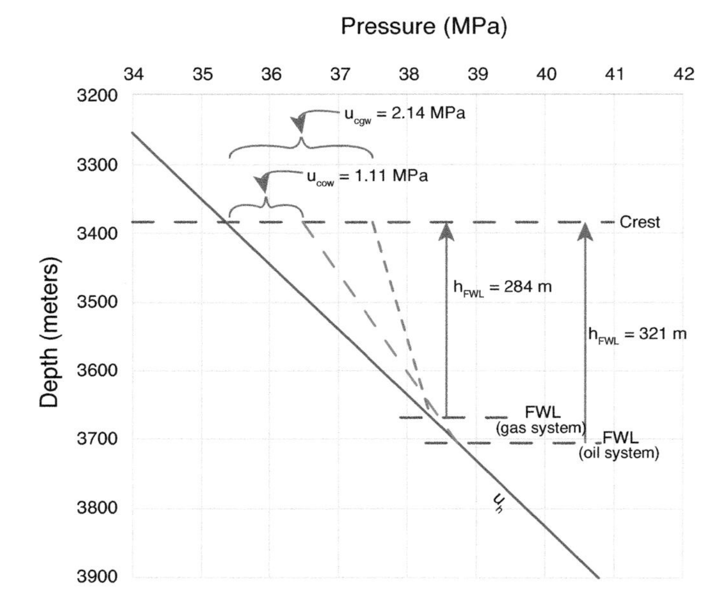

Figure 9.5 Column heights trapped by capillary sealing for oil-water (green) and gas-water (red) systems. This is calculated from the displacement pressure shown in Figure 9.4 and the fluid properties described in the text. The migration pressure in the reservoir is assumed to be zero and, thus, the fluid contact is coincident with the free water level.

I apply Equation 9.1 and find an oil column of 321 m (1 053 ft) can be trapped while only 284 m (931 ft) of gas can be trapped (Fig. 9.5). I equate the depth of the fluid contact to the free water level, because I have assumed the migration pressure in the reservoir is small (Chapter 2). Even though the capillary pressure at migration is lower for the oil-water case (red line, Fig. 9.5), it has a larger column. This is because the oil is denser than the gas.

Blunt (2017) discusses capillary and gravitational equilibrium for a three phase system (e.g., Fig. 2.16). When a trap is continually supplied with oil and gas at a low rate, and there is some permeability present, then the oil pressure at the crest is controlled by the oil-water migration pressure whereas the gas pressure is controlled by the gas-water migration pressure (Fig. 9.6). For this example, the total column is 321 m thick, the same height as predicted for an oil-water system. The three phase region (gas, water, and oil) is 251 m (974 ft) thick.

The fraction of the gas column (h_{FOL}) to the total column (h_{FWL}) at seal failure can be calculated directly by combining Equations 2.22 and 9.1:

$$\frac{h_{FOL}}{h_{FWL}} = \left[\frac{\gamma_{gw} cos\theta_{gw} - \gamma_{ow} cos\theta_{ow}}{\gamma_{ow} cos\theta_{ow}}\right]\left[\frac{\rho_w - \rho_o}{\rho_o - \rho_g}\right], \quad \text{Eq. 9.2}$$

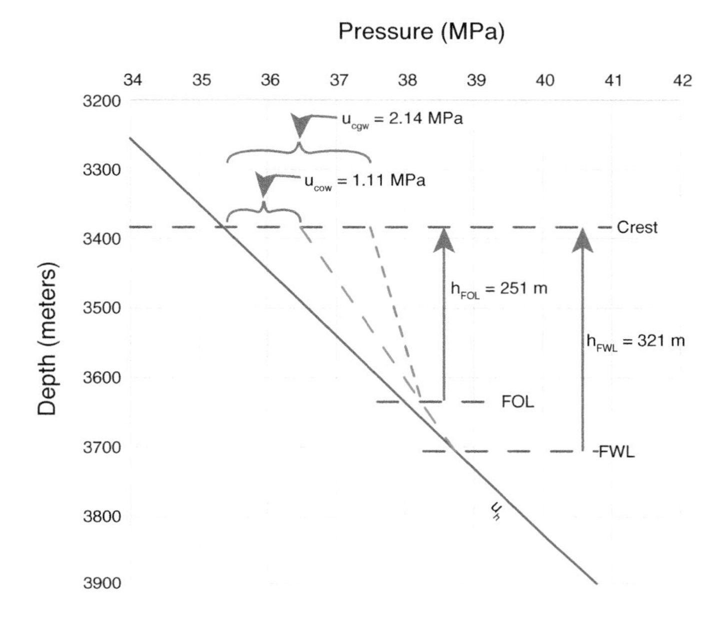

Figure 9.6 Column height for three phase system limited by capillary sealing. The gas pressure is controlled by the gas-water threshold pressure, and the oil pressure is controlled by the oil-water threshold pressure. The total column height is controlled by the oil-water migration pressure (1.11 MPa).

which for this example is 0.78: gas fills 78% of the total column as is observed in Figure 9.6. The total column height does not change relative to that predicted for the oil-water system (Fig. 9.5). This result contrasts that of Watts (1987), who argues that at the gas-oil contact, the oil-water capillary pressure equals the gas-water capillary pressure. Watts (1987) argues that because gas-water migration pressures are greater than oil-water migration pressures, this results in a larger oil column if a thin gas cap is present relative to when no gas is present.

9.2.3 Controls on Displacement Behavior

The study of capillary sealing capacity is a cottage industry. It has long been studied by the petroleum industry for the purposes of fault and top seal analysis. Relevant studies include Dawson and Almon (2006) and Schowalter (1979). There is renewed interest in the topic as we grapple with the challenge of CO_2 sequestration (Espinoza & Santamarina, 2017).

As a mudrock compacts, its pore throats shrink and its migration pressure increases. I illustrate this behavior for intact shallow mudrocks in the Gulf of Mexico (Fig. 9.7). I compare specimens of similar composition and grain size

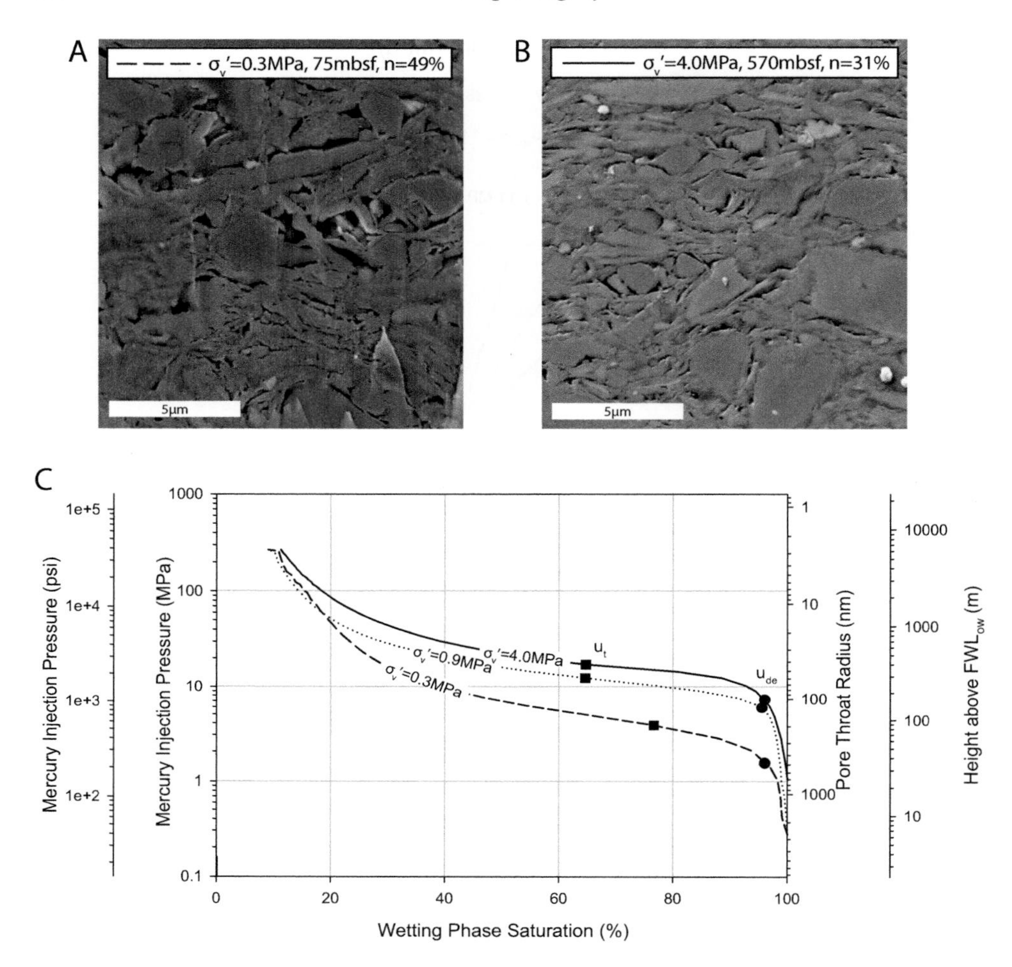

Figure 9.7 (a) Image of mudstone in the Ursa Basin buried 75 meters below seafloor. (b) Image of mudstone in the Ursa Basin buried 570 mbsf. (c) Mercury-air injection pressure versus wetting phase saturation for three samples in the Ursa Basin. These data are described in detail in Day-Stirrat et al. (2012). (a) and (b) are also presented and discussed in Chapter 3 (Fig. 3.3). Adapted from Day-Stirrat et al. (2012) with permission of Elsevier.

distribution from depths ranging from 75 meters below seafloor (mbsf) to 570 mbsf (see also Chapter 3, Fig. 3.3). The images clearly show the decrease in pore size and increase in alignment that occurs with increasing effective stress (Fig. 9.7a, b). This behavior is recorded in the capillary curves (Fig. 9.7c). With increasing stress, the median pore size decreases from 156 nm to 37 nm (delineated by squares in Figure 9.7c).

The fraction of clay versus silt in mudrock also plays a major role in its capillary behavior. I illustrate this with an example for mudrock from the Nankai Trough,

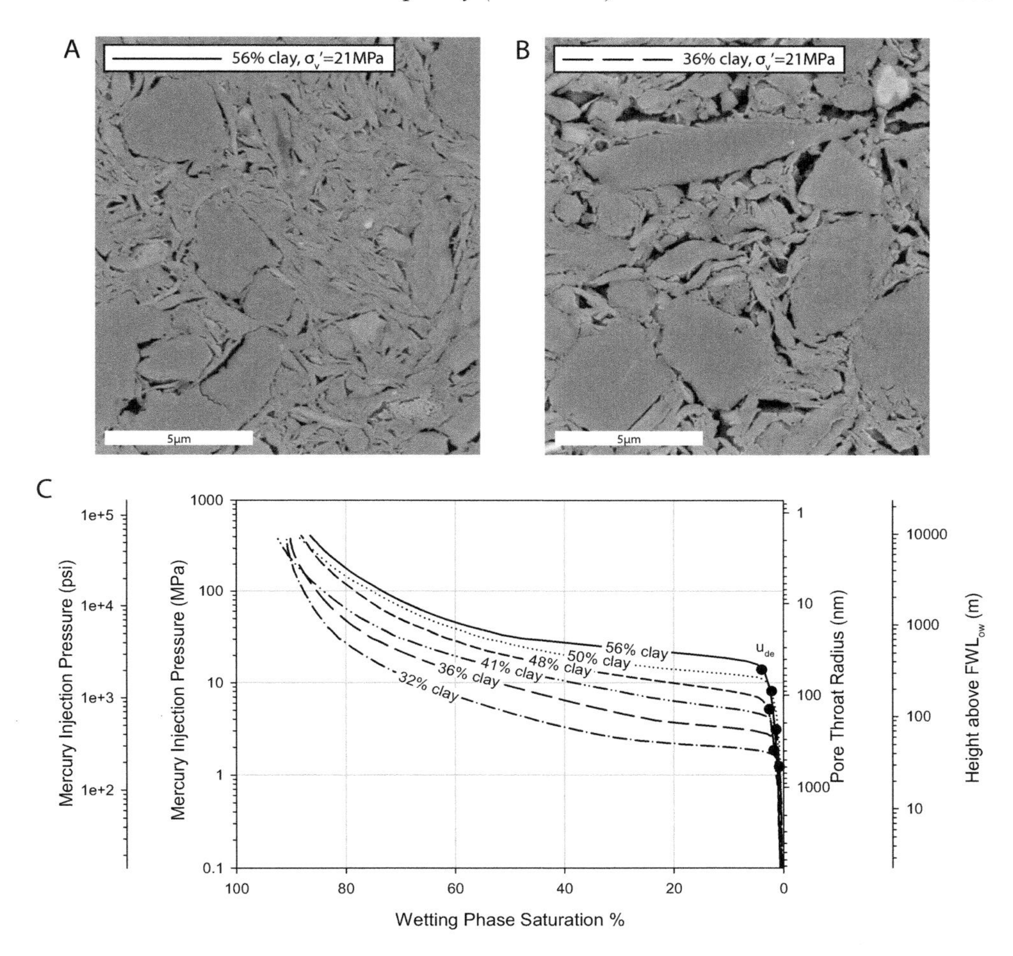

Figure 9.8 Impact of clay fraction on capillary behavior. (a) Resedimented Nankai Trough mudrock. (b) Resedimented Nankai Trough mudrock with admixed silt. (c) Mercury-air capillary measurements for Nankai mudrock with different fractions of silt admixed. Modified from Reece (2013). Reprinted with permission by Julia S. Reece.

offshore Japan (see also Fig. 3.4). As silt is admixed, the displacement pressure systematically decreases (Fig. 9.8c). As the silt fraction increases, silt grains start to touch and form an interconnected framework that shields larger pores from compaction (Fig. 9.8a, b) (Schneider et al., 2011).

9.2.4 Impact of Water Phase Pressure and Seal Heterogeneity on Trapping

Most estimates of trap integrity based on capillary sealing assume that the water pressure follows the hydrostatic gradient upward from the reservoir through the seal and hence that overpressure in the seal is the same as within the reservoir (Fig. 9.9b). However, as discussed in Chapter 10, the overpressure may decrease (Fig. 9.9c) or increase (Fig. 9.9d) above the reservoir. If the overpressure is lower in the overlying seal than in the reservoir (Fig. 9.9c), then the trapped column height may be the same as if the overpressure was the same in the reservoir and the overlying seal. In this case, there is only a thin interface with a high capillary sealing potential (Fig. 9.9c). If the overpressure in the seal is greater than within the reservoir, then a much larger column can be trapped in the reservoir (Fig. 9.9d). The intervening zone beneath the point of maximum pressure will be invaded by hydrocarbons. In summary, if the overpressure in the overlying seal is greater than that in the reservoir, the column height must be calculated from the depth of the maximum pressure (Fig. 9.9d).

Reservoirs that are underpressured relative to their bounding seals (e.g., Fig. 9.9d) are common exploration targets. An example is the Mad Dog field, where a pressure regression is present and an extraordinary (4 700 ft) hydrocarbon column is trapped (Merrell, 2012). The Mad Dog field is part of an interconnected underpressured aquifer that contains the Pony field (GC 468) approximately 29 miles (26.6 km) northwest of Mad Dog, the K2 (GC 562) approximately 18 miles (29 km) north of Mad Dog, Knotty Head (GC 512) approximately 28 miles (45 km) northwest of Mad Dog, Spa Prospect (Walker Ridge 285) approximately 53 miles (85.3 km) southwest of Mad Dog, and Shenzi (GC 609) approximately 14 miles (22.5 km) northeast of Mad Dog (Fredrich et al., 2007; Rohleder et al., 2003; Sanford et al., 2006; Weatherl, 2010; Williams et al., 2008). Another striking pressure regression is present at the Macondo Prospect (Figures 1.3 and 6.12). While one mechanism for successful trapping of exceptional columns is that illustrated in Figure 9.9d, it is equally true that a pressure regression from the overlying seal to the reservoir also results in an increasing effective stress as the reservoir is approached. This will also increase the capillary sealing capacity as the rock compacts in response (Fig. 9.7).

Finally, seal heterogeneity can impact seal capacity. It is common in many stratigraphic sequences to have either a fining-upward succession (e.g., Fig. 9.10a) or lateral-fining (Fig. 9.10b). In this case, there will be a zone of low quality reservoir above the high quality reservoir (commonly called waste rock) that has a low migration pressure relative to the ultimate seal. In these cases, the location of the seal may lie within the sealing rock that has the highest percolation threshold.

It should be clear from the above discussion that we must consider both the quality of the sealing section that is entrapping the hydrocarbons and the pore pressure within that sealing section. This is an area that is only beginning to be addressed.

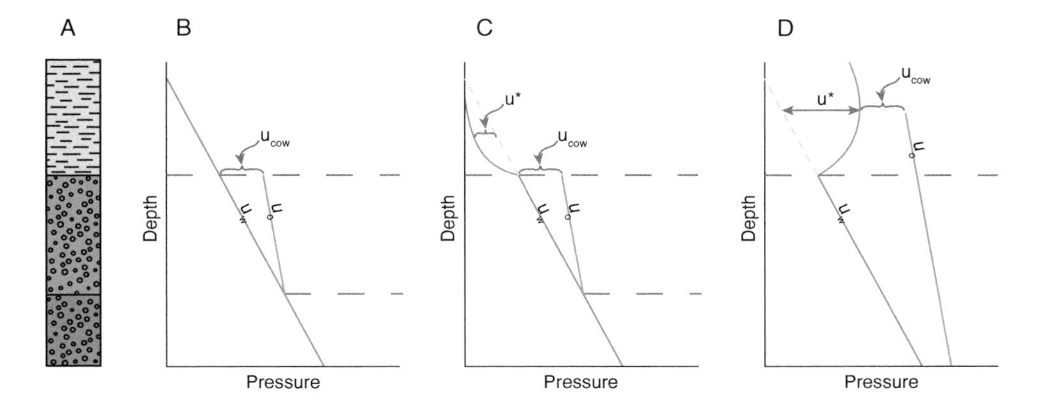

Figure 9.9 Impact of water pressure on capillary sealing. (a) The reservoir (blue) contains a column of hydrocarbons (green) trapped beneath the seal (grey). (b) The water phase pressure (blue line) follows the hydrostatic gradient from the reservoir into the seal, and the overpressure does not change. (c) The overpressure is less within the seal. (d) The overpressure is more in the seal.

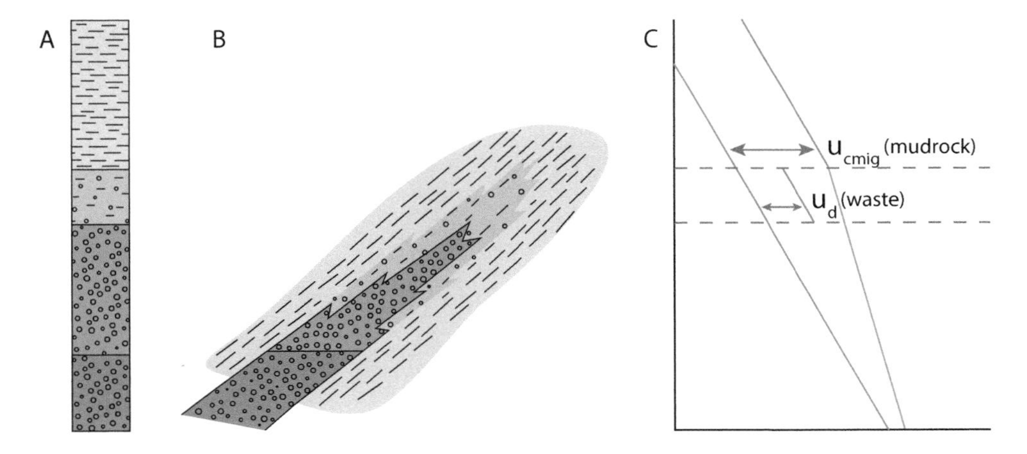

Figure 9.10 (a) Some reservoir systems have an upward-fining succession resulting in a vertical transition to high quality seal rock (grey zone). (b) Other reservoirs can have lateral fining resulting in similar behavior. (c) In these types of reservoirs, there can be a zone of waste rock above the primary reservoir. The waste rock is of an intermediate lithology (e.g., siltstone) that has a fairly low capillary entry pressure. The actual seal is above the waste rock.

9.3 Mechanical Seal

The quality of a seal can also be limited by the stress state. Watts (1987) presented the concept of hydrocarbon entrapment limited by hydraulic fracturing, and Finkbeiner et al. (2001) described quantitatively how hydrocarbons are limited by either hydraulic fracturing or shear failure of the cap rock.

9.3.1 Seal Limited by Hydraulic Fracturing

In this model, buoyant fluids fill a trap until the pressure $\left(u_{crit}^{res}\right)$ equals the least principal stress of the overlying seal $\left(\sigma_3^{seal}\right)$ (Fig. 9.11a):

$$u_{crit}^{res} = \sigma_3^{seal}. \qquad \text{Eq. 9.3}$$

The maximum column height (h_{FWL}) that can be trapped is proportional to the difference between the least principal stress in the overlying seal ($\sigma_3^{seal} = u_{crit}^{res}$) and the water pressure in the reservoir immediately beneath $\left(u_w^{res}\right)$:

$$h_{FWL} = \frac{u_{crit}^{res} - u_w^{res}}{\Delta\rho g}. \qquad \text{Eq. 9.4}$$

This process is illustrated with the aquarium example presented in Chapter 2 (Fig. 2.14). In this case, air fills the structure to a critical height (h_{FWL}) of 7.4 cm

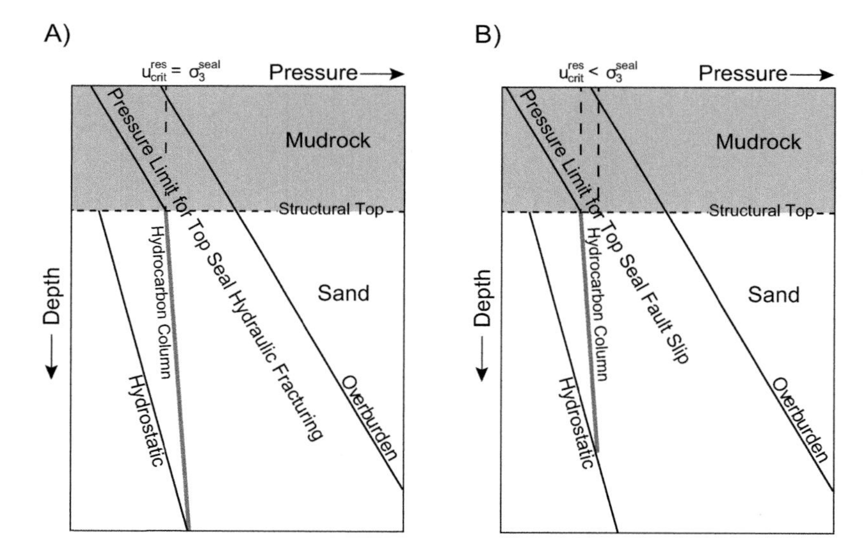

Figure 9.11 Conceptual model of how hydrocarbon column height is limited by stress. (a) When seal is limited by hydraulic fracturing, the maximum column height is achieved when the pressure at the reservoir crest (u_{crit}^{res}) equals the least principal stress in the overlying seal (σ_3^{seal}). (b) When seal is limited by fault slip, the maximum column height is limited by frictional failure and this occurs at a lower pressure than for hydraulic fracturing. Modified from Finkbeiner et al. (2001). Reprinted by permission of the AAPG whose permission is required for further use. AAPG© 2001.

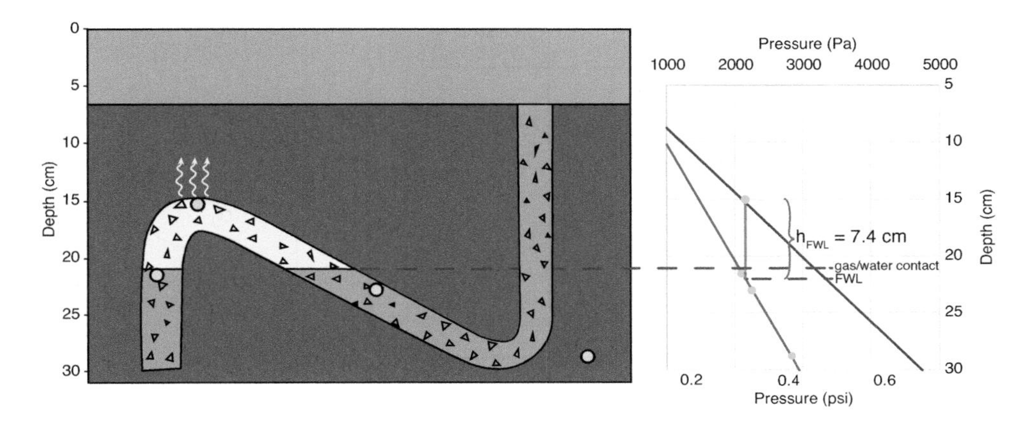

Figure 9.12 Aquarium example of stress-controlled column height. Gas is trapped beneath the seal until the gas pressure equals the weight of the overburden. At that point, it hydraulically fractures the material and gas escapes vertically.

(Fig. 9.12). At that point, the gas pressure equals the overburden stress and the gas hydraulically fractures into the overlying material. It is not obvious why failure occurs at the overburden stress. However, the presence of the aquarium walls may cause the horizontal stress to approximately equal the overburden stress. As in the capillary sealing example, we observed that this process is cyclical; the reservoir fills to the threshold pressure, and then the gas is released through the seal whereupon the column builds again. Fauria and Rempel (2011) describe these processes with a similar example.

9.3.2 Shear Failure

Finkbeiner et al. (2001) described an alternative model for stress-controlled trapping. In this model, reservoir pore pressure increases to the point where it causes slip along favorably oriented faults in the overlying seal. These faults are interpreted to dilate during shear (Figures 9.1c and 9.13), which allows flow through the seal. Barton et al. (1995) showed that fractures optimally oriented for shear failure (i.e., critically stressed) conduct fluids while fractures at other orientations do not. They interpreted that optimally oriented fractures at shear failure have enhanced permeability.

This critical pressure is described in Chapter 8 and can be calculated by rearranging Equation 8.7:

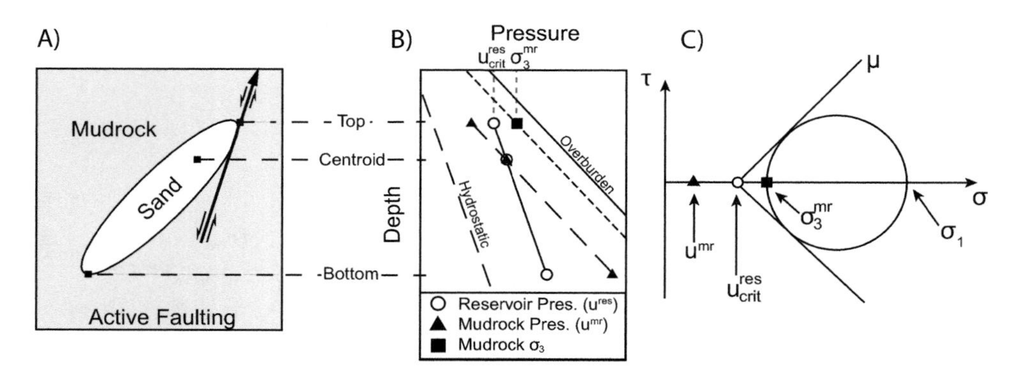

Figure 9.13 Hydrocarbon entrapment limited by fault slip. (a) A dipping overpressured sandstone reservoir abuts a favorably oriented fault. (b) The pore pressure in the reservoir is limited by the pressure at which fault slippage starts to occur (u^{res}_{crit}). (c) Mohr Coulomb diagram. The Mohr circle describes the stress state in the mudrock seal. From Finkbeiner et al. (2001). Reprinted by permission of the AAPG whose permission is required for further use. AAPG© 2001.

$$u^{res}_{crit} = \frac{\sigma^{seal}_h - \left(\frac{1-\sin\phi'}{1+\sin\phi'}\right)\sigma_v}{\left[1 - \left(\frac{1-\sin\phi'}{1+\sin\phi'}\right)\right]}. \quad \text{Eq. 9.5}$$

Equation 9.4 is then applied with this value of u^{res}_{crit} to determine the maximum trapped column height. This approach predicts smaller column heights than the hydraulic fracture model.

9.3.3 Demonstration of Mechanical Seal Capacity

We illustrate these two models of hydrocarbon trapping using the geometry and water pressure of the J3 sand example presented in Chapter 2 (Fig. 9.14). In this location, the least principal stress (σ^{seal}_h) within the seal is assumed to equal 9 100 psi (62.74 MPa) (σ^{seal}_h, black square, Fig. 9.14).

2 602 m of oil or 1 010 m of gas will be trapped before hydraulic fracture (Equations 9.3 and 9.4) (dashed green and dashed red lines, Fig. 9.14a). With a friction angle of 20 degrees, 1 206 m of oil and 499 m of gas can be trapped (solid red and green lines, Fig. 9.14a) before shear failure occurs. Unlike the case of capillary sealing (Fig. 9.5), the capillary pressure at the reservoir crest is the same whether oil or gas is present: the critical pressure is controlled only by the least principal stress and not the fluid properties. As the friction angle is reduced from 30

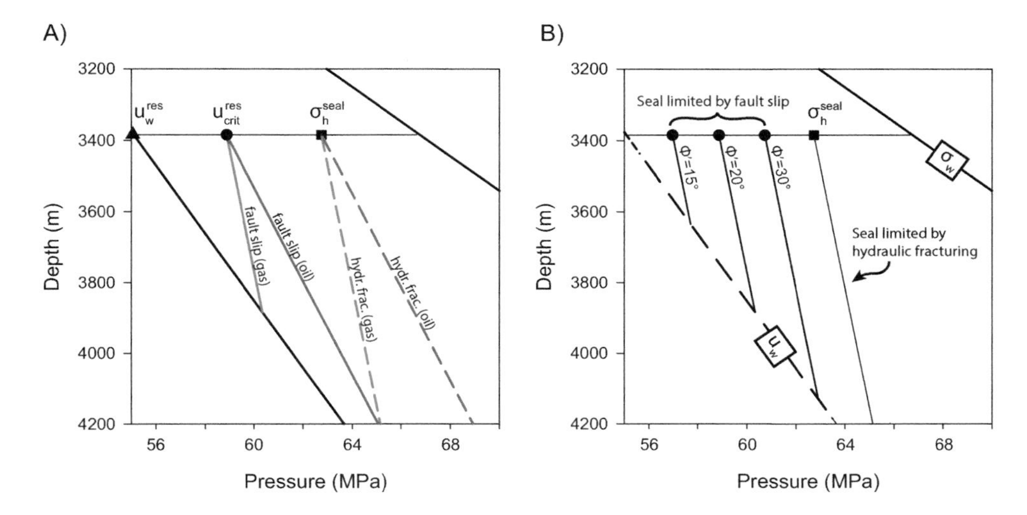

Figure 9.14 Illustration of mechanical trap integrity for the J3 reservoir given the water pressure estimated in Chapter 2 (e.g., Fig. 2.6). (a) The dashed lines show the maximum gas column (red) and the maximum oil column (green) that can be trapped before hydraulic fracturing (Equations 9.3 and 9.4). The solid lines illustrate the maximum column that can be trapped before the overlying seal starts to slip along favorably oriented fractures for a friction angle of 20 degrees (Equations 9.3 and 9.5). (b) Illustration of the effect of friction angle on the amount of gas trapped for shear failure (Equations 9.4 and 9.5). For these examples, a gas density of 0.13 psi/ft, an oil gradient of 0.31 psi/ft (Table 2.2), and a water phase overpressure of 19.5 MPa (Table 2.3) are assumed. The least principal stress (black square, σ_h^{seal}) in the seal above the reservoir is assumed equal to 9 100 psi (62.74 MPa).

to 15 degrees, it takes a lower and lower reservoir pressure (u_{crit}^{res}) to initiate slippage by shear in the seal, and hence a smaller and smaller column is trapped (Fig. 9.14b).

9.3.4 Mechanical Seal Examples

Gaarenstroom et al. (1993) showed that reservoir pressures are limited by the least principal stress in the North Sea. They found that if the difference between the reservoir pressure and the least principal stress is less than ~1 000 psi (6.9 MPa), there is a high probability that no hydrocarbons will be trapped. Perhaps the most obvious example of stress-limited sealing is the case where the water pressure converges on the least principal stress. For example, Reilly and Flemings (2010) demonstrated in the Auger Basin that the reservoir pressure equals the least principal stress at the shallowest reservoir location ("vent," Fig. 9.15a). In this location, drilling documented pore pressures that converged on the least principal stress and

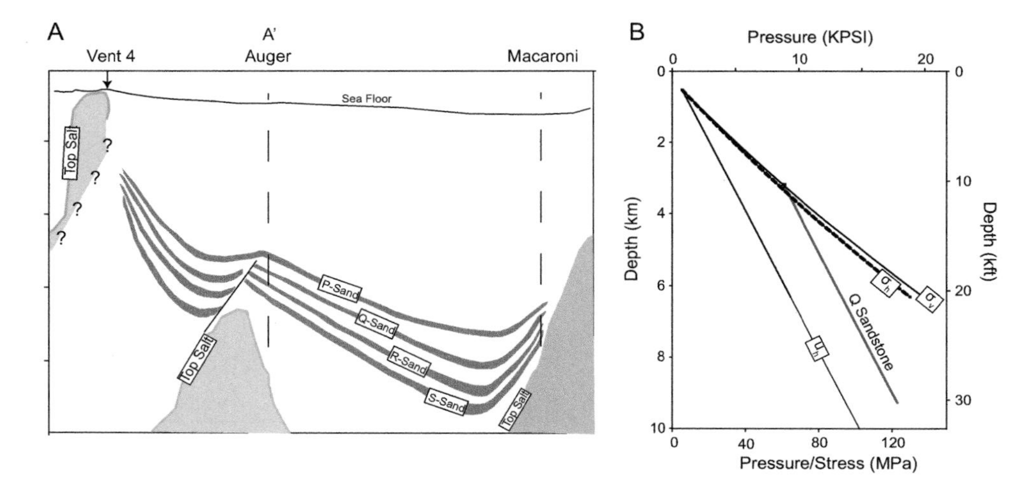

Figure 9.15 Stress limited hydrocarbon entrapment in the Auger Basin. (a) Cross-section through the Auger Basin. On the western (left) side of the basin, the crests of the P, Q, R, and S sandstones thin laterally toward the salt-bounded basin margin and underlie seafloor vents ("Vent 4"). (b) Pressure versus depth of the Q sandstone. At the crest of the Q sandstone, pore pressures equal both the overburden stress and least principal stress and hydraulic fracturing is interpreted to be occurring. Reprinted from Reilly and Flemings (2010) with permission by John Wiley and Sons.

the presence of only trace hydrocarbons. Furthermore, hydrocarbons of the same composition as those vented at the seafloor are found in the deeper reservoir at Auger.

Naruk et al. (2019) describe geysers in the Paradox Basin that are sourced by CO_2 reservoirs. The column height of the CO_2 in these reservoirs is limited by the least principal stress (Eq. 9.3). CO_2 leaks where the fluid pressure equals or exceeds the least principal (horizontal) stress and pre-existing fractures are present. This can be viewed with a map of the least principal effective stress (Fig. 9.16): hydraulic fracturing occurs where the gas pressure equals the least principal stress, and the effective stress is zero. Naruk describes a second fascinating example where overpressures generated by water injection in the Frade field (offshore Brazil) drive the hydrocarbon pressure to the least principal stress, causing hydraulic fracturing and driving hydrocarbons to the seafloor (Fig. 9.17). Tréhu et al. (2004) also show that the gas pressure equals the overburden stress in a zone of leakage on Hydrate Ridge, offshore Oregon.

Finally, Finkbeiner et al. (2001) interprets the hydrocarbon column in the OI-1 reservoir in Eugene Island 330 to be inducing shear failure in the overlying caprock and limiting the hydrocarbon column (Fig. 9.18). In this case, the pore pressure at leakage is less than the least principal stress present but within the range of expected pressure for shear failure.

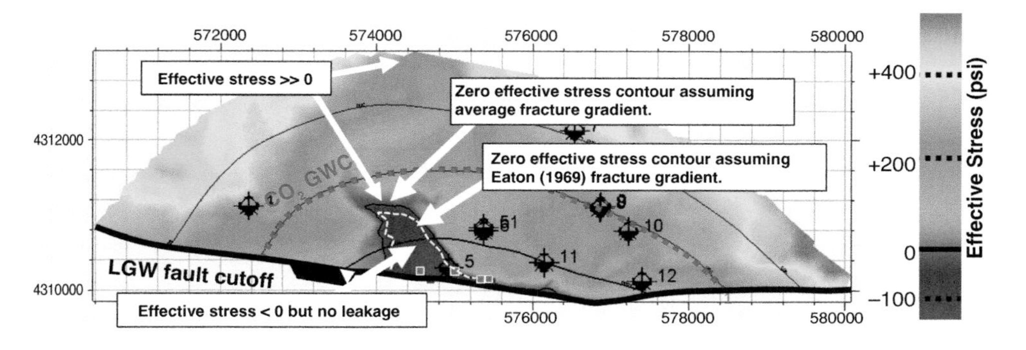

Figure 9.16 Structure contour map of top Navajo Sandstone reservoir at Crystal Geyser, colored by least principal effective stress at the top of the Navajo. The solid black line represents the zero effective stress contour. The seeps and geysers (red squares) are restricted to the fault damage zone where the minimum horizontal is less than zero and there are extensive fractures. North of the damage zone, there is an area where the minimum horizontal is less than zero, but there are no seeps or geysers because of the absence of through-going fault fracture damage zones. GWC = gas-water contact; LGW = Little Grand Wash. From Naruk et al. (2019). Reprinted by permission of the AAPG whose permission is required for further use. AAPG© 2019.

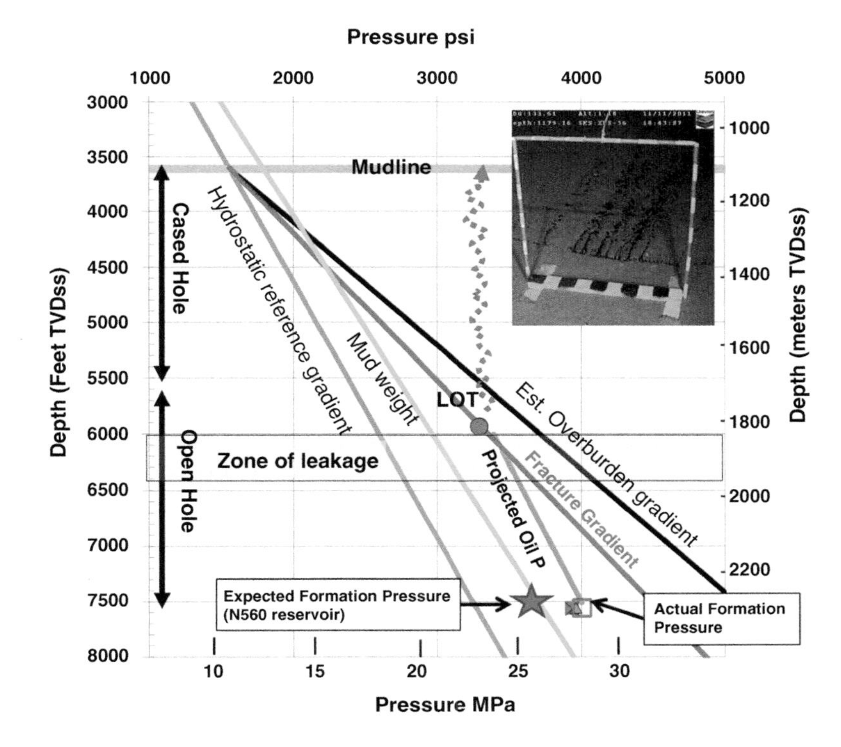

Figure 9.17 Pressure-depth graph for the deepwater Brazil Frade well 9-FR-50DP-RJS, showing that the pressure (P) of the oil that escaped into the wellbore equaled the fracture gradient in the zone of leakage. Inset shows the active seep from the fault at the seafloor. Est. = estimated; LOT = leak-off test; TVDss = true vertical depth subsea. From Naruk et al. (2019). Reprinted by permission of the AAPG whose permission is required for further use. AAPG© 2019.

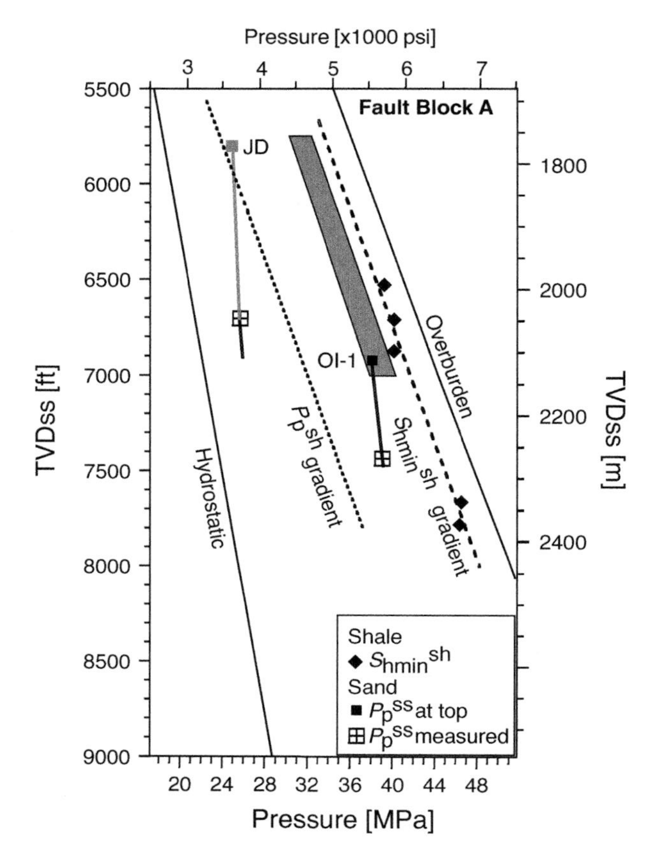

Figure 9.18 Pressure and stress state of the OI-1 sand in the Eugene Island 330 field. The oil pressure at the crest of the OI-1 sand (black square) lies within the range that will induce shear failure in the overlying seal (the grey box defines the critical pressure for shear failure over a range of 17 to 30 degrees). Least principal stresses in shales (σ_h^{seal}) are from leak-off tests (LOTs) and formation integrity tests (FITs). The estimated mudrock pressure is shown with short dashes. From Finkbeiner et al. (2001). Reprinted by permission of the AAPG whose permission is required for further use. AAPG© 2001.

9.4 Summary

Hydrocarbons or CO_2 are trapped by capillary forces or stress. To interpret which of these processes limit the trap behavior, the capacity for capillary sealing and for mechanical sealing must be evaluated. The trap will be limited by the lesser of these two estimates. Mechanical seal is often found to limit seal capability in over-pressured basins.

10
Flow Focusing and Centroid Prediction

10.1 Introduction

I have focused on interpreting pressure in one dimension (e.g., in a well profile). However, one of the most fundamental advances in the field of pore pressure prediction in recent years is the recognition that pressure distribution varies systematically and predictably in three and even four (time) dimensions as a result of the presence of permeable and laterally continuous beds (commonly sandstones) that are encased in low permeability mudrocks. In overpressured basins, these permeable bodies provide pathways that focus flow and perturb the pore pressure distribution. This process has been termed "flow focusing," "lateral transfer," and the "centroid effect."

10.2 Flow Focusing

10.2.1 Physical Example

I illustrate flow focusing with a simple physical example (Fig. 10.1). A dipping permeable body (composed of gravel) is encased in lower permeability material (composed of sand). A constant basal overpressure is generated by pumping fluid into the base. The upper water surface has a constant and zero overpressure. The overpressures at specific points are recorded by the height of the water surface within manometers (tubes) relative to the water surface. For example, the manometer level at the base of the fish tank is approximately 10 cm (dl) above the water surface, which is an overpressure equal to 980 Pa $(u^* = \rho_w g dl)$. Overpressure contours (white lines, Fig. 10.1) are elevated above the crest of the permeable layer (the 0.6 kPa contour, Fig. 10.1) and depressed below the base of the permeable sand (the 0.9 kPa contour, Fig. 10.1). Red dye is injected into the five ports at the base. The dye tracks the water flow and is focused into the base of the aquifer and then along the aquifer itself. The flow direction is orthogonal to the pressure contours.

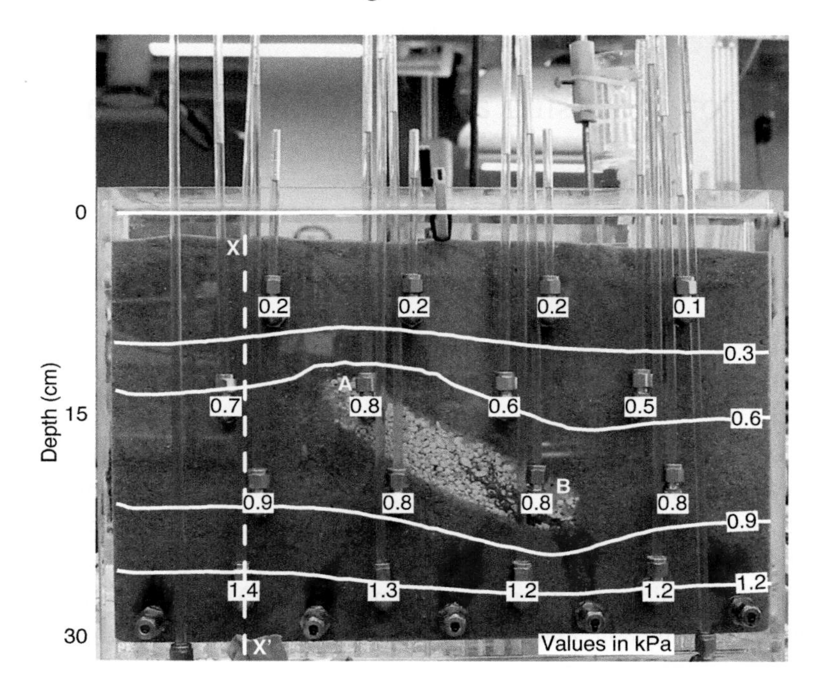

Figure 10.1 Laboratory example of flow focusing. Fluid is supplied to the base at a constant rate. Each vertical tube is a manometer used to describe pressure at the base of each tube. The overpressures are contoured (white lines). Red dye is injected at the base. The dye flows orthogonal to the overpressure contours and moves rapidly along the permeable layer.

The overpressure on a vertical profile to the left, away from the sand (located as X-X', Fig. 10.1), rises approximately linearly with depth (dashed line, Fig. 10.2b). In contrast, the overpressure within the permeable layer is constant (A-B, vertical solid line, Fig. 10.2b). The equivalent mud weight plot (Fig. 10.2c) is striking because the scale is characteristic of those encountered in drilling. This is because the mud weight plot records the pressure gradients encountered. Thus, although the total pressures are small, the gradients are characteristic of those encountered at any scale.

10.2.2 Flow Focusing Model

Flow focusing can also be illustrated with an analytical solution of Laplace's equation $\left(\nabla^2 u^* = 0\right)$ for two-dimensional flow past a thin isolated lens (Fig. 10.3). This solution assumes flow is steady, mudrock and reservoir properties are homogenous and isotropic, and the reservoir permeability is much greater than that of the bounding mudrock (see Flemings et al. (2002) for discussion). Flow is

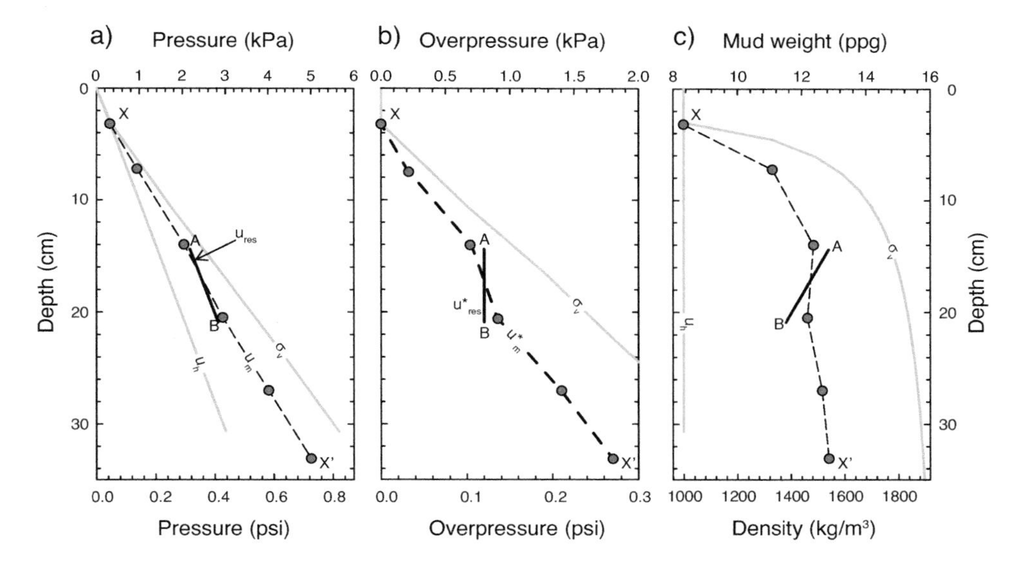

Figure 10.2 Pressure distribution within the fish tank shown in Figure 10.1. Pressure in the low permeability layer along X-X' (Fig. 10.1) is shown with dashed lines. Pressure in the high permeability along A-B (Fig. 10.1) is shown with solid black line. (a) Pressure versus depth. (b) Overpressure versus depth. (c) Equivalent mud weight versus depth.

focused into the reservoir at its base and out of the reservoir at its crest (Fig. 10.3a). At the reservoir base, overpressure contours are depressed and at the crest, they are elevated (Fig. 10.3b). Pressure within the reservoir follows the hydrostatic gradient (Fig. 10.3c and d). The mudrock pressure gradient along a vertical profile that penetrates the midpoint of the reservoir is constant and equal to the far-field pressure gradient (dash-dot line, Fig. 10.3c, d). At its crest, reservoir pressure is greater than the adjacent mudrock pressure (Fig. 10.3c). In a vertical profile through the reservoir crest, the mudrock pressure (dashed line) rises above the far-field pressure (dash-dot line) to equal reservoir pressure (Fig. 10.3c); beneath the reservoir, the mudrock pressure returns to the far-field pressure. Just beneath the crest, the mudrock overpressure decreases with depth, which records downward flow (Fig. 10.3d). At the structural low-point (dotted line), mudrock pressure drops below far-field pressure (dash-dot line) to converge on the reservoir pressure. Beneath the reservoir, the pressure rises toward the far-field pressure (Fig. 10.3c).

The magnitude of flow focusing is described by the flow focusing ratio, G, which is the ratio of the flow velocity in the reservoir relative to the velocity far from the reservoir (Phillips, 1991). G is controlled by the contrast in permeability between the reservoir and the mudrock (k_{res}/k_{mr}) and the aspect ratio of the reservoir $\left(\frac{l}{h}\right)$. For the case of a reservoir oriented parallel to the flow direction (Fig. 10.4a), when

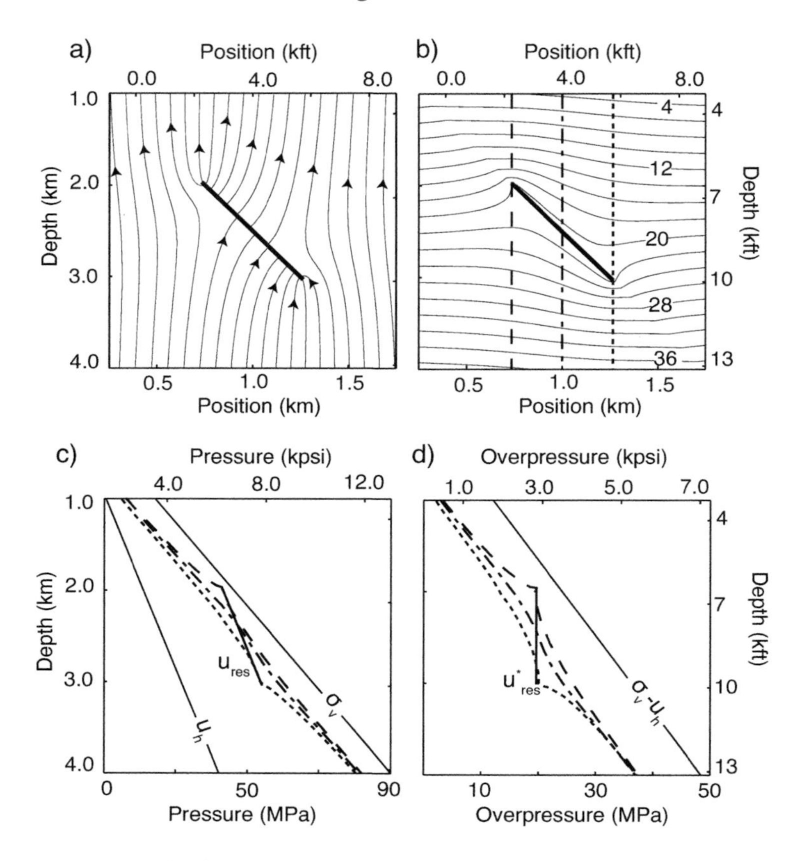

Figure 10.3 Steady-flow solution for a dipping sand within overpressured mudrock. (a) Streamlines illustrate focused flow into the base of the reservoir and out of its crest. (b) Mudrock overpressure contours (MPa) are elevated at the reservoir's crest and depressed near its base. Vertical lines locate pressure profiles shown in (c) and (d). (c) Mudrock pressures converge onto the reservoir pressure at the crest (long dashes), center (dash-dots), and base (dots) of the reservoir body. (d) Overpressure in the reservoir is constant (solid black line), overpressure in mudrock rises to meet the pressure at the crest of the reservoir (dashed line). Modified from Flemings et al. (2002). Reprinted by permission of the American Journal of Science.

$\frac{k_{res}}{k_{mr}} > \frac{l}{h}$, flow is fully developed, the flow focusing ratio equals the aspect ratio of the sand $\left(\frac{l}{h}\right)$. Thus, for a sand 100 times longer (l) than it is thick (h), the flow velocity within the sand is 100 times greater than the flow velocity in the far-field mudrock. An alternative way to view this effect is that flow focusing captures all flow from the bounding mudrock beneath it over a width equal to the sand length (Fig. 10.4a). For flow parallel to the reservoir, when $\frac{k_{res}}{k_{mr}} \leq \frac{l}{h}$, flow focusing is not fully developed, and the flow focusing ratio equals the permeability ratio (k_{res}/k_{mr}).

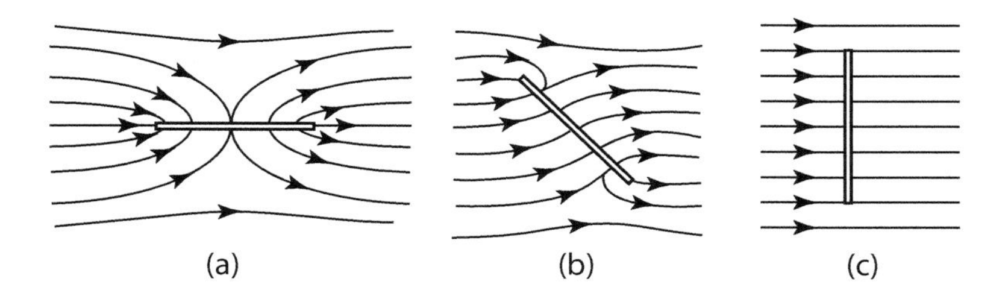

Figure 10.4 Streamline patterns near thin isolated lens that is much more permeable than the surrounding material. (a) Permeable bed is parallel to flow and captures flow over a width equal to its length. (b) Permeable bed dipping at 45 degrees. (c) Permeable bed perpendicular to the flow. From Phillips (1991). Reprinted with permission by Cambridge University Press.

When the flow is parallel to the sandstone, the flow focusing is greatest (Fig. 10.4a). However, the flow focusing decreases as the angle between the flow and the sandstone increases (Fig. 10.4b), and there is no flow focusing when the sandstone is perpendicular to the flow direction (Fig. 10.4c).

It is important in these calculations to recognize that it is not the absolute permeability, but the contrast in permeability between the mudrock and sandstone and the reservoir geometry, that controls the flow. Thus, the effect will be present both in conventional reservoirs and in unconventional (low permeability) reservoirs where there is sufficient permeability contrast between lithofacies.

The analytical solution just described gives tremendous insight. However, it is now common to use numerical models to simulate the flow focusing effect in more complicated geometries that evolve with time (Yardley & Swarbrick, 2000). I present a simple example of this for the burial of a permeable sand within mudrock (Fig. 10.5). This is a two-dimensional basin model that assumes uniaxial strain (Eq. 4.31) and only considers the effect of sediment loading on pore pressure. The modeling approach is described in Section 4.3.2. For 20 million years, mudrock is deposited to an average thickness of 11 km, reflecting an average sedimentation rate of 0.95 mm/year (Fig. 10.5). Thereafter, a permeable sandstone 6 km long and 0.3 km thick is deposited at 0.1 mm/year (Fig. 10.5c). Mudrock of 0.3 km thick is concurrently deposited adjacent to the sand deposition. From 23 million years to 43 million years, mudrock is deposited asymmetrically on top of the sandstone (Fig. 10.5d). As a result, the sand body dips at approximately 7.5° at the end of the simulation (43 million years).

At the end of the simulation, the pressure distribution around the sandstone is similar but not identical to the analytical model (Fig. 10.6 versus Fig. 10.3). Far from the sandstone, the mudrock overpressure increases at a rate almost equal to the overburden stress (compare red to black line, Fig. 10.6b), whereas the sandstone overpressure is approximately constant. However, the sandstone

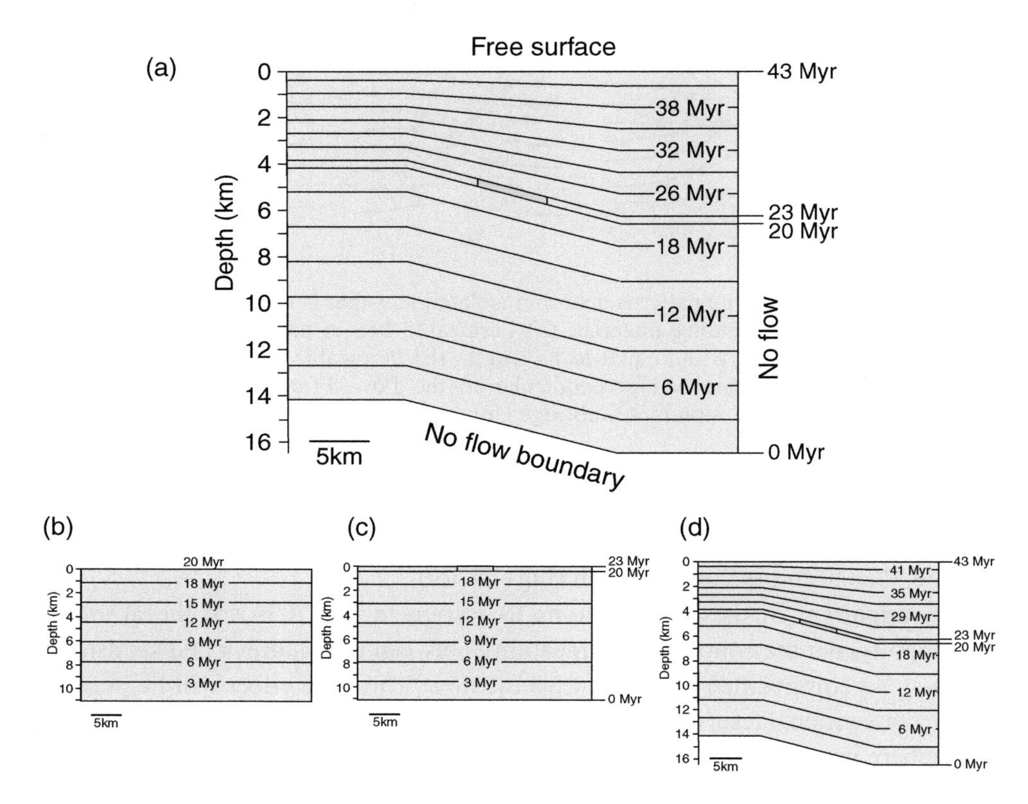

Figure 10.5 Basin model of flow focusing. (a) Boundary conditions, basin geometry, and timelines at end of simulation. (b) Basin at 20 million years. (c) Basin at 23 million years. (d) Basin at final state, 43 million years. Reprinted from Gao and Flemings (2017) with permission of Elsevier.

pressure equals the far-field pressure at roughly one third of the way down from the top of the structure (red circle, Fig. 10.6b) in contrast to the simple physical example (Fig. 10.2) and the analytical model (Fig. 10.3), where this depth is at the midpoint of the structure (green dot, Fig. 10.6b). This is because the numerical model includes the effect of changing mudrock permeability. As a result of flow focusing, the effective stress is higher near the base of the reservoir and lower near the crest of the reservoir. Consequently, the mudrock is more compacted near the base and less compacted near the crest. This difference in compaction results in much lower permeability around the base of the structure and much higher permeability near the crest (Fig. 10.6a). As a result of the lower permeability at the base of the reservoir, the pressure in the reservoir is lower and the point where the reservoir pressure equals the mudrock pressure is above the midpoint (Fig. 10.6b). I discuss the impact of mudrock permeability in more detail below.

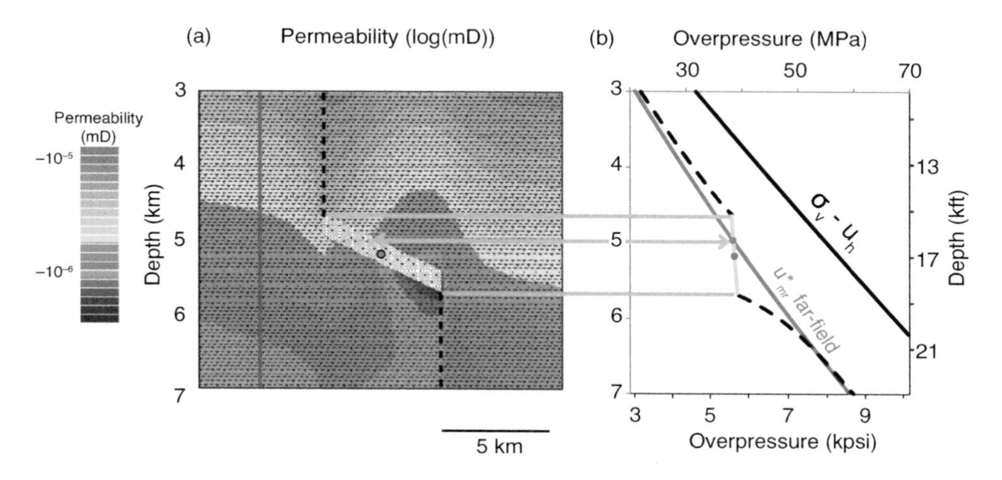

Figure 10.6 Basin model simulation of flow focusing (see Fig. 10.5). (a) Permeability variation at end of simulation. Mudrock permeability is decreased (darker blue) around the base of the reservoir where the effective stress is high and increased (yellow) around the crest of the reservoir where effective stress is reduced. The green dot marks the midpoint of the reservoir structure. (b) Overpressure plot. The equal-pressure or centroid depth (marked as red dot) is about one third depth of the reservoir structure. Rock properties for this simulation are shown in Figure 10.13. Reprinted from Gao and Flemings (2017) with permission of Elsevier.

10.3 Field Examples

Flow focusing is mapped in the field by comparing estimated mudrock pressures to measured reservoir pressures. I compare the predicted mudrock overpressure with the reservoir overpressure in two wells in the Bullwinkle Basin (Fig. 10.7). Near the top of the structure (Fig. 10.7b), u_{mr}^* increases linearly into the sandstone, yet is lower than the sandstone pressure at the sandstone-mudrock interface (black dashed line intersects sand at lower pressure than measured water pressure [square symbol]). At the bottom of the structure, the opposite behavior is present: u_{mr}^* is greater than the reservoir overpressure at the sandstone-mudrock interface (Fig. 10.7c). In the wells that penetrate the J sandstone in the deeper locations, there is a 200 meter thick zone where the porosity and predicted overpressure in the mudrock decrease toward the J sandstone (Fig. 10.7c).

The mudrock overpressure (u_{mr}^*) at the sandstone-mudrock interface is plotted for all penetrations (diamonds, Fig. 10.8). The overpressure gradient in the mudrock is 9.6 MPa/km (dashed line, Fig. 10.8); this equals an absolute pore pressure gradient of 19.4 MPa/km, which is the lithostatic gradient. For this to be the case, the effective stress ($\sigma_v^{'} = \sigma_v - u$) must be constant. This implies that whatever effective stress

a)

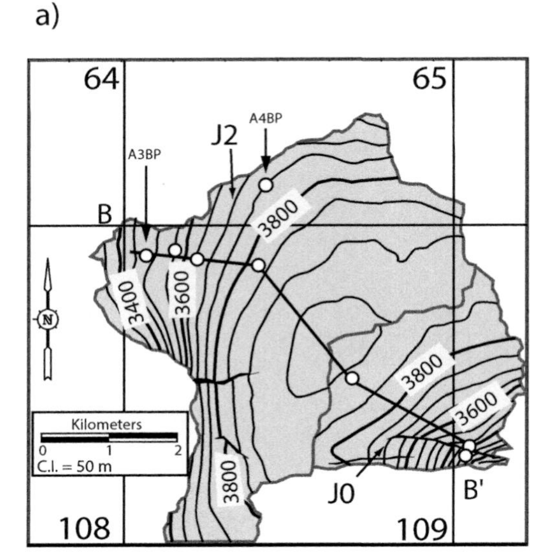

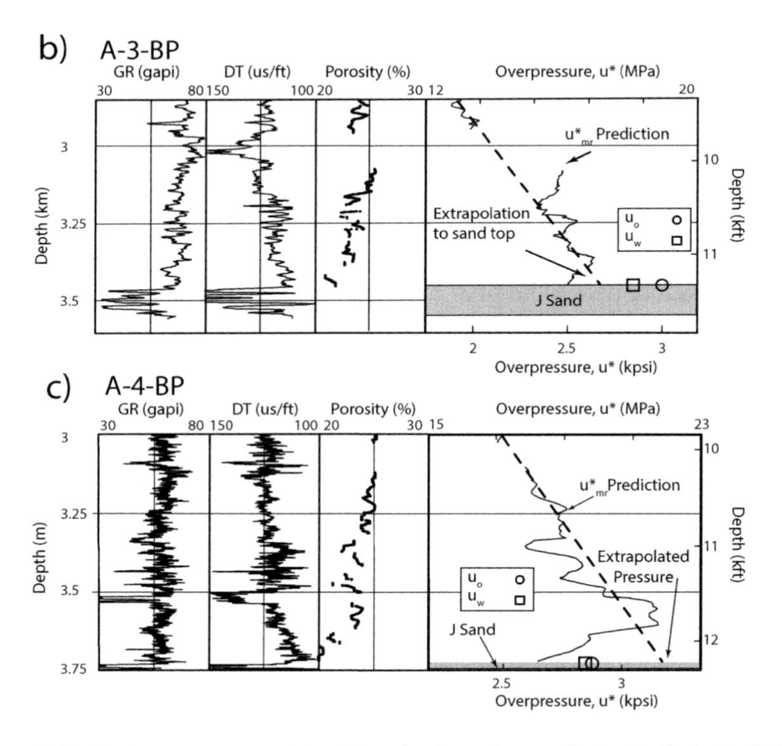

Figure 10.7 (a) Structure map of the J-2 and J-0 sandstones in the Bullwinkle field in meters beneath the sea surface. Mudrock overpressure at (b) the A-3-BP well (near the crest of the reservoir) and (c) the A-4-BP well (in the syncline) (dashed lines in (b) and (c)). To estimate the mudrock overpressure (u^*_{mr}) at the sandstone-mudrock interface, mudrock overpressure is projected to the sandstone top (dashed line in (b) and (c)). Water (square) and oil (circle) overpressures for the sandstone are indicated at the sandstone-mudrock boundary. Modified from Flemings and Lupa (2004) with permission of Elsevier.

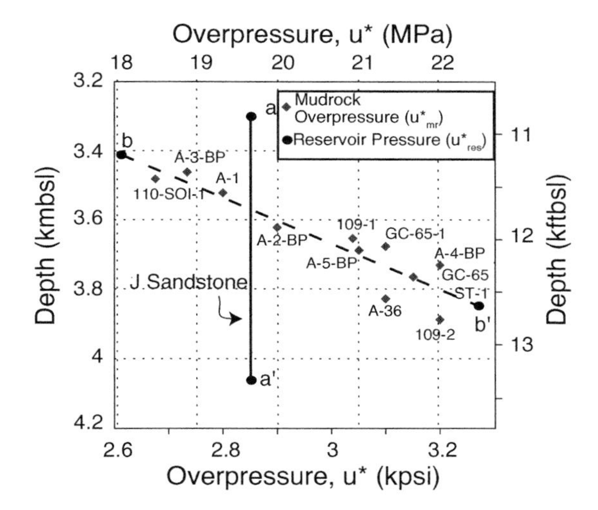

Figure 10.8 Predicted overpressure (u^*_{mr}) in mudrocks immediately above the J-2 and J-0 sandstones (black diamonds) and water phase overpressure (u^*_w) of the J-2 and J-0 sandstones (a-a'). The mudrock overpressure gradient (slope of b-b') is 9.6 MPa/km; this is equal to the lithostatic gradient less the hydrostatic gradient. Reprinted from Flemings and Lupa (2004) with permission of Elsevier.

dependent proxy is used to predict mudrock pressure (e.g., velocity or resistivity), this value does not change as a function of structural depth. While the predicted mudrock pressure follows the lithostatic gradient, the water pressure in the permeable reservoir must follow the hydrostatic gradient (vertical line a-a', Fig. 10.8). As a result, any well that penetrates the J sandstone above 3.575 km has a mudrock pressure that is less than the reservoir pressure, whereas any well that penetrates beneath this point has a mudrock pressure that is greater than the reservoir pressure (Fig. 10.8).

Flemings et al. (2002) introduced the parameter Z, the dimensionless depth along the structure where the sandstone and mudrock pressure are equal:

$$Z = \frac{\bar{z} - z_{crest}}{R}. \qquad \text{Eq. 10.1}$$

$\bar{z}$ is the depth where the mudrock pressure and the reservoir pressure are equal, z_{crest} is the depth of the crest of the structure, and R is the total relief of the structure. In the Bullwinkle example (Fig. 10.8), $Z = \sim 1/3$: the reservoir pressure equals the mudrock pressure one third of the way down the structure. In contrast, in the fish tank example (Fig. 10.1) and in the analytical model presented (Fig. 10.3), $Z = 1/2$: the mudrock pressure and the reservoir pressure are equal at the midpoint of the structure.

In the OI-1 Fault Block B and C reservoirs of the Eugene Island 330 field, the mudrock pressure once again follows the lithostatic gradient and the reservoir water pressure follows the hydrostatic pressure (Fig. 10.9). u_w (short dashed line) exceeds u_{mr} (long dashed line, black circles) above 2 500 m (8 200 ft) and $u_w < u_{mr}$ below that depth (Fig. 10.9a). Z is approximately 1/2. Figure 10.9 emphasizes that this analysis is done based on the water-phase pressures. The buoyant effect of the hydrocarbons is in addition to the water phase pressures. At the crest, the gas phase pressure (u_g) is 5 MPa greater than the water pressure (u_w) because of the buoyant effect of the 518 m gas column.

10.4 Predicting the Centroid Depth

Flow focusing is often conceptually simplified to the image shown in Figure 10.10. The reservoir has a crest and a base, and its water pressure follows the hydrostatic gradient. The mudrock pressure follows a gradient that is higher than hydrostatic and is often lithostatic. A challenge for both exploration and drilling is to know what the reservoir pressure is if the mudrock pressure is known. For example, if drilling into the crest of the reservoir, the reservoir pressure encountered could be much larger than that in the bounding mudrock. If this pressure is underestimated, there could be rapid inflow into the borehole (i.e., a "kick"). In contrast, if drilling on the downdip side, the reservoir pressure will be low and the drilling mud pressure may be so high it fractures the formation and losses occur.

Commonly, but not always, the mudrock pressure is known in advance, for example by prediction from seismic velocity or offset logs; however, the reservoir pressure is not known. To characterize the reservoir pressure, the depth where the mudrock and reservoir pressure are equal must be determined and then the water-phase reservoir pressure can be projected along the hydrostatic gradient to determine its pressure at any depth. This depth, which we parameterize with the variable $\bar{z}$ (Eq. 10.1), has become known as the "centroid" depth. In Figure 10.10b and c, three successively deeper centroid depths result in successively larger reservoir pressures. The fundamental challenge is to predict this depth in advance. As described below, the centroid depth is controlled by the mudrock pressure, the geometry of the reservoir, and the mudrock permeability.

As is clear from the physical and theoretical examples (Figures 10.1, 10.3, and 10.6), the reservoir pressure results from the flow field present. The quantitative approaches presented earlier (e.g., Figures 10.3 and 10.6) capture this two-dimensional flow behavior. Unfortunately, it is seldom possible to pursue detailed numerical simulations for every exploration well. However, it is possible to develop

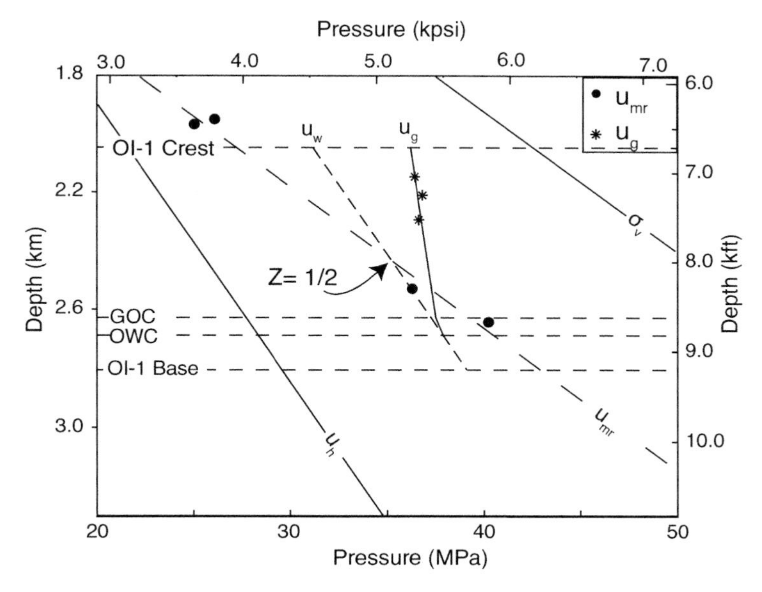

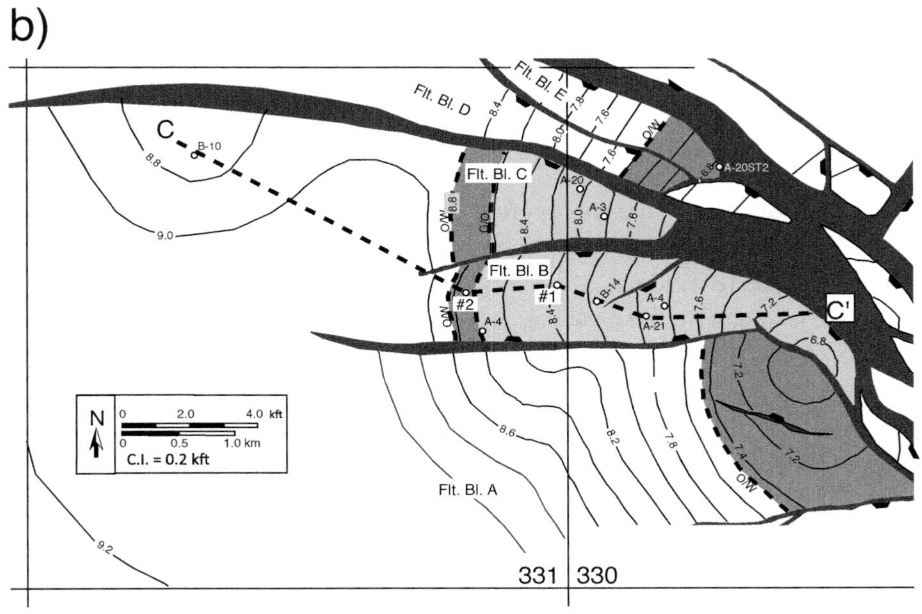

Figure 10.9 (a) Pressure-depth plot for the OI-1 FB B reservoir in the Eugene Island 330 field. Black circles illustrate velocity-predicted mudrock pressures (u_{mr}) immediately above the OI-1 at four penetrations. At the crest, the water phase pressure (u_w) and the gas phase pressure (u_g) are significantly greater than u_{mr}. u_w equals u_{mr} at 2 500 m (8 200 ft) and $Z = \sim 1/2$. u_{mr} line is a linear regression of predicted mudrock pressure values (circles). Reservoir pressures are calculated by extrapolating known pressure gradients from measured values (stars) in the OI-1 FB-B reservoir. (b) Structure map of the OI-1 sandstone. Depth contours are in kilo-feet (TVD_{ss}) (1 km = 3.28 kilo-feet). Reservoir penetrations marked with circles. Oil-water (O/W) and gas-water (G/W) fluid contacts are labeled for each fault block. The depth intervals filled with oil and gas are shown in dark and light grey. Modified from Flemings et al. (2002). Reprinted by permission of the American Journal of Science.

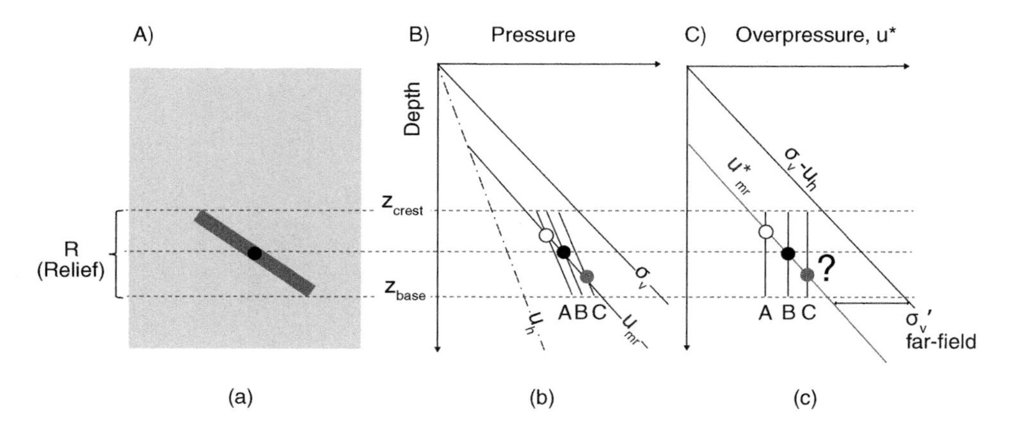

Figure 10.10 Conceptual view of the pressure system around a dipping reservoir within overpressured mudrock. (a) A dipping reservoir is encased in overpressured mudrock. (b) In the reservoir, the pressure gradient (black lines) is equal to the hydrostatic gradient (u_h dash-dot line), whereas the mudrock has a higher pressure gradient (u_{mr}, dotted line). The depth at which the pressure within the reservoir equals that of the far-field mudrock is unknown. The three circles mark possible depths where the reservoir pressure equals the far-field mudrock pressure. (c) Overpressure plot of the same system. Modified from Gao and Flemings (2017) with permission of Elsevier.

simple approximations that incorporate the effect of mudrock permeability and reservoir geometry to predict reservoir pressure.

We assume the reservoir is at a steady state: the flow into the reservoir below the centroid is balanced by the flow out of the reservoir above it. The overpressure in the mudrock $\left(u^*_{mr}\right)$ far from the reservoir is known, and the reservoir water overpressure $\left(u^*_{res}\right)$ is unknown, but assumed constant. Flow between the reservoir and mudrock occurs only across the top and bottom surfaces (and not the ends), is normal to the reservoir-mudrock interface, and the drainage distance from the far-field pressure to the reservoir pressure is constant. Under these conditions, the flow balance can be expressed as:

$$\int \frac{-k_{mr}(A)}{L\mu}\left(u^*_{res} - u^*_{mr}(A)\right)dA = 0. \qquad \text{Eq. 10.2}$$

k_{mr} varies spatially, μ is the water viscosity, L is the drainage length (assumed constant), and dA is the differential reservoir area facing the mudrock. Equation 10.2 can be simplified, and in the following two sections we consider first the effect of reservoir area and then mudrock permeability.

10.4.1 Predicting Centroid Depth from Reservoir Geometry

For the case of constant mudrock permeability, Eq. 10.2 simplifies to:

$$u^*_{res} = \int_{z_{base}}^{z_{crest}} \frac{u^*_{mr}(z)dA(z)}{A}. \qquad \text{Eq. 10.3}$$

If the mudrock overpressure gradient is constant (e.g., it follows the lithostatic gradient, Fig. 10.9a) then the reservoir pressure equals the mudrock pressure at the depth where half the area of the reservoir is above and half is below:

$$\bar{z} = \frac{\sum_{z_{base}}^{z_{crest}} z_i dA_i}{A}. \qquad \text{Eq. 10.4}$$

$\bar{z}$ is the depth where the reservoir pressure and the mudrock pressure are equal, Z_i is the depth at i, and A_i is the reservoir area at depth i. In fact, $\bar{z}$ is the centroid, or the depth of the center of mass of the reservoir body for a reservoir that is of constant thickness. For this reason, the depth where the reservoir pressure equals the mudrock pressure has been termed the "centroid" depth (Traugott & Heppard, 1994). Flemings and Lupa (2004) showed that Eq. 10.4 successfully predicts the reservoir pressures that result from more complicated two- and three-dimensional geometries.

The effect of reservoir geometry is illustrated in Figure 10.11. A synclinal structure generates higher reservoir pressure, a deeper centroid ($\bar{z}$), and a larger relative depth (Z) because more of the reservoir is at deeper depths where it is exposed to higher mudrock pressures (Fig. 10.11a). In contrast, an anticlinal structure generates a lower reservoir pressure, a shallower centroid ($\bar{z}$), and a smaller relative depth (Z) because more of the reservoir is at shallow depths where it is exposed to lower mudrock pressures (Fig. 10.11a). A reservoir with a three-dimensional geometry amplifies this effect. In Figure 10.11b, for example, a reservoir that is cone shaped, expanding downward, has a higher pressure than one that is cone shaped but expanding upward. Iliffe et al. (1999) describe this as the "Whoopee Cushion" effect for turbidite reservoirs that expand downdip and thus have the majority of the reservoir exposed to deeper, higher pressures.

Equations 10.3 and 10.4 can easily be applied to a reservoir structure map. For example, a structure map of the sand (Fig. 10.12a) is gridded into equal area bins and the average depth is determined to be 12 820 ft (Fig. 10.12c). This is the centroid depth. The reservoir pressure is assumed to equal the mudrock pressure (black dot, Fig. 10.12b) at the centroid depth, and the pressure is extrapolated over the reservoir (blue dotted line, Fig. 10.12b). This compares closely with the actual

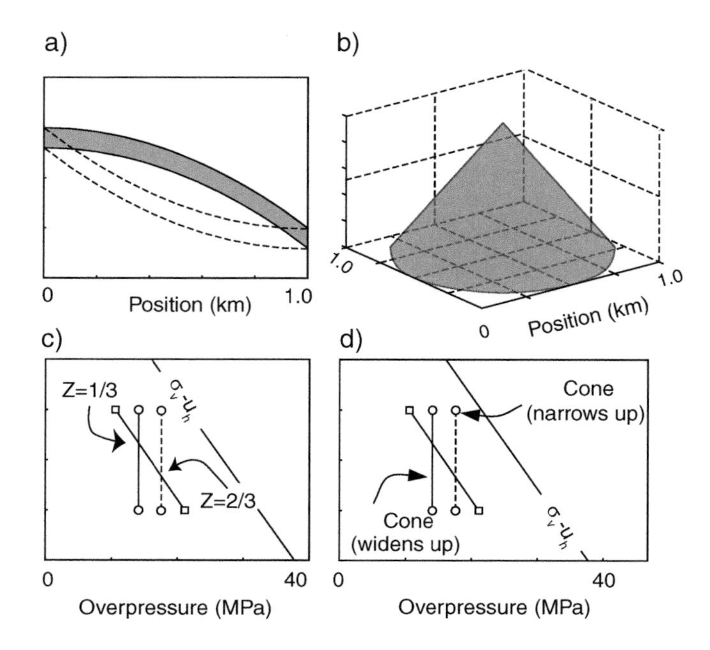

Figure 10.11 Overpressure for different reservoir geometries. (a) Synclinal geometry (dashed line) increases sandstone overpressures relative to anticlinal geometry (solid line). (b) A three-dimensional, cone-shaped, anticlinal structure (dashed line) elevates sandstone overpressure, whereas a three-dimensional synclinal geometry reduces sandstone overpressure (solid line). Modified from Flemings et al. (2002). Reprinted by permission of the American Journal of Science.

water pressure of the reservoir (blue solid line, Fig. 10.12b). These types of calculations are now routinely made in many seismic interpretation packages. They can also be easily extended to incorporate Eq. 10.3 where the value of the pressure at every depth is included (i.e., a linear increase in mudrock pressure is not assumed). The approach presented is slightly simplified because the mapped bin area should be divided by the cosine of the dip angle to calculate the area facing the reservoir. Often the biggest challenge is to have a clear map of the reservoir and its boundaries. For example, sometimes the reservoir and its aquifer may extend off the mapped area or may not be mappable throughout the region.

10.4.2 Dependence of Centroid Depth on Mudrock Permeability

Mudrock permeability also impacts the centroid depth (Fig. 10.6) (Gao & Flemings, 2017). The vertical effective stress (the difference between the overburden stress and the reservoir pressure, σ_v') is lower at the crest than at the base of these systems (e.g., Fig. 10.10c). Because rocks compact with increasing effective stress,

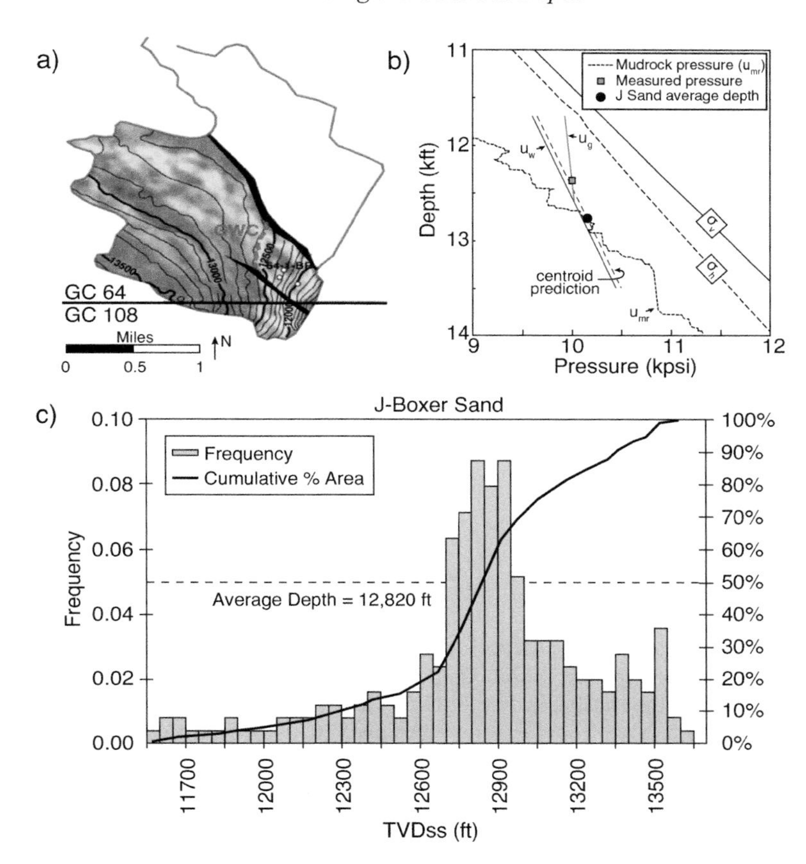

Figure 10.12 a) Structure map with seismic amplitude overlain of the Boxer sand, Bullwinkle Basin. The gas-water contact (GWC) is interpreted with the red dashed line. (b) Pore pressure versus depth in the J-Boxer sand interval. The mudrock pressure predicted from the sonic log (u_{mr}) parallels the overburden stress (σ_v). Pore pressure was measured directly at the green square, and the gas-water contact (GWC) is known from seismic image (see part a); these data were used to estimate the water phase pressure in the reservoir (u_w, solid blue line) (e.g., Fig. 2.6). The black dot marks the average depth of the reservoir (see part c). The centroid-predicted water pressure assumes that the reservoir water phase pressure equals the mudrock pore pressure at the average depth, which is interpreted to be approximately 12 820 ft. The centroid-predicted pressure is shown with the blue dashed line that extends along the water gradient from the black dot. The centroid-predicted reservoir pressure (blue dashed line) is very close to the measured reservoir water pressure (solid blue line). (c) Frequency histogram illustrates that 50% of the sand area is above ~12 820 feet and 50% is below. This is the depth of the centroid. See Lupa et al. (2002) for a discussion of the Boxer sand. Republished with permission of the Society of Exploration Geophysicists (SEG), with modifications from Lupa et al. (2002); permission conveyed through Copyright Clearance Center, Inc.

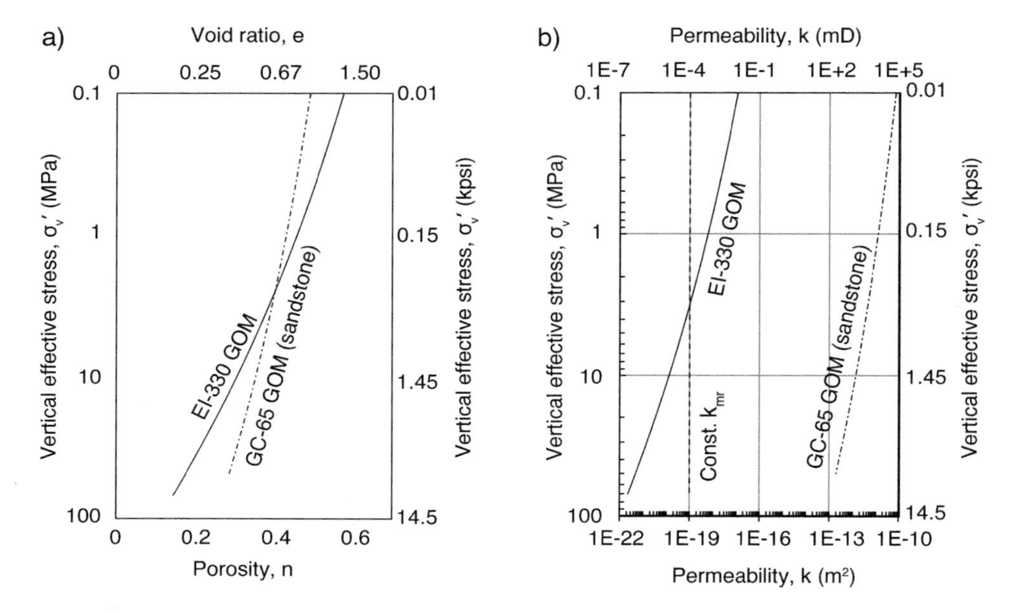

Figure 10.13 Typical (a) compaction behavior and (b) permeability behavior for Gulf of Mexico mudrock (solid line, RGoM-EI, [Casey et al., 2013]) and for sandstone reservoir (dash-dot line, GC-65) at Bullwinkle (Best, 2002; Flemings et al., 2001). Reprinted from Gao and Flemings (2017) with permission of Elsevier.

the mudrock is less compacted near the crest, where effective stress is low, and more compacted near the base of the reservoir, where effective stress is high. Permeability scales with porosity (Fig. 10.13b); thus, the mudrock permeability at the crest will be higher than at the base (e.g., Fig. 10.6).

If we consider a two-dimensional reservoir that dips linearly, then Equation 10.2 can be discretized and simplified to:

$$\sum\nolimits_{z_{base}}^{z_{crest}} k_{mr}(z)\left(u_{res}^{*} - u_{mr}^{*}(z)\right)dz = 0. \qquad \text{Eq. 10.5}$$

u_{res}^{*} depends on how mudrock permeability varies with depth ($k_{mr}(z)$). If the mudrock permeability is constant with depth, then the flow field is symmetric above and below the reservoir (Fig. 10.14a). The magnitude of the flow into the reservoir at the base equals the magnitude of the flow out of the reservoir at the crest, and the centroid lies at the midpoint depth. In contrast, if the mudrock permeability declines with increasing effective stress (Fig. 10.14b), then permeability is greatest at the crest, where the effective stress (defined by the difference in the overburden stress and the reservoir pressure) is least. In contrast, permeability is least at the base, where the effective stress is highest. To have an equal flux into and

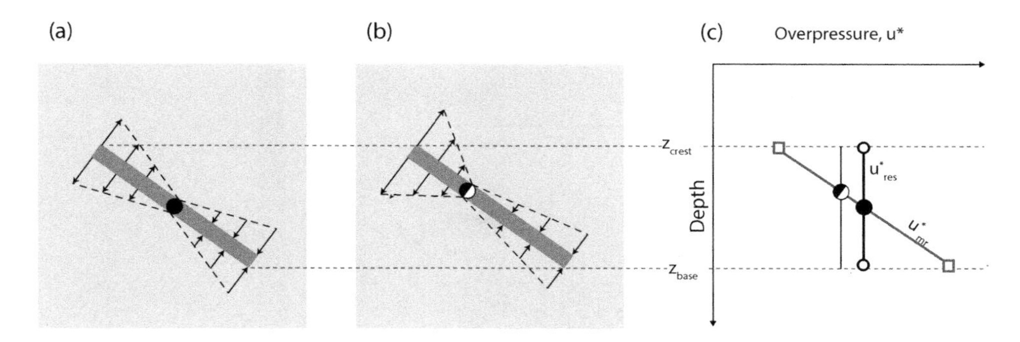

Figure 10.14 The effect of mudrock permeability on centroid depth. (a) Dipping reservoir encased in constant permeability mudrock. The equal-pressure depth (centroid, black dot) marks the depth where the reservoir overpressure equals the mudrock overpressure. Below this depth, fluid flows into the reservoir; above it, fluid flows out of the reservoir. With constant mudrock permeability, the flow into and out of the reservoir is symmetric and the centroid is at the midpoint. (b) Dipping reservoir encased in mudrock with permeability that decreases with depth. The flow velocity into the base is less than (a) because the permeability is lower, and the flow velocity out of the crest is higher than (a) because the permeability is higher. To achieve mass balance, the total flux in (the area under the dashed line) must equal the total flux out (the area under the dash-dot line). (c) Reservoir pressure and mudrock overpressure for (a) and (b).

out of the reservoir, it is necessary for the centroid depth to shift upwards (Fig. 10.14b), with the result that the reservoir pressure is lower (Fig. 10.14c).

In practice, Equation 10.5 is solved with an iterative approach to arrive at the exact centroid depth (Gao & Flemings, 2017). Figure 10.15 is a nomogram that illustrates that the centroid depth moves from the midpoint depth to shallower relative depths as the structural relief increases or the average effective stress decreases for a simple, linear, two-dimensional structure with mudrock permeability described as in Figure 10.14. An increase in structural relief results in a greater difference in permeability between the top and the bottom of the structure, which drives the centroid depth upwards. A greater effective stress means that the change in permeability from the top to the bottom of the structure is reduced. This is because of the nature of the compaction curve. At low effective stresses there is a large change in porosity (and hence, permeability) with effective stress (Fig. 10.13). However, at higher effective stress, this effect is not as dramatic.

10.4.3 Centroid Prediction as Function of Both Permeability and Reservoir Geometry

I close with an idealized trapezoidal geometry to illustrate the impact of the reservoir geometry and permeability on reservoir overpressure (Fig. 10.16). If we

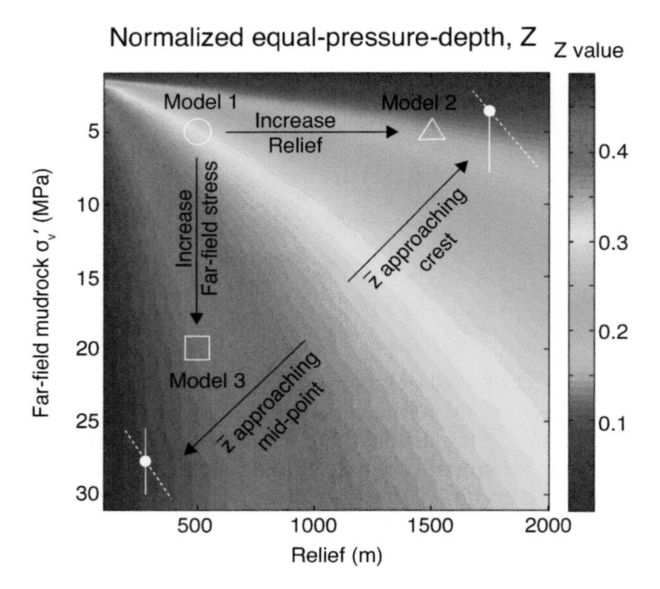

Figure 10.15 Effect of structural relief and effective stress on reservoir pressure for linear two-dimensional structures. Normalized equal-pressure depth (centroid), Z, as a function of the reservoir structural relief and the far-field vertical effective stress. Warmer colors indicate a higher Z value and a higher reservoir pressure. The circle marker shows that for the case with 500 m relief and 5 MPa far-field vertical effective stress, the equal-pressure depth is at about one third of the reservoir structure ($Z =$ 0.33). The triangle shows the high relief case with 1 500 m relief and 5 MPa far-field vertical effective stress ($Z = 0.17$). The square marker shows the high far-field effective stress case with 500 m relief and 20 MPa far-field vertical effective stress ($Z = 0.43$). Reprinted from Gao and Flemings (2017) with permission of Elsevier.

considered only a two-dimensional structure with constant permeability, the centroid would be at the midpoint ($Z = 0.5$). However, if the effect of mudrock permeability is included for this simple structure, then the centroid moves upward ($Z = 0.33$) (purple line, Fig. 10.16b). In contrast, if we neglect permeability and consider only the effect of geometry, the centroid shifts downward ($Z = 0.61$) (green line, Fig. 10.16b). Finally, when the effect of area and decrease in permeability with depth are combined, the centroid moves upward (compare green line to red line, Fig. 10.16b).

10.5 The Impact of Reservoir Hydrocarbons on Centroid Position

A long-standing question is how the presence of hydrocarbons in the reservoir impacts the centroid depth. A hydrocarbon column will have a very low relative

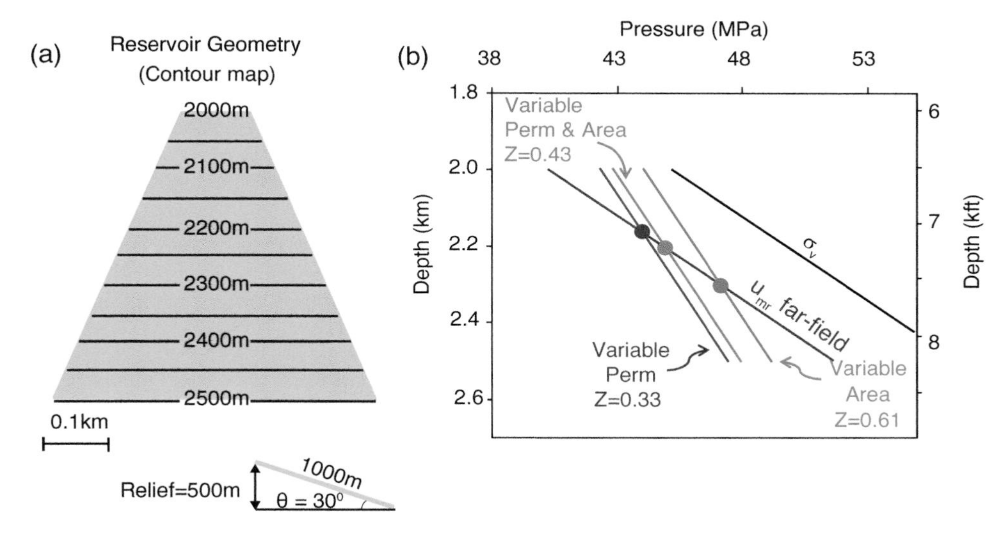

Figure 10.16 Effect of reservoir geometry on the depth where the mudrock and reservoir pressure are equal. (a) Reservoir geometry labeled with depth contour. (b) Predicted pressure using three approaches: 1) area-weighted method with constant permeability (green dot), 2) constant area with variable permeability method (purple dot), 3) combining both area-weighted method and effective stress dependent permeability method (red dot). A greater reservoir area at depth results in a higher pore pressure, because the reservoir is exposed to more of the overpressured mudrocks at depth (e.g., Iliffe et al., 1999). Reprinted from Gao and Flemings (2017) with permission of Elsevier.

permeability to water (Fig. 10.17). Thus, where the hydrocarbon column is present, there will be no (or little) water flow in the reservoir. As a result, it may be appropriate for the centroid depth to be calculated by the portion of the reservoir where only brine and no hydrocarbons are present. In the simple, two-dimensional example in Figure 10.17, the centroid depth lies at the midpoint of the water leg in the reservoir (Fig. 10.17b). The net result is to elevate the reservoir pressure, because it is exposed only to lower and higher mudrock pressures (Fig. 10.17).

10.6 Reservoir Compaction and the Centroid Model

I have focused on understanding reservoir pressure based on the pressure of the bounding mudrock. However, reservoir properties will also change systematically because of flow focusing. Effective stress is lowest at the crest and increases with depth (Fig. 10.18b). As a result, there will be more reservoir compaction and lower reservoir permeability at the deeper points of the reservoir. This systematic change in porosity will impact how these reservoirs are imaged. In addition, the deeper

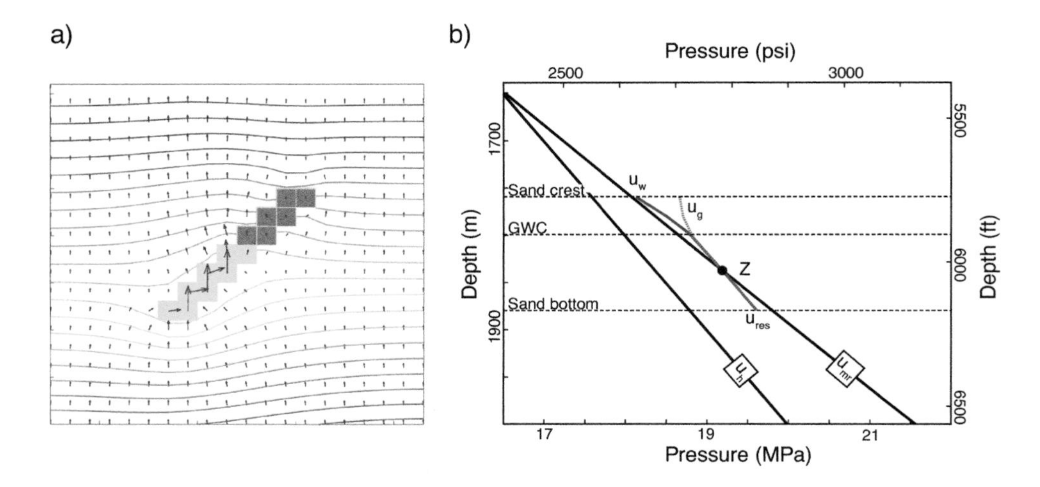

Figure 10.17 Multiphase flow simulation of centroid development. (a) Gas and water are introduced from below. Gas is trapped in the sand body (red) with water beneath it (yellow), and there is little water flow in the gas interval (blue arrows). (b) The far-field mudrock pressure is shown with the solid black line. The water phase pore pressure is shown with the blue line. The reservoir gas pressure is shown with the red dotted line. The centroid (where the gas pressure and water pressure are equal) is at the midpoint of the water leg. From Liu (2003). Reprinted with permission by Xiaoli Liu.

parts of these structures are commonly hotter. Thus, higher effective stresses and higher temperatures may drive more reservoir diagenesis in the deeper section of the reservoir. A better predictive understanding of this behavior may allow more accurate predictions of reserves, strengthen reservoir models, and assist exploitation decisions.

Flemings et al. (2001) explored this behavior in the Bullwinkle J3 sand. Porosity declined from 0.36 to 0.30 in the J3 sand (GC 65), which resulted in a 58% decline in permeability (3.3 to 1.4 darcies). In turn, this resulted in a 75% decrease in acoustic reflectivity over a depth range of only 500 ft. This provided a mechanistic explanation for why the sands deep in the Bullwinkle Basin could not be imaged seismically.

10.7 Summary

The process of flow focusing within a thin, isolated, and permeable lens was presented by Lamb (1932), and the resultant pressure distribution was recognized in the field many decades ago (e.g., Hubbert & Rubey, 1959). Phillips (1991) presented an elegant analysis of these solutions with application to geological

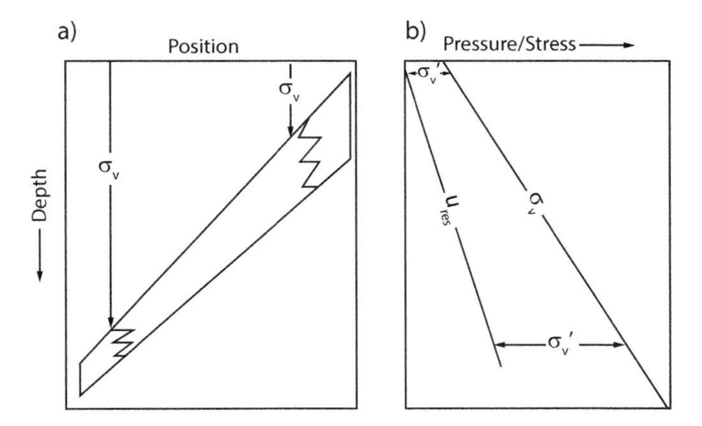

Figure 10.18 Diagram illustrating increased reservoir compaction at greater depth due to higher effective stress in reservoir systems. (a) Sketch illustrating greater compaction in down dip location of the reservoir relative to up dip location due to higher stress. (b) Illustration that in any reservoir the effective stress increases with depth leading to more compaction in the deeper parts of the reservoir. This model was used to describe the observed compaction in the J3 reservoir at Bullwinkle. Reprinted with permission of Offshore Technology Conference, from Flemings et al. (2001); permission conveyed through Copyright Clearance Center, Inc.

systems. England et al. (1987) were the first to tie this problem directly to the study of overpressure in sedimentary basins. Flemings et al. (2002) further explored flow focusing in sedimentary basins and the implications for pore pressure prediction, slope stability, and well design. Yardley and Swarbrick (2000) used basin modeling to describe flow focusing and fluid lateral transfer. In this chapter, I described flow focusing, illustrated how to map this effect in the field, and then provided simple approaches to predict reservoir pressure in these systems.

I close by describing a workflow to predict reservoir pressures for a prospective well that will penetrate three reservoirs (S1, S2, and S3) before reaching the target reservoir S4 (Fig. 10.19a). Each reservoir has a distinct geometry, and the well will penetrate each reservoir at different relative structural positions. The far-field mudrock pressure profile is known by either offset wells or through seismic velocity analysis. To predict the reservoir pressures, the process described above must be applied to each reservoir. For this example, we base the calculation only on the relief in the reservoir using the mudrock permeability model presented in Figure 10.14. The corresponding centroid depths are $Z1 = 0.43$, $Z2 = 0.37$, $Z3 = 0.40$, and $Z4 = 0.31$. These values are used to locate the absolute depth of where each reservoir's pressure equals the far-field pressure. We then plot the pore pressure everywhere in the reservoir assuming the pressure follows the hydrostatic gradient.

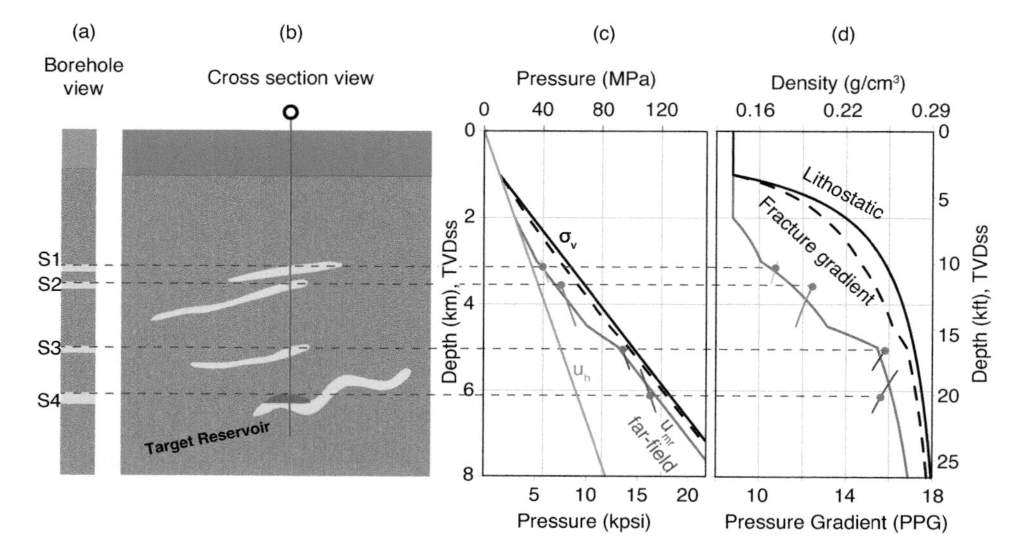

Figure 10.19 Schematic illustration of reservoir pressure prediction along a well path intersecting permeable dipping structures. (a) Sandstone layers (S1, S2, S3, S4) in a borehole view. Yellow layers represent sandstone, and grey layers represent mudrock. (b) Cross-section showing the two-dimensional extension of the layers. Hydrocarbon accumulation is shown in green. The red line shows the well path. (c) Pressure plot. The cyan line, grey line, and black line are the hydrostatic pressure, far-field mudrock pressure, and lithostatic pressure respectively. The yellow, green, purple, and dark blue lines show the predicted reservoir pore pressures in S1, S2, S3, and S4. The red dots show the predicted pore pressures at the well penetration locations in (b). The black dashed line plots the least principal stress (assuming $K_0 = 0.8$). (d) Pressure gradient (mud weight) plot. The grey line is the far-field mudrock pressure gradient, and the black line is the lithostatic pressure gradient. The black dashed line plots the corresponding fracture pressure gradient. Reprinted from Gao and Flemings (2017) with permission of Elsevier.

We then mark the actual pressure that will be encountered by the well when it penetrates each reservoir (red dots, Fig. 10.19c, d). The S1 reservoir is penetrated at a depth close to where the mudrock pressure equals reservoir pressure, thus the encountered reservoir pore pressure (yellow, Fig. 10.19c) is close to the far-field mudrock pressure (grey line, Fig. 10.19c). The S2 and S3 reservoirs are penetrated near their crests and, therefore, the reservoir pore pressures are higher than the far-field mudrock pressure. In contrast, the penetration depth of the target reservoir S4 is about 6 200 m, which is far below the equal-pressure depth of 5 810 m (Fig. 10.19c). Thus, at the target prospect, a pressure regression is expected.

11

Flow Focusing, Fluid Expulsion, and the Protected Trap

11.1 Introduction

As described in Chapter 10, the water phase pressure within a laterally extensive aquifer bounded by overpressured mudrocks follows the hydrostatic gradient. As a result, at structural crests, the effective stress is low because the pore pressure in the reservoir follows the hydrostatic gradient but the overburden stress follows the lithostatic gradient. There is a limiting case where the structural relief of the permeable body is so large that the aquifer pressure converges on the least principal stress and mechanical failure of the overlying seal occurs.

This behavior is illustrated with a simple cartoon (Fig. 11.1). A permeable sand body is initially deposited and buried along the hydrostatic gradient (Fig. 11.1a, b). At a critical depth, either because the mudrock permeability has reduced sufficiently or because the sedimentation rate has increased, the mudrock can no longer drain its pore fluid at the rate it is loaded (see Section 4.3.2.2). Thereafter, the pore fluid bears all of the incremental load and its pressure then follows the lithostatic gradient (Fig. 11.1c). Finally, the sandstone is structurally rotated (Fig. 11.1d). As the relief increases, the pore pressure converges toward the least principal stress at the crest of the structure. This simple model has enormous implications. When the pore pressure converges on the least principal stress, the caprock mechanically fails and fluids are driven vertically (Chapter 9). This process impacts hydrocarbon entrapment, slope stability, and borehole stability.

11.2 The Popeye-Genesis Protected Trap

I review a spectacular example of this behavior from the Popeye minibasin in the deepwater Gulf of Mexico that is described in detail by Seldon and Flemings (2005).

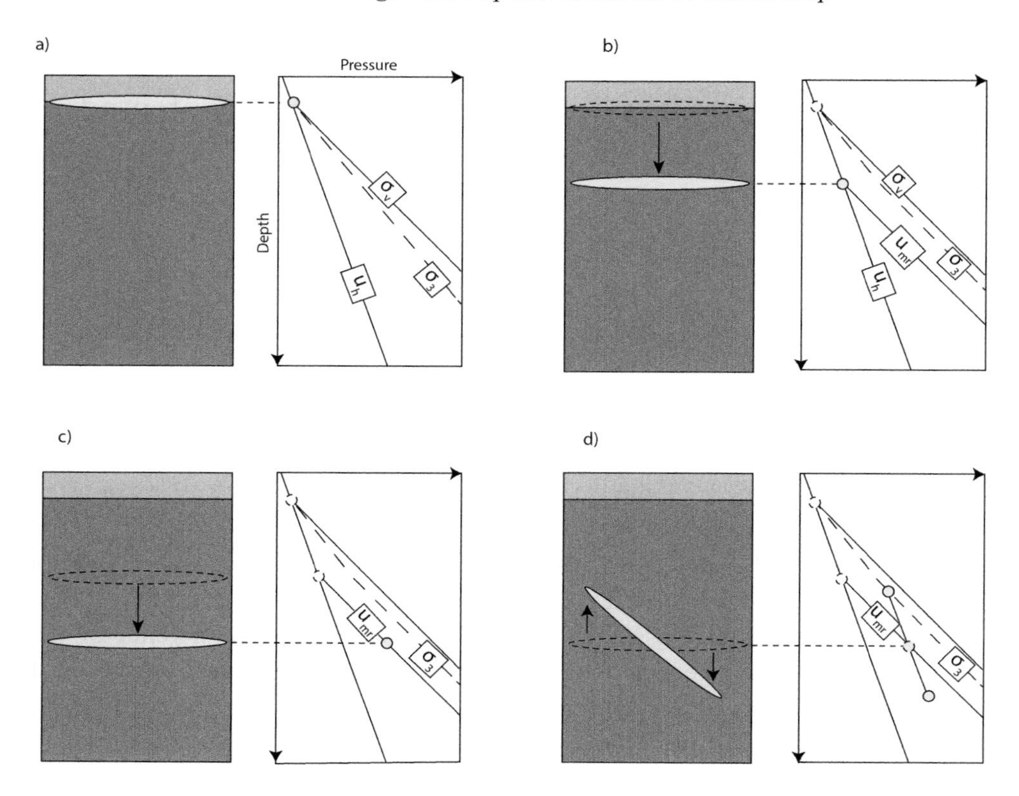

Figure 11.1 Flow focusing can generate reservoir pressures (circles) that converge on the least principal stress (σ_3). (a) A permeable body is deposited at the seafloor. (b) The permeable body is buried by mudrock along the hydrostatic gradient. (c) At a certain depth, the sediment becomes overpressured and its pressure follows the lithostatic gradient with further burial. (d) The sand is rotated, the overpressure of the sandstone is the average of the mudrock overpressure around it (Chapter 10), and the pressure follows the hydrostatic gradient. As a result, the reservoir pressure converges on the least principal stress at the peak of the structure.

11.2.1 The Popeye Basin: Gulf of Mexico

The Popeye minibasin lies in 2 000 ft to 2 900 ft (610 m to 880 m) of water on the Gulf of Mexico continental slope (located in Fig. 1.2). Its reservoirs are bounded to the west and south by a salt-cored high, to the north by a regional growth fault, and to the east by a salt diapir (Fig. 11.2a, b). The sands have bowl-like geometries that are longer in the north-south direction (10.6 miles or 17 km) than in the east-west direction (5 miles or 8 km) (Fig. 11.2a). As is characteristic of minibasin deposition, the sands are ponded in the center of the basin and stratigraphically thin and ultimately pinch at its margins (e.g., Fig. 11.2b). Gas is trapped in the G sand at

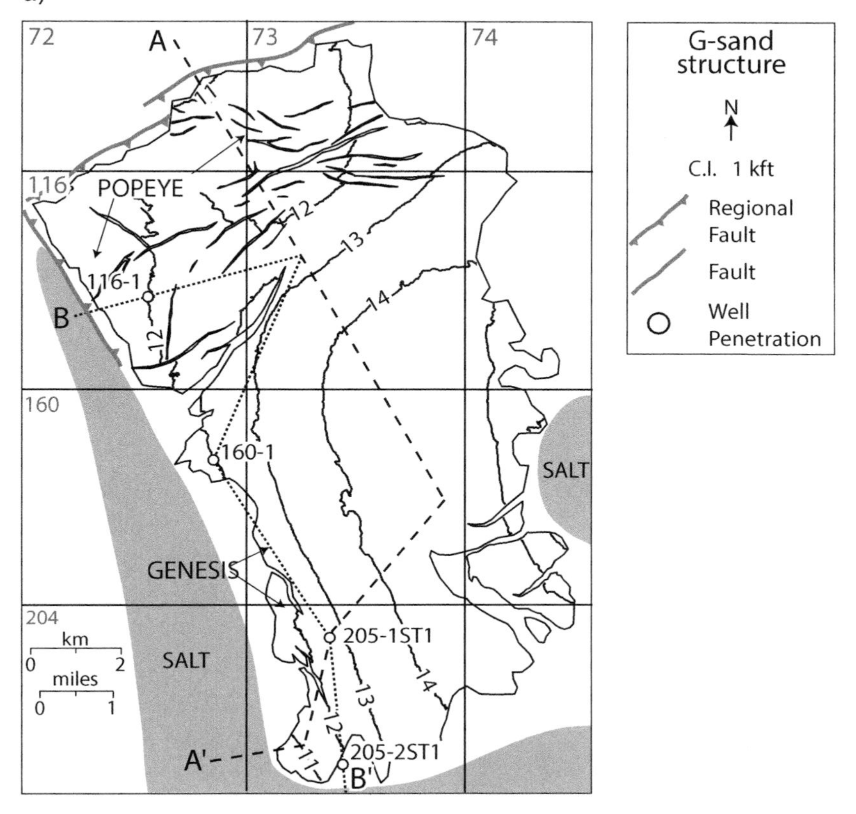

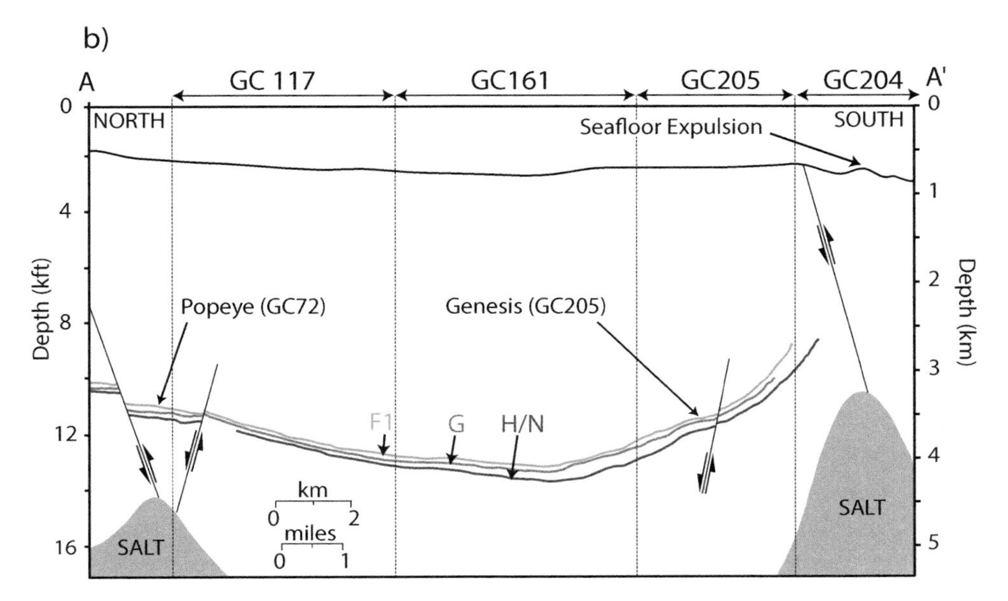

Figure 11.2 (a) Depth to the top of the G sand in thousands of feet in the Popeye Genesis minibasin. Cross-section AA' and the Popeye and Genesis Fields are annotated. (b) Structural cross-section AA' illustrates the synclinal geometry of the reservoir sands and the structural traps of the two fields. From Seldon and Flemings (2005). Reprinted by permission of the AAPG whose permission is required for further use. AAPG© 2005.

the Popeye field (GC72, 73, 116, and 117), and oil is trapped in the underlying H/N sands at the Genesis field (GC160, 161, and 205) (Figures 11.2a, 11.2b and 11.3).

The N sands are the most significant producing units of the Genesis field and the deepest in this study. They have the broadest spatial distribution of the reservoirs, extending from Popeye in the north, where they correlate to the H sands (Yuvancic Strickland et al., 2003), over the Genesis high, to the southern limit of the seismic data. The G sand overlies the N sands. It is thickest at Popeye, thins to the southwest, and is not present at Genesis. The crest of the G sand is significantly deeper than the crest of the underlying H/N sand (Fig. 11.2b). The F1 sand is also present across the minibasin. Its crest is only slightly deeper than the underlying N1 sand.

The pressures and stresses in the Popeye-Genesis minibasin are illustrated in Figure 11.3. The overburden stress was determined by integrating the density log (Chapter 8), and the least principal stress is estimated from leak-off data (Chapter 8). A linear regression of the leak-off data versus depth is used to estimate the least principal stress at any depth (Fig. 11.3a). The least principal stress is extremely close to the overburden stress ($\frac{\sigma_h}{\sigma_v} \geq 0.95$). The water phase reservoir pore pressures are extrapolated from measurements of the gas or oil pressure assuming the hydrocarbon phase pressure and water phase pressure are equal at the fluid contact (Fig. 11.3a) (Chapter 2).

The water phase pressures within each reservoir are assumed to follow the hydrostatic gradient where stratigraphic and structural interpretations suggest the reservoirs are continuous (Fig. 11.3a). This is consistent with the 235 to 3 000 millidarcies core-sampled permeabilities from the Popeye and Genesis reservoir sands (Rafalowski et al., 1994; Yuvancic Strickland et al., 2003). The shallowest location for each of the sands is in the southwest corner of the minibasin in Blocks 204 or 205 (Fig. 11.2a). At this crestal location, the projection of the hydrostatic pressure gradient shows that the F1 and N1-aquifer pressures (u_w) approximately equal the least principal stress (σ_h) (Fig. 11.3a); the F1 pressure is slightly higher and the N1 pressure is slightly lower than the least principal stress at the crest. In general, we find that in these crestal locations, the least principal stress, the overburden stress, and the pore pressure converge toward the overburden stress as is suggested by Equation 8.4.

Immediately above this crestal location, spectacular seafloor vents overlie a zone of chaotic, low amplitude reflectors (Fig. 11.4). Two symmetric mounds (locations 1 and 3, Fig. 11.4a) overlie discrete, vertical, seismically transparent zones (Fig. 11.4b, c) and are interpreted to be mud volcanoes. The seafloor seismic amplitudes over these mounds are four to eight times the magnitude of the surrounding seafloor muds. Flowlike features, delineated by slightly dimmer

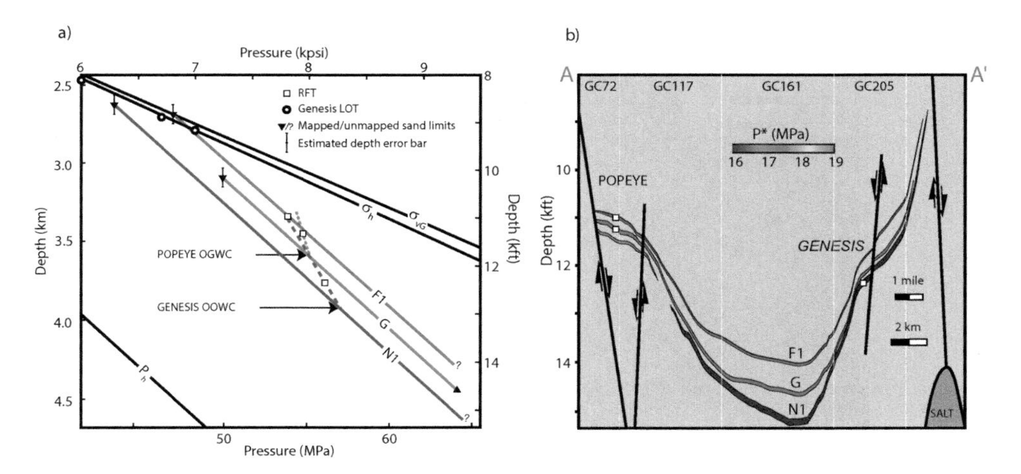

Figure 11.3 (a) Pore pressures in the F1 and N1 sands converge on the least principal stresses (σ_h). Vertical bars at the sand crest are the estimated errors resulting from the conversion of time to depth in the seismic data. (b) Structural cross-section AA' (located in Figure 11.2a) with sands colored by their overpressure. This emphasizes that the overpressure in the F1 sand is greater than that in the underlying N1 sand. Sand thicknesses are exaggerated for display purposes (OGWC = original gas-water contact and OOWC = original oil-water contact). From Seldon and Flemings (2005). Reprinted by permission of the AAPG whose permission is required for further use. AAPG© 2005.

amplitudes (location 2, Fig. 11.4), emanate from these conical mounds in a radial pattern. Away from the mounds, they rotate down the bathymetric slope to the south, extending laterally to the limit of the seismic data set (more than 6 km [3.6 mi]) (Fig. 11.4a). Abrams and Boettcher (2000) sampled the Genesis expulsion structures with a submersible. They found areas of bacterial mats, live clams, and oil and gas saturation and relict carbonate blocks, dead clams, and mud mounds. Oil samples had a similar chemistry to oils in the N sand series at Genesis (Abrams & Boettcher, 2000).

At the crestal position, the sand pressure is generating fracture permeability in the overlying mudrocks either by hydraulically fracturing or by critically stressing favorably oriented fractures (Chapter 8). This location is the minibasin leak point. At this point, projection of the pore pressures in the N1 and F1 sand to the crestal location shows that the pore pressure converges on the least principal stress, which results in an effective horizontal stress of approximately zero at the crest of the N1 and the F1 sand (Fig. 11.3). The exact relation between pore pressure and failure stress is difficult to determine for several reasons. First, it is difficult to map the exact location of the crest of the sand. Second, it is impossible to know exactly the

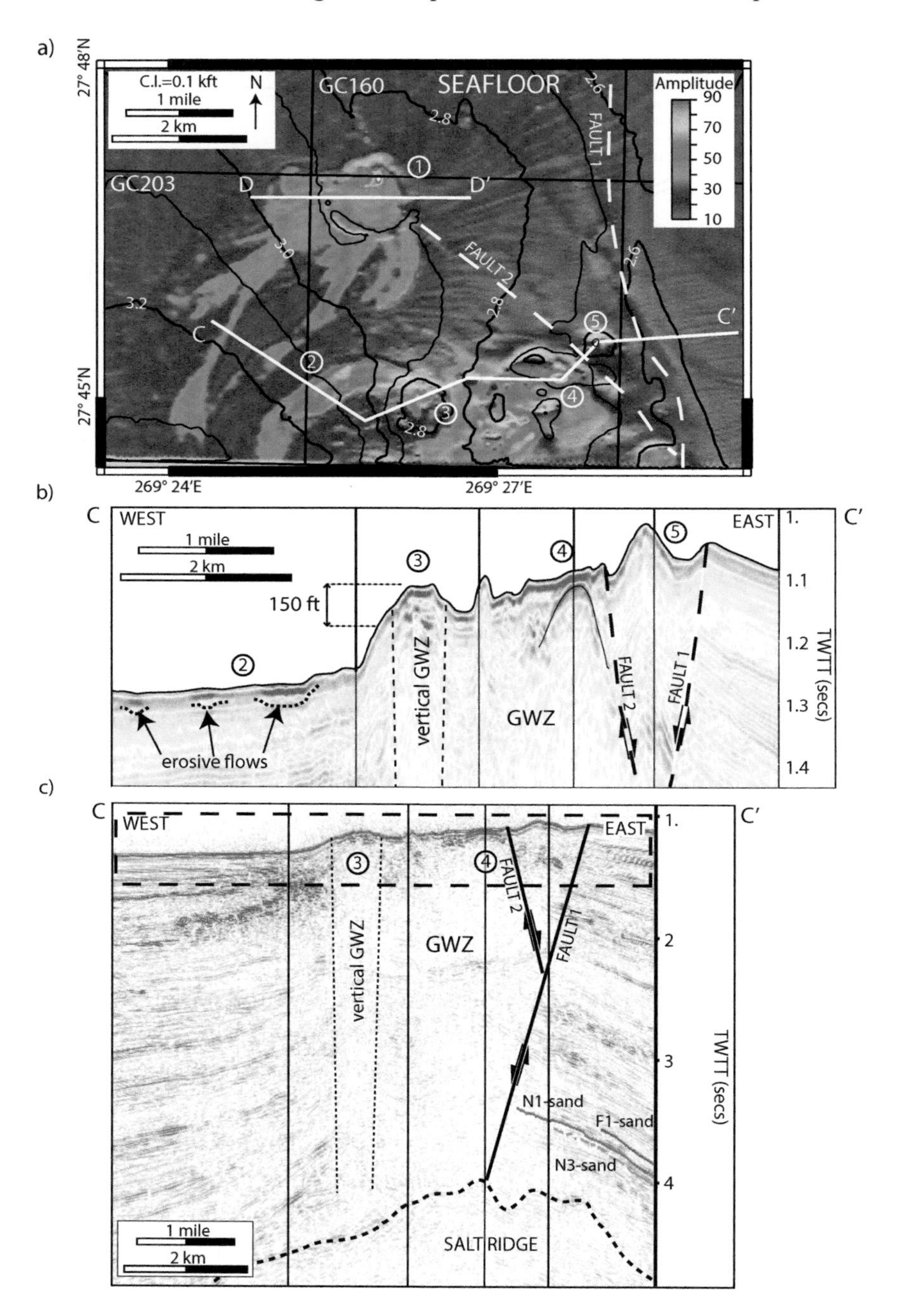

Figure 11.4 (a) Seismic amplitude of the seafloor at Block GC204 with bathymetric contours (in thousands of feet) overlain. Fluid expulsion features (circles 1–5) are annotated. (b) and (c) Seismic cross-sections illustrate wipe-out zone feeding fluid expulsion features. From Seldon and Flemings (2005). Reprinted by permission of the AAPG whose permission is required for further use. AAPG© 2005.

least principal stress in the crestal location. Ultimately, we observe that at the leak point locations the pore pressure, least principal stress (σ_3), and overburden stress (σ_v) all converge.

The seismic amplitudes of the N1 horizon provide a striking illustration of the expulsion pathway (Fig. 11.5). A broad zone of moderate amplitudes (less than those of the Genesis reservoir but greater than the low amplitudes of the aquifer region) is present to the southwest of the Genesis reservoir (Fig. 11.5). These amplitudes extend upward toward the leak point and expulsion features 4 and 5 (Fig. 11.4); we infer that they result from residual hydrocarbon saturation in the pore spaces. These amplitudes may record ongoing migration or multiple charge and expulsion events through the history of the Popeye-Genesis minibasin (e.g., Sassen et al., 2003). The downdip limit of the residual amplitudes records a relict hydrocarbon-water contact (dotted line, delineated as ROWC, Fig. 11.5). The relict contact merges with the present oil-water contact to the north, and it continues beneath the oil-water contact to the south (Fig. 11.5). We interpret that in the past, the Genesis field was significantly larger than it is today (as delineated by the relict hydrocarbon-water contact). Elevation of pore pressures to the least principal stress and resultant expulsion at the leak point may have resulted from uplift of the basin margin. This uplift may also have created the Genesis trapping fault structure and inclined the relict contact. Ultimately, the hydrocarbons not trapped by the Genesis fault were expelled through the leak point.

An oil slick is commonly observed above the leak point. This analysis was not done at the Popeye minibasin. However, a similar study at the Auger Basin identified a leak point where the reservoir pressure exceeded the least principal stress, resulting in oil and gas venting to the seafloor (Reilly & Flemings, 2010). At Auger, oil slicks are present that are clearly sourced at distinct vent locations (Fig. 11.6).

The minibasin leak point causes unusual overpressures within the basin sands (Fig. 11.3): the N1 has lower overpressure than the overlying F1 sand. Both the F1 sand and the underlying N1 sand overpressure (Fig. 11.3) converge on the least principal stress and are interpreted to be leaking. However, the F1 sand crests at a deeper level where the least principal stress is higher: thus it has a higher overpressure than the N1 sand. The situation is akin to a kettle boiling: pressure builds up beneath the crest until it reaches a critical pressure that equals the least principal stress of the overlying rock; thereafter fluids migrate vertically through the caprock precluding further pressure increases. The G sand is not leaking fluid at its crest because its pressure is significantly lower than the least principal stress. However, it is sandwiched between the overlying, higher pressured, F1 sand and the underlying, lower pressured, N1 sand; as a result it has an intermediate overpressure.

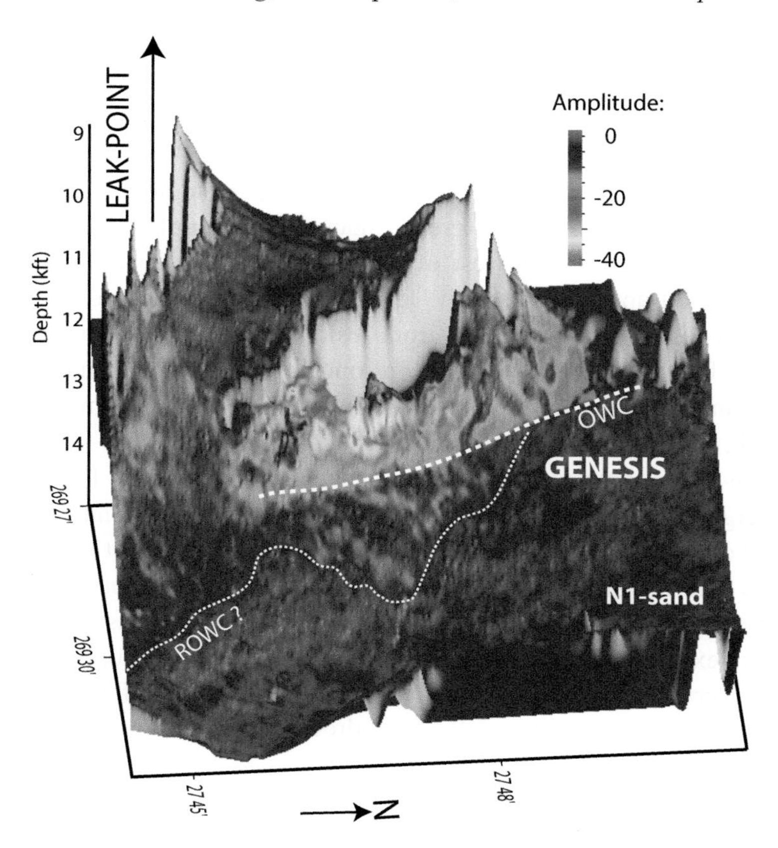

Figure 11.5 Three-dimensional image of the N1 sand structure with its amplitudes overlain. The oil-water contact (OWC) meets the southern (left) edge of the Genesis fault, which is the reservoir spill point. The dotted line delineates a relict hydrocarbon-water contact (ROWC). Above this dotted line, extending up to the leak point, there are significant amplitudes that are interpreted to record residual hydrocarbon saturation. We interpret that in the past, the Genesis oil field extended to this relict contact. Subsequently, the system leaked fluids at the leak point, perhaps because of structural rotation. From Seldon and Flemings (2005). Reprinted by permission of the AAPG whose permission is required for further use. AAPG© 2005.

11.2.2 The Protected Trap Concept

If a leak point is present, it fixes the overpressure everywhere in the connected reservoir to the overpressure at the leak point. The reservoir water pressure and effective stress everywhere else in the minibasin can then be estimated (Fig. 11.7a). For example, the Genesis field is approximately 3 000 ft below the leak point. From the leak point, the water pressure increases downward along the hydrostatic

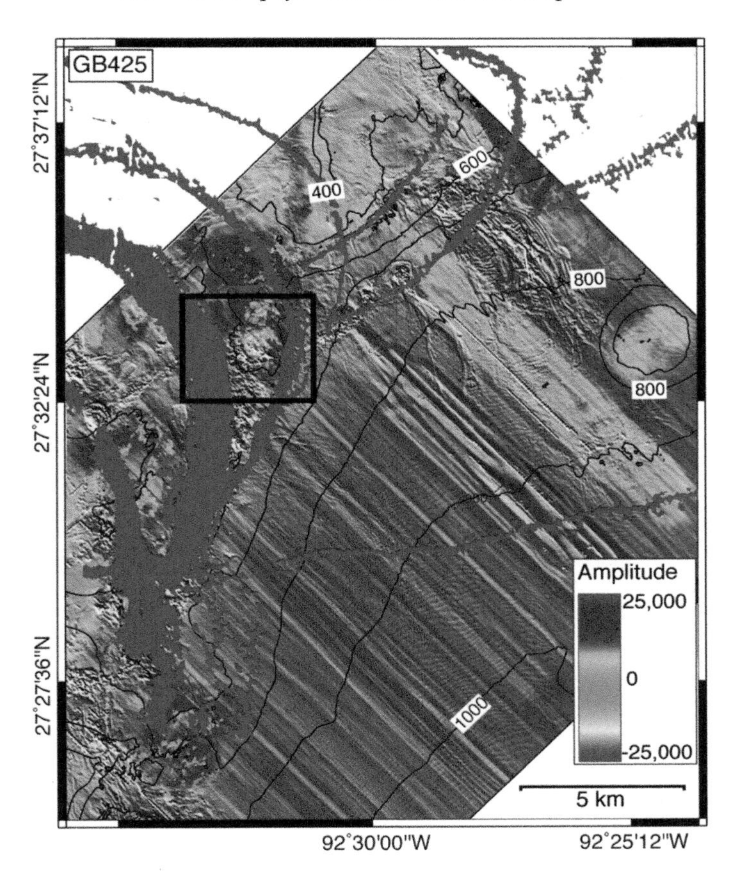

Figure 11.6 Example of oil slicks observed at sea surface above leak point in the Auger basin, Gulf of Mexico. Seafloor amplitude (color) and seafloor bathymetry (meters) are shown. Higher amplitudes (purple) record the presence of authigenic carbonates or hydrates. Lower amplitudes (yellow–red) are interpreted to record locations where there is active gas venting from the seafloor. Maps are overlain with backscatter anomalies (red) detected from synthetic Aperture Radar (SAR) images taken in May 2006. Backscatter anomalies are interpreted as oil slicks on the ocean surface. Contours are in meters below sea level. Reprinted from Smith et al. (2014) with permission of Elsevier.

gradient, but the least principal stress increases approximately along the lithostatic gradient. As a result, the horizontal effective stress, even after accounting for the 1 400 ft oil column that is present (Beeunas et al., 1999), is significant (≈1 000 psi, 6.9 MPa) (Fig. 11.7b). The critical concept is that the effective stress at Genesis is high because the fluid pressures throughout the sand are limited to the least principal stress at the leak point. As one moves to structurally deeper positions,

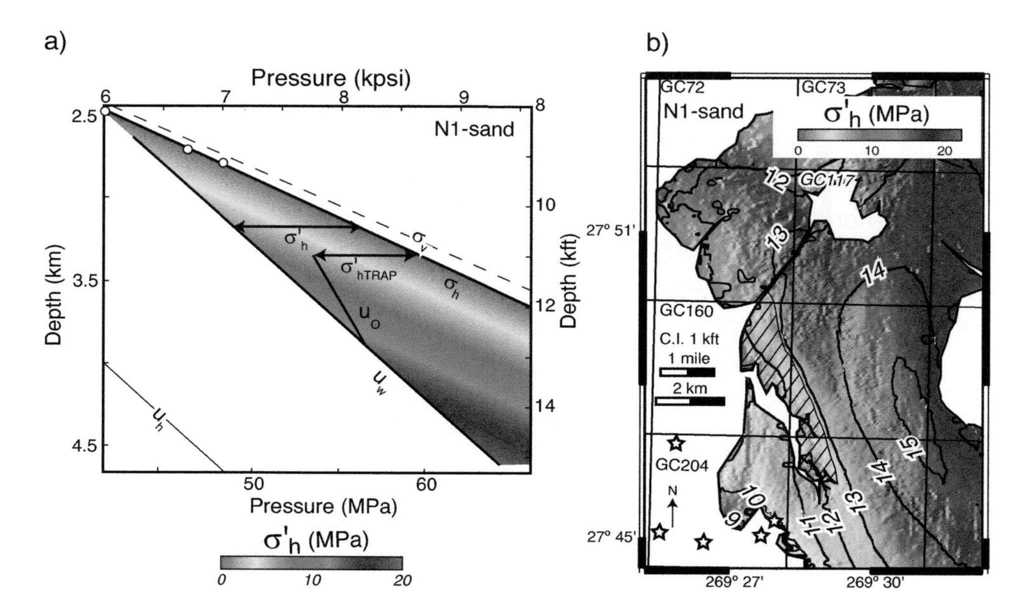

Figure 11.7 The mechanical trap integrity for the N1 sand in the Popeye-Genesis minibasin. (a) Pressure-depth plot of the N1 sand. The minimum horizontal effective stress ($\sigma_h{}'$) is mapped in color. (b) Map view of effective stress. Very low effective stresses (orange) are present at the leak point in the southwest. However, the effective stress increases linearly with depth, resulting in substantial effective stresses at Genesis. Where effective stresses are low, the seal is likely to be breached by mechanical failure, whereas where effective stresses are high and there is structural closure, protected traps are present (e.g., the Genesis oil pool is drawn with the diagonal pattern). From Seldon and Flemings (2005). Reprinted by permission of the AAPG whose permission is required for further use. AAPG© 2005.

the effective stress increases. The Genesis and Popeye fields are thus protected traps: they trap hydrocarbons because the pore pressure is relieved at the leak point.

The protected trap concept is used to construct effective stress maps of potential reservoir horizons, where the water pressure is subtracted from the least principal stress $\left(\sigma_h' = \sigma_h - u_w\right)$ (Fig. 11.7b). These maps predict the effective stress state of the caprock above the reservoir and are a proxy for trap integrity. In the N1 horizon, we see that the Genesis field is located in a zone of moderate effective stress ($\sigma_h' \approx$ 1 000 to 1 700 psi, 7 to 12 MPa), which allows for the entrapment of a significant column of hydrocarbons behind the normal fault spill point (hatched region, Fig. 11.7b).

In the Popeye-Genesis example, we began by mapping the pressures and showing that they converged on the stress at the leak point. However, in basins with significant structural relief where venting is present, the upper bound of pore pressures within reservoirs can be estimated before drilling by assuming the pressure at the crest of the structure is at the least principal stress. Generally, we find that the least principal stress is approximately equal to the vertical stress in these locations. Once this prediction is made, it is possible to estimate the minimum-horizontal effective stress everywhere within the basin by assuming that the sand is hydraulically connected (e.g., Fig. 11.7b). The trap risk in any structure within the interconnected system can be estimated. This allows one to examine subsidiary structures, risking the trap integrity and potential column heights.

The conceptual model that pore pressures at the crest of dipping reservoirs can converge on the overburden stress and cause seal failure has been developing since at least the early 1990s. Gaarenstroom et al. (1993) documented small hydrocarbon columns where pore pressures converge on the least principal stress in the North Sea. Darby et al. (1996) described how high pore pressures at structural crests are driven by flow focusing along the permeable reservoir from deeper overpressured mudrocks, and Darby et al. (1998) modeled this behavior. Iliffe et al. (1999) described how permeable systems with relief create leak points at the crests of structures. Finkbeiner et al. (2001) show that some reservoirs in the Gulf of Mexico Eugene Island 330 (EI-330) field have pore pressures that equal the least principal stress and interpret that hydraulic fracturing is occurring; other EI-330 reservoirs have pore pressures near, but not equal to, the least principal stress and these are interpreted to be leaking by shear failure along favorably oriented faults. Couzens-Schultz et al. (2006) described how pore pressures converge on the least principal stress in the Baja well, the first well into the Wilcox formation in the deepwater Gulf of Mexico. More recently, Reilly and Flemings (2010) documented the presence of a single reservoir at a nearly constant overpressure across the 18 km Auger Basin. They predicted that the pore pressure at the crestal location equaled the overburden stress and showed the expulsion of hydrocarbons and mud volcanoes above this leak point. A well drilled at the leak point confirmed the presence of lithostatic pore pressures and residual hydrocarbons.

11.3 Shallow Water Flow and Submarine Landslides

Near the seafloor, it is common for a near-horizontal and permeable horizon to be loaded asymmetrically by low permeability mudrocks (Fig. 11.8). As in the case of a dipping reservoir, asymmetric sediment loading drives flow laterally along the permeable layer away from where the overburden is thick and toward where it is thin: low effective stresses result where overburden is thin (Fig. 11.8). This process

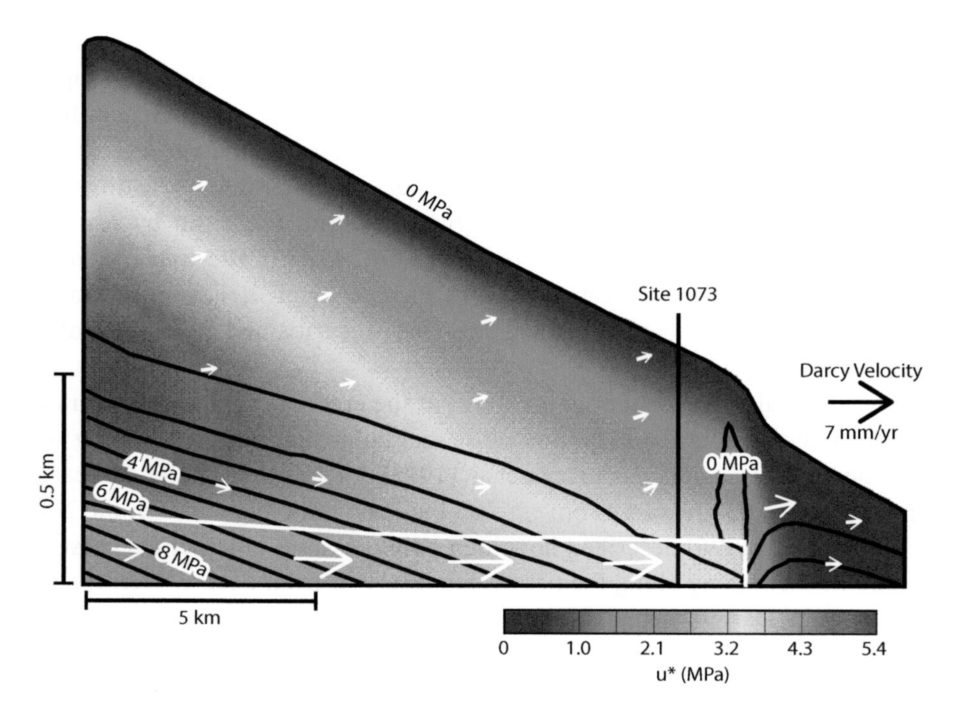

Figure 11.8 Simulated vertical effective stress (contour interval = 1 MPa), overpressure (color contours), and flow fields for the New Jersey slope after one million years of deposition. The left edge (upper slope) is a no-flow boundary, and the right edge (lower slope) is a constant-pressure boundary. The model geometry is constrained from regional seismic data. The white surface is the Miocene-Pliocene boundary. Vertical effective stress is less than 1 MPa for much of the section and is <0 MPa above the toe of the Miocene bed. The low vertical effective stress indicates the lower slope is at failure conditions. From Dugan and Flemings (2000). Reprinted with permission from AAAS.

can drive slope instability on the downslope portions of continental margins (Dugan & Flemings, 2000b; Stigall & Dugan, 2010; Terzaghi, 1950). There is a growing understanding that this process may have driven some of the world's largest submarine landslides such as the Storegga Slides (Bryn et al., 2005; Kvalstad et al., 2005).

In many cases, flow focusing is the root cause for a problem encountered while drilling that is termed shallow waterflow. Alberty et al. (1999) define shallow waterflow as water flowing on the outside of structural casing to the ocean floor. This can erode the structural support of the well, which may lead to buckling of the casing and subsequent casing failure. One of the most common types of shallow waterflow is when overpressured permeable sands are penetrated near the seafloor (before running the 20-inch conductor and the high pressure well head). In this case,

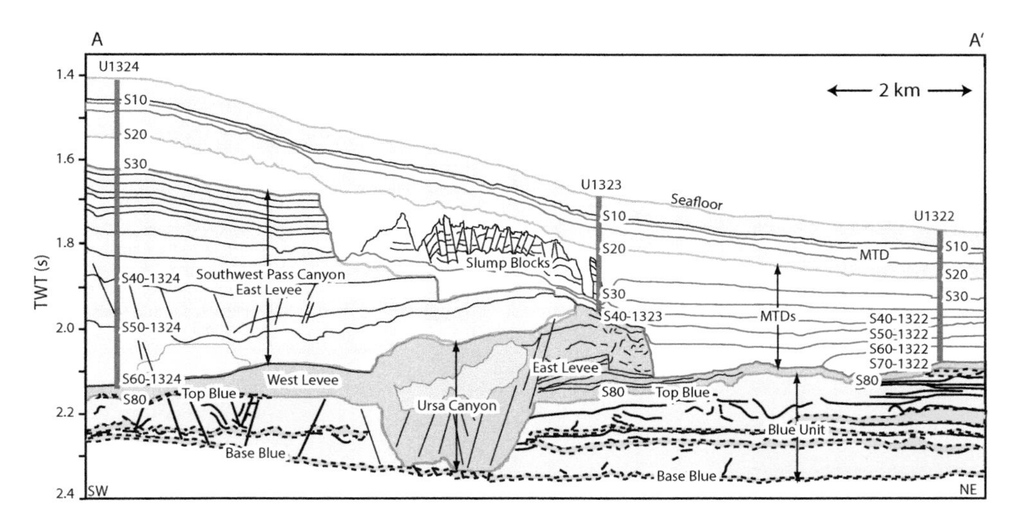

Figure 11.9 East-West interpreted cross-section in the Ursa region. Light and dark grey represent mud-rich levee, rotated channel-margin slides, and hemipelagic drape; yellow represents sand-rich channel fill. The Blue Unit (light blue) is composed of sand (yellow) and mudrock (blue). Seismically mapped surfaces (e.g., 'S10') are shown. Reprinted from Flemings et al. (2008) with permission of Elsevier.

there is no riser and hence no conventional control of wellbore pressure. Sediment loading and flow focusing such as is illustrated in Figure 11.8 can result in surprisingly high pressures in unconsolidated sands when they are drilled where the overburden is thin.

Perhaps the most spectacular example of this was the Ursa shallow water-flow event (Winker & Stancliffe, 2007a, 2007b). In this case, the permeable Blue Unit is overlain asymmetrically by mudrock that is significantly overpressured (Flemings et al., 2008) (Fig. 11.9). The Ursa wells were drilled where the overburden was thin (e.g., Site 1322, Fig. 11.9). Shallow waterflow from the Blue Unit during the initial drilling ultimately led to the abandonment of the original tension leg platform location and numerous wells drilled for that location.

11.4 Outcrop Examples

There are fascinating outcrop examples that record how reservoir overpressure has fractured the overburden and injected reservoir sand into dikes and sills in the overlying caprock. A spectacular example is found at Yellow Bank Beach, near

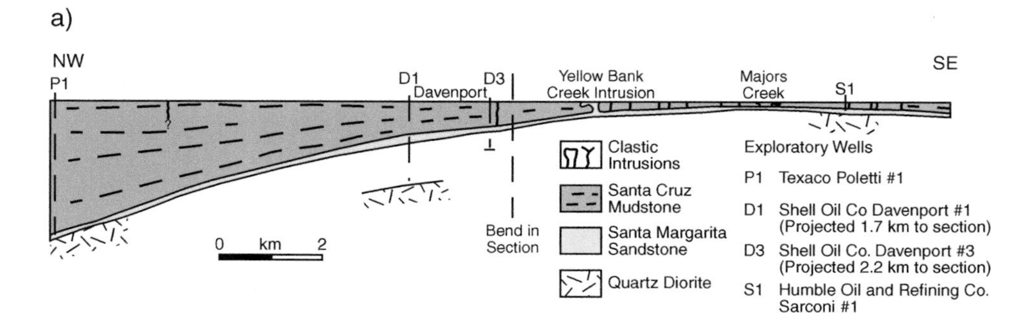

Figure 11.10 (a) Cross-section that parallels the coastline near Santa Cruz, California. The Santa Cruz mudstone thins dramatically to the crest where the Yellow Bank Creek Intrusion is present. (b) The Yellow Bank Intrusion just south of Yellow Bank Beach, Santa Cruz, California. Reprinted from Boehm and Moore (2002) with permission of John Wiley and Sons.

Santa Cruz, California (Fig. 11.10) (Boehm & Moore, 2002). In this location, the Santa Margarita Sandstone is overlain asymmetrically by the upper Miocene Santa Cruz mudrock (Fig. 11.10a). At the crest of this structure, huge sandstone intrusions composed of sand from the Santa Margarita Sandstone are injected to form dikes and sills within the Santa Cruz Mudstone (Fig. 11.10b). In places, these injections vent at the paleo-seafloor, thus confirming that injection occurred during deposition of the Santa Cruz Mudstone (9–7 Ma).

Cosgrove (2001) also documented the presence of sedimentary dikes recording hydraulic fracturing where overpressured and unconsolidated sands have been injected into more cohesive overlying mudrocks at the crests of structures. Finally, numerous authors have also described these types of features in the Panoche Hills of California (Scott et al., 2013; Vétel & Cartwright, 2010; Vigorito & Hurst, 2010; Vigorito et al., 2008).

11.5 Summary

Flow focusing within permeable aquifers encased in low permeability overpressured mudrocks can drive pore pressures to converge on the least principal stress, which produces dramatic results. Understanding this two- and three-dimensional process is critical for oil exploration, for the design of stable wellbores, and for predicting geohazards such as submarine slope failure.

When the reservoir pressure converges on the least principal stress, the seal mechanically fails and no hydrocarbons can be trapped. Instead, the water phase pore pressure at the leak point is limited to the least principal stress (Fig. 11.11a). If subsidiary structures are present, these structures are "protected" because the overpressure is fixed at the crest of the structure (e.g., the right limb of Figure 11.11a). Thus, flow focusing within overpressured sandstones controls both the amount of hydrocarbons that can be trapped and the migration pathway for the hydrocarbons.

To drill the crest of the shallow structure in Figure 11.11a, a borehole fluid density of 2.1 g/cc must be used in order for the borehole pressure to exceed the reservoir pressure (Fig. 11.11e). However, with this fluid density, the borehole pressure will exceed the least principal stress (σ_3) in the mudrock at depths shallower than 4.7 km (Fig. 11.11e). If the borehole is open to the formation in this zone, the drilling fluids will fracture and enter the formation instead of returning to the drill rig. For this reason, casing is periodically set to protect the shallow borehole from the pressures necessary to drill the deeper horizons.

Flow is not necessarily focused toward the crests of sandstone bodies when there is significant seafloor relief (Fig. 11.11f, g). For example, on progradational continental margins, it is common for mudrocks to asymmetrically load underlying horizontal sandstones (Fig. 11.1f). Sediment loading drives flow laterally along the permeable layer, and low effective stresses are generated at the toe of the slope where the overburden is thin. This process may cause instability on the downslope portions of continental margins even though sedimentation rates are low.

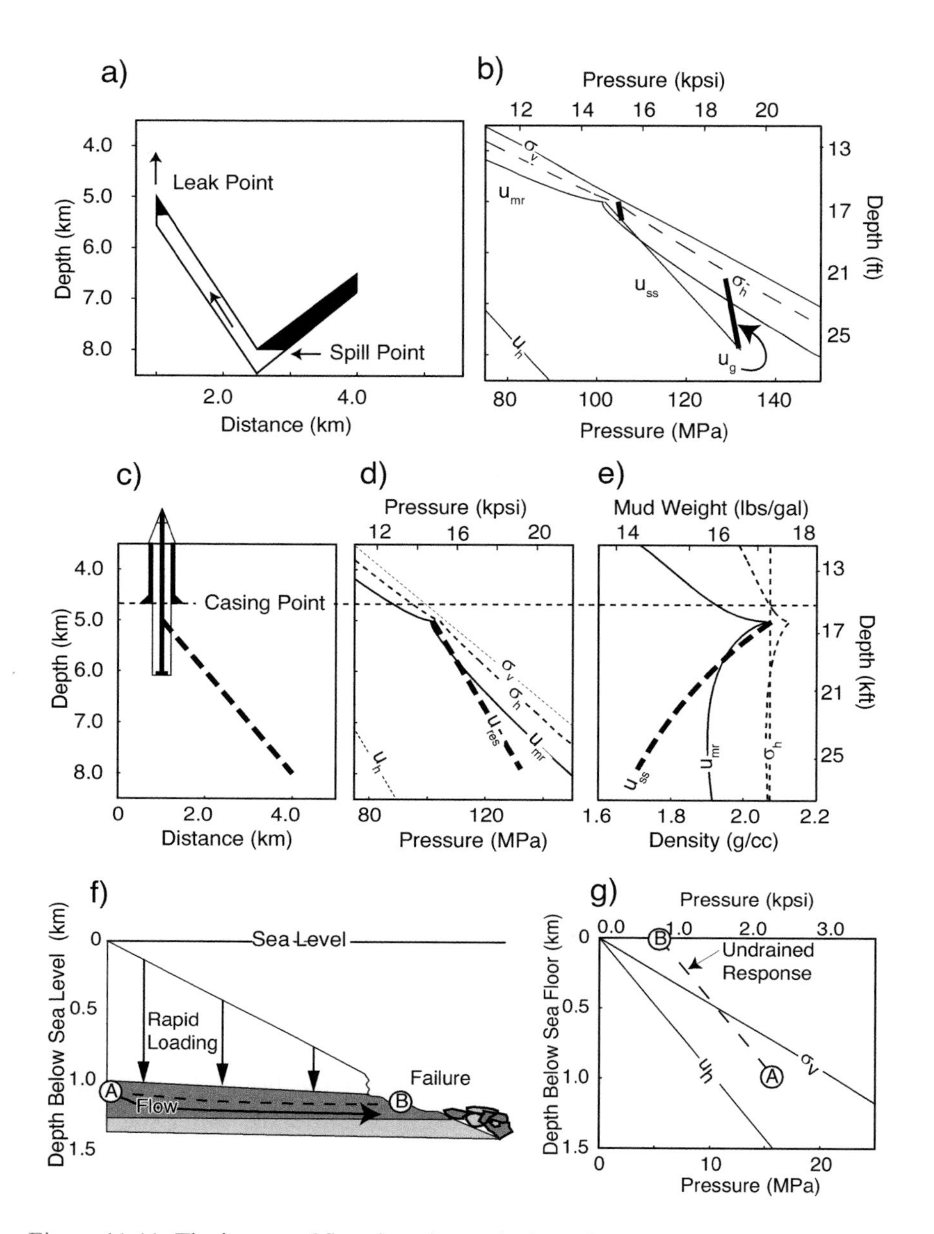

Figure 11.11 The impact of flow focusing on hydrocarbon migration and entrapment, well design, and slope stability. (a) A synclinal reservoir with one shallow limb and one deep limb. Gas is shaded in black; hydrocarbons are trapped in the deeper limb. (b) At the crest of the shallow limb, the reservoir pressure (u_{ss}) is very close to the least principal stress (σ_h) and little or no gas is trapped. At the crest of the lower limb, the reservoir pressure is much less than σ_h and a large gas column is trapped to the spill point of the structure. (c) Pressure and stresses encountered during drilling the shallow limb of A. (d) Reservoir pressure is nearly at the least principal stress. (e) Equivalent mud weight versus depth. Casing point must be set at the horizontal dashed line in order to be able to have a mud weight sufficient to balance the reservoir pressure at the crest. (f) Rapid sediment-loading at continental margins focuses flow along permeable aquifers and can generate slope instability. (g) Where overburden is thin, pore pressures converge on the overburden stress and slope failure is likely to occur. From Flemings et al. (2002). Reprinted by permission of the American Journal of Science.

References

Abrams, M. A., & Boettcher, S. S. (2000). *Mapping migration pathways using geophysical data, seabed core geochemistry and submersible observations in the central Gulf of Mexico*. Paper presented at the AAPG Annual Meeting Expanded Abstracts, New Orleans, Louisiana.

Alberty, M. W., Hafle, M. E., Minge, J. C., Byrd, T. M., & BP Exploration. (1999). Mechanisms of Shallow Waterflows and Drilling Practices for Intervention. *SPE Drilling & Completion*, *14*(2), 123–129.

Alberty, M. W., & McLean, M. R. (2001). *Fracture Gradients in Depleted Reservoirs - Drilling Wells in Late Reservoir Life*. Paper presented at the SPE/IADC Drilling Conference, Amsterdam, The Netherlands.

Alberty, M. W., & McLean, M. R. (2004). *A Physical Model for Stress Cages*. Paper presented at the SPE Annual Technical Conference and Exhibition, 26–29 September, Houston, Texas.

Alberty, M. W., & McLean, M. R. (2014). *The Use of Modeling to Enhance the Analysis of Formation Pressure Integrity Tests*. Paper presented at the IADC/SPE Drilling Conference and Exhibition, Fort Worth, Texas.

Albertz, M., Beaumont, C., & Ings, S. J. (2011). Geodynamic Modeling of Sedimentation-induced Overpressure, Gravitational Spreading, and Deformation of Passive Margin Mobile Shale Basins. In *Shale Tectonics* (pp. 29–62).

Alexander, L. L., & Flemings, P. B. (1994). *Architecture and Evolution of a Plio-Pleistocene Salt Withdrawal Mini-Basin: Eugene Island, Block 330, Offshore Louisiana*. AAPG Annual Meeting. Denver, CO.

Alexander, L. L., & Flemings, P. B. (1995). Geologic Evolution of a Pliocene-Pleistocene Salt-Withdrawal Minibasin: Eugene Island Block 330, Offshore Louisiana. *AAPG Bulletin*, *79*(12), 1737–1756.

Allen, D. F., Best, D. L., Evans, M., & Holenka, J. M. (1993). The Effect of Wellbore Condition on Wireline and MWD Neutron Density Logs. *SPE Formation Evaluation*, *8*(01), 50–56.

Anderson, E. M. (1951). *The dynamics of faulting and dyke formation with applications to Britain*. Edinburgh: Oliver and Boyd.

Anderson, R. L., Ratcliffe, I., Greenwell, H. C., Williams, P. A., Cliffe, S., & Coveney, P. V. (2010). Clay swelling – A challenge in the oilfield. *Earth-Science Reviews*, *98*(3–4), 201–216.

Aplin, A. C., Yang, Y., & Hansen, S. (1995). Assessment of β the compression coefficient of mudstones and its relationship with detailed lithology. *Marine and Petroleum Geology*, *12*(8), 955–963.

Athy, L. F. (1930). Density, Porosity, and Compaction of Sedimentary Rocks. *AAPG Bulletin*, *14*(1), 1–24.

Balk, R. (1949). Structure of Grand Saline Salt Dome, Van Zandt County, Texas. *AAPG Bulletin*, *33*(11), 1791–1829.

Balk, R. (1953). Salt structure of Jefferson Island salt dome, Iberia and Vermilion Parishes, Louisiana. *AAPG Bulletin*, *37*(11), 2455–2474.

Bangs, N. L., Shipley, T. H., Gulick, S. P. S., Moore, G. F., Kuromoto, S., & Nakamura, Y. (2004). Evolution of the Nankai Trough décollement from the trench into the seismogenic zone: Inferences from three-dimensional seismic reflection imaging. *Geology*, *32*(4).

Barker, J. W., & Meeks, W. R. (2003). *Estimating fracture gradient in Gulf of Mexico deepwater, shallow, massive salt sections*. Paper presented at the SPE Annual Technical Conference and Exhibition, Denver, Colorado.

Barton, C. A., Zoback, M. D., & Moos, D. (1995). Fluid flow along potentially active faults in crystalline rock. *Geology*, *23*(8), 683–686.

Becker, D. E., Crooks, J. H. A., Been, K., & Jefferies, M. G. (1987). Work as a criterion for determining in situ and yield stresses in clays. *Canadian Geotechnical Journal*, *24*(4), 549–564.

Beeunas, M. A., Hudson, T. A., Valley, J. A., Clark, W. Y., & Baskin, D. K. (1999). Reservoir continuity and architecture of the Genesis field, Gulf of Mexico (Green Canyon 205); an integration of fluid geochemistry within the geologic and engineering framework. *Gulf Coast Association of Geological Societies Transactions*, *49*, 90–95.

Benson, S. M., & Cole, D. R. (2008). CO2 Sequestration in Deep Sedimentary Formations. *Elements*, *4*(5), 325–331.

Berg, R. R. (1975). Capillary pressures in stratigraphic traps. *AAPG Bulletin*, *59*(6), 939–956.

Best, K. D. (2002). *Development of an Integrated Model for Compaction/Water Driven Reservoirs and its Application to the J1 and J2 Sands at Bullwinkle, Green Canyon Block 65, Deepwater Gulf of Mexico*. Master's thesis, The Pennsylvania State University.

Bethke, C. M., Altaner, S. P., Harrison, W. J., & Upson, C. (1988). Supercomputer Analysis of Sedimentary Basins. *Science*, *239*(4837), 261–267.

Biot, M. A. (1941). General Theory of Three-Dimensional Consolidation. *Journal of Applied Physics*, *12*(2), 155–164.

Birchwood, R. A., & Turcotte, D. L. (1994). A unified approach to geopressuring, low-permeability zone formation, and secondary porosity generation in sedimentary basins. *Journal of Geophysical Research: Solid Earth*, *99*(B10), 20051–20058.

Bird, D. E., Burke, K., Hall, S. A., & Casey, J. F. (2005). Gulf of Mexico tectonic history: Hotspot tracks, crustal boundaries, and early salt distribution. *AAPG Bulletin*, *89*(3), 311–328.

Blunt, M. J. (2017). *Multiphase Flow in Permeable Media: A Pore-Scale Perspective*. Cambridge, United Kingdom: Cambridge University Press.

Boebert, E., & Blossom, J. M. (2016). *Deepwater Horizon*. Cambridge, Massachusetts: Harvard University Press.

Boehm, A., & Moore, J. C. (2002). Fluidized sandstone intrusions as an indicator of Paleostress orientation, Santa Cruz, California. *Geofluids*, *2*(2), 147–161.

Bowers, G. L. (1995). Pore Pressure Estimation From Velocity Data: Accounting for Overpressure Mechanisms Besides Undercompaction. *SPE Drilling & Completion*, *10*(02), 89–95.

Bowers, G. L. (2001). *Determining an Appropriate Pore-Pressure Estimation Strategy.* Paper presented at the Offshore Technology Conference, Houston, Texas.

Braunsdorf, N., & Kittridge, M. (2003). *Overburden Pressure Estimation in Deep Water Settings.* Paper presented at the PSU GeoFluids Consortium Annual Meeting (2003.15), Santa Cruz, California.

Breckels, I. M., & van Eekelen, H. A. M. (1982). Relationship Between Horizontal Stress and Depth in Sedimentary Basins. *Journal of Petroleum Technology, 34*(9), 2191–2199

Brooks, J. M., Kennicutt, M. C., 2nd, Fisher, C. R., Macko, S. A., Cole, K., Childress, J. J., et al. (1987). Deep-sea hydrocarbon seep communities: evidence for energy and nutritional carbon sources. *Science, 238*(4830), 1138–1142.

Bryn, P., Berg, K., Forsberg, C. F., Solheim, A., & Kvalstad, T. J. (2005). Explaining the Storegga Slide. *Marine and Petroleum Geology, 22*(1–2), 11–19.

Burland, J. B. (1990). On the compressibility and shear strength of natural clays. *Geotechnique, 40*(3), 329–378.

Burwicz, E., Reichel, T., Wallmann, K., Rottke, W., Haeckel, M., & Hensen, C. (2017). 3-D basin-scale reconstruction of natural gas hydrate system of the Green Canyon, Gulf of Mexico. *Geochemistry, Geophysics, Geosystems, 18*(5), 1959–1985.

Butterfield, R. (1979). A natural compression law for soils (an advance on e-logp'). *Geotechnique, 29*(4), 469–480.

Casagrande, A. (1936). The determination of the pre-consolidation load and its practical significance. In A. Casagrande, P. C. Rutledge, & J. D. Watson (Eds.), *Proceedings of the 1st International Conference on Soil Mechanics and Foundation Engineering* (Vol. 3, pp. 60–64): American Society of Civil Engineers.

Casey, B. (2014). *The Consolidation and Strength Behavior of Mechanically Compressed Fine-Grained Sediments.* Doctoral thesis, Massachusetts Institute of Technology. Retrieved from http://hdl.handle.net/1721.1/90039

Casey, B., Germaine, J. T., Flemings, P. B., & Fahy, B. P. (2015). Estimating horizontal stresses for mudrocks under one-dimensional compression. *Marine and Petroleum Geology, 65*, 178–186.

Casey, B., Germaine, J. T., Flemings, P. B., & Fahy, B. P. (2016). In situ stress state and strength in mudrocks. *Journal of Geophysical Research-Solid Earth, 121*(8), 5611–5623.

Casey, B., Germaine, J. T., Flemings, P. B., Reece, J. S., Gao, B., & Betts, W. (2013). Liquid limit as a predictor of mudrock permeability. *Marine and Petroleum Geology, 44*, 256–263.

Casey, B., Reece, J. S., & Germaine, J. T. (2019). One-dimensional normal compression laws for resedimented mudrocks. *Marine and Petroleum Geology, 103*, 397–403.

Chopra, S., & Huffman, A. R. (2006). Velocity determination for pore-pressure prediction. *The Leading Edge, 25*(12), 1502–1515.

Christman, S. A. (1973). Offshore Fracture Gradients. *Journal of Petroleum Technology, 25* (08), 910–914.

Comisky, J. T. (2002). *Petrophysical Analysis and Geologic Model for the Bullwinkle J Sands with Implications for Time-Lapsed Reservoir Monitoring, Green Canyon Block 65, Offshore Louisiana.* Master's thesis, The Pennsylvania State University.

Comisky, J. T., Santiago, M., McCollom, B., Buddhala, A., & Newsham, K. E. (2011). *Sample Size Effects on the Application of Mercury Injection Capillary Pressure for Determining the Storage Capacity of Tight Gas and Oil Shales.* Paper presented at the Canadian Unconventional Resources Conference, Calgary, Canada.

Cosgrove, J. W. (2001). Hydraulic fracturing during the formation and deformation of a basin: A factor in the dewatering of low-permeability sediments. *AAPG Bulletin*, *85* (4), 737–748.

Couzens-Schultz, B. A., Axon, A., Azbel, K., Hansen, K. S., Haugland, M., Sarker, R., et al. (2013). *Pore Pressure Prediction in Unconventional Resources*. Paper presented at the International Petroleum Technology Conference, Beijing, China.

Couzens-Schultz, B. A., Hedlund, C. A., & Guzman, C. E. (2006). *Integrating Geology and Velocity Data to Constrain Pressure Prediction in Foldbelts*. Paper presented at the International Oil Conference and Exhibition in Mexico, Cancun, Mexico.

Craig, R. F. (2004). *Craig's Soil Mechanics* (7th ed.). London; New York: Spon Press.

Crook, A. J. L., Willson, S. M., Yu, J. G., & Owen, D. R. J. (2006). Predictive modelling of structure evolution in sandbox experiments. *Journal of Structural Geology*, *28*(5), 729–744.

Cruz-Atienza, V. M., Villafuerte, C., & Bhat, H. S. (2018). Rapid tremor migration and pore-pressure waves in subduction zones. *Nature Communications*, *9*(1), 2900.

Daines, S. R. (1982). Prediction of fracture pressures for wildcat wells. *Journal of Petroleum Technology*, *34*(4), 863–872.

Dake, L. P. (1978a). Chapter 1 Some Basic Concepts in Reservoir Engineering. In L. P. Dake (Ed.), *Fundamentals of Reservoir Engineering* (Vol. 8, pp. 1–43): Elsevier.

Dake, L. P. (1978b). Chapter 10 Immiscible Displacement. In L. P. Dake (Ed.), *Fundamentals of Reservoir Engineering* (Vol. 8, pp. 343–430): Elsevier.

Darby, D., Haszeldine, R. S., & Couples, G. D. (1996). Pressure cells and pressure seals in the UK Central Graben. *Marine and Petroleum Geology*, *13*(8), 865–878.

Darby, D., Haszeldine, R. S., & Couples, G. D. (1998). Central North Sea overpressures: insights into fluid flow from one- and two-dimensional basin modelling. In S. J. Duppenbecker & J. E. Iliffe (Eds.), *Basin Modelling: Practice and Progress* (Vol. 141, pp. 95–107). London: Geological Society, Special Publications.

Dawson, W. C., & Almon, W. R. (2006). *Shale Facies and Seal Variability in Deepwater Depositional Systems*. Poster presented at the AAPG Annual Convention (Search and Discovery Article #40199), Houston, TX.

Day-Stirrat, R. J., Aplin, A. C., Środoń, J., & van der Pluijm, B. A. (2008). Diagenetic Reorientation of Phyllosilicate Minerals in Paleogene Mudstones of the Podhale Basin, Southern Poland. *Clays and Clay Minerals*, *56*(1), 100–111.

Day-Stirrat, R. J., Flemings, P. B., You, Y., Aplin, A. C., & van der Pluijm, B. A. (2012). The fabric of consolidation in Gulf of Mexico mudstones. *Marine Geology*, *295–298*, 77–85.

de Gennes, P.-G., Brochard-Wyart, F., & Quéré, D. (2004). *Capillarity and Wetting Phenomena* (A. Reisinger, Trans.). New York: Springer Science+Business Media.

de Marsily, G. (1986). *Quantitative Hydrogeology: Groundwater Hydrology for Engineers*. Orlando, FL: Academic Press.

Deepwater Horizon Study Group. (2011). *Final Report on the Investigation of the Macondo Well Blowout*. Retrieved from http://ccrm.berkeley.edu/index.shtml

Dewhurst, D. N., Brown, K. M., Clennell, M. B., & Westbrook, G. K. (1996). A comparison of the fabric and permeability anisotropy of consolidated and sheared silty clay. *Engineering Geology*, *42*(4), 253–267.

Dickinson, G. (1953). Geological aspects of abnormal reservoir pressures in Gulf Coast Louisiana. *AAPG Bulletin*, *37*, 410–432.

Dugan, B., & Flemings, P. B. (2000a). Overpressure and Fluid Flow in the New Jersey Continental Slope: Implications for Slope Failure and Cold Seeps. *Science*, *289* (5477), 288–291.

Dugan, B., & Flemings, P. B. (2000b). *Rapid Sediment Loading, Lateral Fluid Flow, and Slope Stability on the US East Coast Margin*. Paper presented at the AGU Fall Meeting, San Francisco, CA.

Dusseault, M. B., Maury, V., & Santarelli, F. J. (2004). *Drilling Through Salt: Constitutive Behavior and Drilling Strategies*. Paper presented at the Gulf Rocks 2004, the 6th North America Rock Mechanics Symposium (NARMS), Houston, Texas.

Dutta, N. C. (1983). *Shale compaction and abnormal pore-pressures: A model of geopressures in the Gulf Coast basin*. Paper presented at the SEG Technical Program Expanded Abstracts 1983, Las Vegas.

Dutta, N. C. (2002a). Deepwater geohazard prediction using prestack inversion of large offset P-wave data and rock model. *The Leading Edge*, *21*(2), 193–198.

Dutta, N. C. (2002b). Geopressure prediction using seismic data: current status and the road ahead. *Geophysics*, *67*(6), 2012–2041.

Dutta, N. C. (Ed.) (1987). *Geopressure* (Vol. 7). Tulsa, Oklahoma: Society of Exploration Geophysicists.

Eaton, B. A. (1969). Fracture gradient prediction and its application in oil field operations. *Journal of Petroleum Technology*, *21*(10), 1353–1360.

Eaton, B. A. (1975). *The Equation for Geopressure Prediction from Well Logs*. Paper presented at the Fall Meeting of the Society of Petroleum Engineers of AIME.

Economides, M. J., Nolte, K. G., & Ahmed, U. (Eds.). (1989). *Reservoir stimulation* (2nd ed.). Englewood Cliffs, New Jersey: Prentice Hall.

Ellis, S., Ghisetti, F., Barnes, P. M., Boulton, C., Fagereng, Å., & Buiter, S. (2019). The contemporary force balance in a wide accretionary wedge: numerical models of the southcentral Hikurangi margin of New Zealand. *Geophysical Journal International*, *219*(2), 776–795.

Ellsworth, W. L. (2013). Injection-induced earthquakes. *Science*, *341*(6142), 1225942.

England, A. H., & Green, A. E. (1963). Some two-dimensional punch and crack problems in classical elasticity. *Mathematical Proceedings of the Cambridge Philosophical Society*, *59*(02), 489–500.

England, W. A., Mackenzie, A. S., Mann, D. M., & Quigley, T. M. (1987). The movement and entrapment of petroleum fluids in the subsurface. *Journal of the Geological Society*, *144*(2), 327–347.

Espinoza, D. N., & Santamarina, J. C. (2017). CO_2 breakthrough—Caprock sealing efficiency and integrity for carbon geological storage. *International Journal of Greenhouse Gas Control*, *66*, 218–229.

Evans, K. F. (1989). Appalachian Stress Study: 3. Regional Scale Stress Variations and Their Relation to Structure and Contemporary Tectonics. *Journal of Geophysical Research:Solid Earth and Planets*, *94*(B12), 17619–17645.

Fang, Y., Flemings, P. B., Daigle, H., Phillips, S. C., Meazell, P. K., & You, K. (2020). Petrophysical properties of the Green Canyon block 955 hydrate reservoir inferred from reconstituted sediments: Implications for hydrate formation and production. *AAPG Bulletin*, *104*(9), 1997–2028.

Fauria, K. E., & Rempel, A. W. (2011). Gas invasion into water-saturated, unconsolidated porous media: Implications for gas hydrate reservoirs. *Earth and Planetary Science Letters*, *312*(1–2), 188–193.

Fertl, W. H. (1976). *Abnormal Formation Pressures: Implications to Exploration, Drilling, and Production of Oil and Gas Resources*. Amsterdam: Elsevier Scientific Publishing Company.

Finkbeiner, T., Zoback, M., Flemings, P., & Stump, B. (2001). Stress, Pore Pressure, and Dynamically Constrained Hydrocarbon Columns in the South Eugene Island 330 Field, Northern Gulf of Mexico. *AAPG Bulletin*, *85*(6), 1007–1031.

Flemings, P. B., Comisky, J. T., Liu, X., & Lupa, J. A. (2001). *Stress-Controlled Porosity in Overpressured Sands at Bullwinkle (GC65), Deepwater Gulf of Mexico*. Paper presented at the Offshore Technology Conference, Houston, Texas.

Flemings, P. B., Long, H., Dugan, B., Germaine, J., John, C. M., Behrmann, J. H., & Sawyer, D. (2008). Erratum to "Pore pressure penetrometers document high overpressure near the seafloor where multiple submarine landslides have occurred on the continental slope, offshore Louisiana, Gulf of Mexico" [Earth and Planetary Science Letters 269/3–4 (2008) 309–32]. *Earth and Planetary Science Letters*, *274*(1–2), 269–283.

Flemings, P. B., & Lupa, J. A. (2004). Pressure prediction in the Bullwinkle Basin through petrophysics and flow modeling (Green Canyon 65, Gulf of Mexico). *Marine and Petroleum Geology*, *21*(10), 1311–1322.

Flemings, P. B., & Saffer, D. M. (2018). Pressure and Stress Prediction in the Nankai Accretionary Prism: A Critical State Soil Mechanics Porosity-Based Approach. *Journal of Geophysical Research: Solid Earth*, *123*(2), 1089–1115.

Flemings, P. B., Stump, B. B., Finkbeiner, T., & Zoback, M. (2002). Flow focusing in overpressured sandstones: theory, observations, and applications. *American Journal of Science*, *302*(10), 827–855.

Fowler, A. C., & Yang, X. (1999). Pressure solution and viscous compaction in sedimentary basins. *Journal of Geophysical Research: Solid Earth*, *104*(B6), 12989–12997.

Fredrich, J. T., Fossum, A. F., & Hickman, R. J. (2007). Mineralogy of deepwater Gulf of Mexico salt formations and implications for constitutive behavior. *Journal of Petroleum Science and Engineering*, *57*(3–4), 354–374.

Gaarenstroom, L., Tromp, R. A. J., de Jong, M. C., & Brandenburg, A. M. (1993). *Overpressures in the Central North Sea: Implications for Trap Integrity and Drilling Safety*. Paper presented at the Petroleum Geology of Northwest Europe: 4th Conference, London, UK.

Gao, B. (2018). *Stress, Porosity, and Pore Pressure in Fold-and-Thrust Belt Systems*. Doctoral thesis, The University of Texas at Austin. Retrieved from https://hdl.handle.net/2152/73911

Gao, B., & Flemings, P. B. (2017). Pore pressure within dipping reservoirs in overpressured basins. *Marine and Petroleum Geology*, *80*, 94–111.

Gardner, G. H. F., Gardner, L. W., & Gregory, A. R. (1974). Formation velocity and density; the diagnostic basics for stratigraphic traps. *Geophysics*, *39*(6), 770–780.

Gera, F. (1972). Review of Salt Tectonics in Relation to the Disposal of Radioactive Wastes in Salt Formations. *Geological Society of America Bulletin*, *83*(12), 3551–3574.

Gibson, R. E. (1958). The Progress of Consolidation in a Clay Layer Increasing in Thickness with Time. *Geotechnique*, *8*(4), 171–182.

Gordon, D. S., & Flemings, P. B. (1998). Generation of overpressure and compaction-driven fluid flow in a Plio-Pleistocene growth-faulted basin, Eugene Island 330, offshore Louisiana. *Basin Research*, *10*(2), 177–196.

Gradmann, S., & Beaumont, C. (2012). Coupled fluid flow and sediment deformation in margin-scale salt-tectonic systems: 2. Layered sediment models and application to the northwestern Gulf of Mexico. *Tectonics*, *31*(4).

Gradmann, S., Beaumont, C., & Ings, S. J. (2012). Coupled fluid flow and sediment deformation in margin-scale salt-tectonic systems: 1. Development and application of simple, single-lithology models. *Tectonics*, *31*(4).

Green, D. H., & Wang, H. F. (1986). Fluid pressure response to undrained compression in saturated sedimentary rock. *Geophysics*, *51*(4), 948–956.

Growcock, F. B., Kaageson-Loe, N., Friedheim, J., Sanders, M. W., & Bruton, J. (2009). *Wellbore Stability Stabilization and Strengthening*. Paper presented at the Offshore Mediterranean Conference (OMC) 2009, Ravenna, Italy.

Gutierrez, M. A., Braunsdor, N. R., & Couzens, B. A. (2006). Calibration and ranking of pore-pressure prediction models. *The Leading Edge*, *25*(12), 1516–1523.

Hantschel, T., & Kauerauf, A. (2009). *Fundamentals of Basin and Petroleum Systems Modeling* (1 ed.): Springer-Verlag Berlin Heidelberg.

Harrison, W. J., & Summa, L. L. (1991). Paleohydrology of the Gulf of Mexico basin. *American Journal of Science*, *291*(2), 109–176.

Hart, B. S., Flemings, P. B., & Deshpande, A. (1995). Porosity and Pressure – Role of Compaction Disequilibrium in the Development of Geopressures in a Gulf-Coast Pleistocene Basin. *Geology*, *23*(1), 45–48.

Hauser, M. R., Couzens-Schultz, B. A., & Chan, A. W. (2014). Estimating the influence of stress state on compaction behavior. *Geophysics*, *79*(6), D389–D398.

Hauser, M. R., Petitclerc, T., Braunsdorf, N. R., & Winker, C. D. (2013). Pressure prediction implications of a Miocene pressure regression. *The Leading Edge*, *32*(1), 100–109.

Heidari, M., Nikolinakou, M. A., & Flemings, P. B. (2018). Coupling geomechanical modeling with seismic pressure prediction. *Geophysics*, *83*(5), B253–B267.

Hickman, S. H., Hsieh, P. A., Mooney, W. D., Enomoto, C. B., Nelson, P. H., Mayer, L. A., et al. (2012). Scientific basis for safely shutting in the Macondo Well after the April 20, 2010 Deepwater Horizon blowout. *Proceedings of the National Academy of Sciences of the United States of America*, *109*(50), 20268–20273.

Holman, W. E., & Robertson, S. S. (1994). *Field Development, Depositional Model, and Production Performance of the Turbiditic "J" Sands at Prospect Bullwinkle, Green Canyon 65 Field, Outer-Shelf Gulf of Mexico*. Paper presented at the GCSSEPM Foundation 15th Annual Research Conference, Submarine Fans and Turbidite Systems, Houston, Texas.

Hubbert, M. K. (1953). Entrapment of Petroleum under Hydrodynamic Conditions. *AAPG Bulletin*, *37*(8), 1954–2026.

Hubbert, M. K., & Rubey, W. W. (1959). Role of Fluid Pressure in Mechanics of Overthrust Faulting Part I. Mechanics of Fluid-Filled Porous Solids and Its Application to Overthrust Faulting. *Geological Society of America Bulletin*, *70*(2), 115–166.

Hubbert, M. K., & Willis, D. G. (1957). Mechanics of Hydraulic Fracturing. *Transactions of the AIME, 210*(01), 153–168.

Hudec, M. R., Jackson, M. P. A., & Schultz-Ela, D. D. (2009). The paradox of minibasin subsidence into salt: Clues to the evolution of crustal basins. *Geological Society of America Bulletin*, *121*(1–2), 201–221.

Huffman, A. R., & Bowers, G. L. (Eds.). (2001). *Pressure Regimes in Sedimentary Basins and Their Prediction* (Vol. 76). Tulsa, OK: American Association of Petroleum Geologists.

Iliffe, J. E., Robertson, A. G., Wynn, G. H. F., Pead, S. D. M., & Cameron, N. (1999). The importance of fluid pressures and migration to the hydrocarbon prospectivity of the Faeroe-Shetland White Zone. In A. J. Fleet & S. A. R. Boldy (Eds.), *Petroleum Geology of Northwest Europe: Proceedings of the 5th Conference* (pp. 601–611). London: The Geological Society.

Ingebritsen, S. E., Sanford, W. E., & Neuzil, C. E. (2006). *Groundwater in Geologic Processes* (2nd ed.): Cambridge University Press.

Issler, D. R. (1992). A new approach to shale compaction and stratigraphic restoration, Beaufort-Mackenzie Basin and Mackenzie Corridor, northern Canada. *AAPG Bulletin*, *76*(8), 1170–1189.

Jaky, J. (1944). The Coefficient of Earth Pressure at Rest. *Journal of the Society of Hungarian Architects and Engineers*, 355–358.

Karig, D. E., & Ask, M. V. S. (2003). Geological perspectives on consolidation of clay-rich marine sediments. *Journal of Geophysical Research: Solid Earth*, *108*(B4), 2197.

Katahara, K. (2006). *Overpressure and Shale Properties: Stress Unloading Or Smectite-illite Transformation?* In *SEG Technical Program Expanded Abstracts 2006* (pp. 1520–1524). Society of Exploration Geophysicists.

Katz, A. J., & Thompson, A. H. (1986). Quantitative prediction of permeability in porous rock. *Phys Rev B Condens Matter*, *34*(11), 8179–8181.

Katz, A. J., & Thompson, A. H. (1987). Prediction of rock electrical conductivity from mercury injection measurements. *Journal of Geophysical Research: Solid Earth*, *92* (B1), 599–607.

Kirsch, G. (1898). Die Theorie der Elastizität und die Bedürfnisse der Festigkeitslehre. *Z. Ver. dtsch. Ing*, *42*, 707.

Kunze, K. R., & Steiger, R. P. (1992). *Accurate In-Situ Stress Measurements During Drilling Operations*. Paper presented at the SPE Annual Technical Conference and Exhibition 4–7 October, Washington, DC.

Kvalstad, T. J., Andresen, L., Forsberg, C. F., Berg, K., Bryn, P., & Wangen, M. (2005). The Storegga slide: evaluation of triggering sources and slide mechanics. *Marine and Petroleum Geology*, *22*(1–2), 245–256.

Lahann, R. (2002). Impact of Smectite Diagenesis on Compaction Modeling and Compaction Equilibrium. In A. R. Huffman & G. L. Bowers (Eds.), *Pressure regimes in sedimentary basins and their prediction* (pp. 61–72).

Lahann, R. W., & Swarbrick, R. E. (2011). Overpressure generation by load transfer following shale framework weakening due to smectite diagenesis. *Geofluids*, *11*(4), 362–375.

Lamb, H. (1932). *Hydrodynamics* (6th ed.). Cambridge: The University Press.

Lambe, T. W., & Whitman, R. V. (1979). *Soil Mechanics: SI Version*. New York: John Wiley & Sons.

Lerche, I., & Petersen, K. (1995). *Salt and Sediment Dynamics*: Boca Raton, Florida: CRC Press.

Li, B., & Wong, R. C. K. (2016). Quantifying structural states of soft mudrocks. *Journal of Geophysical Research: Solid Earth*, *121*(5), 3324–3347.

Liu, X. (2003). *Multiphase Centroid Model*. Paper presented at the PSU GeoFluids Consortium Annual Meeting (2003–12), Santa Cruz, California.

Long, H., Flemings, P. B., Germaine, J. T., & Saffer, D. M. (2011). Consolidation and overpressure near the seafloor in the Ursa Basin, Deepwater Gulf of Mexico. *Earth and Planetary Science Letters*, *305*(1–2), 11–20.

Lopez, J. L., Rappold, P. M., Ugueto, G. A., Wieseneck, J. B., & Vu, C. K. (2004). Integrated shared earth model: 3D pore-pressure prediction and uncertainty analysis. *The Leading Edge*, *23*(1), 52–59.

Losh, S., Eglinton, L. B., Schoell, M., & Wood, J. R. (1999). Vertical and lateral fluid flow related to a large growth fault, South Eugene Island Block 330 Field, offshore Louisiana. *AAPG Bulletin*, *83*(2), 244–276.

Lupa, J., Flemings, P. B., & Tennant, S. (2002). Pressure and trap integrity in the deepwater Gulf of Mexico. *The Leading Edge*, *21*(2), 184–187.

Ma, X., & Zoback, M. D. (2017). Laboratory experiments simulating poroelastic stress changes associated with depletion and injection in low-porosity sedimentary rocks. *Journal of Geophysical Research: Solid Earth*, *122*(4), 2478–2503.

Mann, D. M., & Mackenzie, A. S. (1990). Prediction of pore fluid pressures in sedimentary basins. *Marine and Petroleum Geology*, *7*(1), 55–65.

Matthews, W. R., & Kelly, J. (1967). How to Predict Formation Pressure and Fracture Gradient. *Oil and Gas Journal*, *20*(February), 92–106.

Maury, V., & Idelovici, J. L. (1995). *Safe Drilling of HP/HT Wells, The Role of the Thermal Regime in Loss and Gain Phenomenon*. Paper presented at the SPE/IADC Drilling Conference, Amsterdam, Netherlands.

McKenzie, D. (1984). The Generation and Compaction of Partially Molten Rock. *Journal of Petrology*, *25*(3), 713–765.

Merrell, M. P. (2012). *Pressure and Stress at Mad Dog Field, Gulf of Mexico*. Master's thesis, The University of Texas at Austin. Retrieved from http://hdl.handle.net/2152/20068

Merrell, M. P., Flemings, P. B., & Bowers, G. L. (2014). Subsalt pressure prediction in the Miocene Mad Dog field, Gulf of Mexico. *AAPG Bulletin*, *98*(2), 315–340.

Mesri, G., & Castro, A. (1987). Cα/Cc Concept and K0 During Secondary Compression. *Journal of Geotechnical Engineering*, *113*(3), 230–247.

Mesri, G., & Godlewski, P. M. (1977). Time- and stress-compressibility interrelationship. *Journal of the Geotechnical Engineering Division*, *103*(GT5), 417–430.

Mesri, G., & Hayat, T. M. (1993). The coefficient of earth pressure at rest. *Canadian Geotechnical Journal*, *30*(4), 647–666.

Mikada, H., Becker, K., Moore, J. C., Klaus, A., & Leg 196 Shipboard Scientific Party. (2002). ODP Leg 196: Logging-While-Drilling and Advanced CORKs at the Nankai Trough Accretionary Prism. *JOIDES Journal*, *28*(2), 8–12.

Milliken, K. L., & Hayman, N. W. (2020). Mudrock Components and the Genesis of Bulk Rock Properties: Review of Current Advances and Challenges. In T. Dewers, J. Heath, & M. Sánchez. (Eds.), *Shale: Subsurface Science and Engineering* (First Edition ed., Vol. 245, pp. 1–25). NJ and Washington, D.C.: American Geophysical Union and John Wiley & Sons, Inc.

Mitchell, J. K., & Soga, K. (2005). *Fundamentals of Soil Behavior* (3rd ed.). Hoboken, New Jersey: John Wiley & Sons.

Mondol, N. H., Bjørlykke, K., Jahren, J., & Høeg, K. (2007). Experimental mechanical compaction of clay mineral aggregates—Changes in physical properties of mudstones during burial. *Marine and Petroleum Geology*, *24*(5), 289–311.

Morency, C., Huismans, R. S., Beaumont, C., & Fullsack, P. (2007). A numerical model for coupled fluid flow and matrix deformation with applications to disequilibrium compaction and delta stability. *Journal of Geophysical Research*, *112*(B10).

Morgan, J. K., & Ask, M. V. S. (2004). Consolidation state and strength of underthrust sediments and evolution of the décollement at the Nankai accretionary margin: Results of uniaxial reconsolidation experiments. *Journal of Geophysical Research: Solid Earth*, *109*(B3), B03102.

Morrow, N. R. (1990). Wettability and Its Effect on Oil Recovery. *Journal of Petroleum Technology*, *42*(12), 1476–1484.

Mouchet, J. P., Mitchell, A., & Boussens, C. d. r. d. (1989). *Abnormal Pressures While Drilling: Origins, Prediction, Detection, Evaluation* (Vol. 2). Boussens, France: Technip Editions.

Munson, D. E., & Dawson, P. R. (1979). *Constitutive model for the low temperature creep of salt (with application to WIPP)* (SAND–79–1853). IAEA: Sandia Labs.,

Albuquerque, NM (USA). Retrieved from: https://inis.iaea.org/search/search.aspx?orig_q=RN:11521637

Nance, R. D., Rovick, J. E., & Wilcox, R. E. (1979). *Lithology of the Vacherie dome core.* (Topical Report E511-02500-5). Prepared for U.S. Department of Energy. Baton Rouge, Louisiana: Institute for Environmental Studies, Louisiana State University.

Naruk, S. J., Solum, J. G., Brandenburg, J. P., Origo, P., & Wolf, D. E. (2019). Effective stress constraints on vertical flow in fault zones: Learnings from natural CO_2 reservoirs. *AAPG Bulletin*, *103*(8), 1979–2008.

Neuzil, C. E. (1995). Abnormal pressures as hydrodynamic phenomena. *American Journal of Science*, *295*(6), 742–786.

Nihei, K. T., Nakagawa, S., Reverdy, F., Myer, L. R., Duranti, L., & Ball, G. (2011). Phased array compaction cell for measurement of the transversely isotropic elastic properties of compacting sediments. *Geophysics, 76*(3), WA113-WA123.

Nikolinakou, M. A., Gao, B., Flemings, P. B., & Saffer, D. M. (in review). Dramatic fluid overpressuring and megathrust weakening initiate at the trench. *Journal of Geophysical Research.*

Nikolinakou, M. A., Heidari, M., Flemings, P. B., & Hudec, M. R. (2018). Geomechanical modeling of pore pressure in evolving salt systems. *Marine and Petroleum Geology*, *93*, 272–286.

Nur, A., & Byerlee, J. D. (1971). An exact effective stress law for elastic deformation of rock with fluids. *Journal of Geophysical Research*, *76*(26), 6419–6419.

O'Connell, J. K., Kohli, M., & Amos, S. (1993). Bullwinkle: A unique 3-D experiment. *Geophysics*, *58*(1), 167–176.

Obradors-Prats, J., Rouainia, M., Aplin, A. C., & Crook, A. J. L. (2017). Hydromechanical Modeling of Stress, Pore Pressure, and Porosity Evolution in Fold-and-Thrust Belt Systems. *Journal of Geophysical Research: Solid Earth*, *122*(11), 9383–9403.

Osborne, M. J., & Swarbrick, R. E. (1997). Mechanisms for generating overpressure in sedimentary basins: a reevaluation. *AAPG Bulletin*, *81*(6), 1023–1041.

Ostermeier, R. M., Pelletier, J. H., Winker, C. D., & Nicholson, J. W. (2001). *Trends in Shallow Sediment Pore Pressures – Deepwater Gulf of Mexico*. Paper presented at the SPE/IADC Drilling Conference, Amsterdam, Netherlands.

Ostermeier, R. M., Pelletier, J. H., Winker, C. D., Nicholson, J. W., Rambow, F. H., & Cowan, K. M. (2002). Dealing with shallow-water flow in the deepwater Gulf of Mexico. *The Leading Edge*, *21*(7), 660–668.

Pennebaker, E. S. (1968). An Engineering Interpretation of Seismic Data. Paper presented at the Fall Meeting of the Society of Petroleum Engineers of AIME, Houston, Texas.

Pepin, G., Gonzalez, M., Bloys, J. B., Lofton, J., Schmidt, J., Naquin, C., & Ellis, S. (2004). *Effect of Drilling Fluid Temperature on Fracture Gradient: Field Measurements and Model Predictions*. Paper presented at the Gulf Rocks 2004, the 6th North America Rock Mechanics Symposium (NARMS): Rock Mechanics Across Borders and Disciplines, Houston, Texas.

Perić, D., & Crook, A. J. L. (2004). Computational strategies for predictive geology with reference to salt tectonics. *Computer Methods in Applied Mechanics and Engineering*, *193*(48–51), 5195–5222.

Petley, D. N. (1999). Failure envelopes of mudrocks at high confining pressures. *Geological Society, London, Special Publications*, *158*(1), 61–71.

Phillips, O. M. (1991). *Flow and reactions in permeable rocks*. New York: Cambridge University Press.

Pinkston, F. W. M. (2017). *Pore Pressure and Stress at the Macondo Well, Mississippi Canyon, Gulf of Mexico*. Master's thesis, The University of Texas at Austin.

Pinkston, F. W. M., & Flemings, P. B. (2019). Overpressure at the Macondo Well and its impact on the Deepwater Horizon blowout. *Sci Rep*, *9*(1), 7047.

Pittman, E. D. (1992). Relationship of Porosity and Permeability to Various Parameters Derived from Mercury Injection-Capillary Pressure Curves for Sandstone. *AAPG Bulletin*, *76*(2), 191–198.

Rafalowski, J. W., Regel, B. W., Jordan, D. L., & Lucidi, D. O. (1994). Green Canyon Block 205 lithofacies, seismic facies, and reservoir architecture. In P. Weimer & T. L. Davis (Eds.), *AAPG Studies in Geology* (pp. 133–142). Tulsa, OK: AAPG/SEG.

Raleigh, C. B., Healy, J. H., & Bredehoeft, J. D. (1976). An experiment in earthquake control at Rangely, Colorado. *Science*, *191*(4233), 1230–1237.

Reece, J. S. (2013). *Seal Capacity in Mudrocks: the Impact of Silt Fraction*. Paper presented at the UT GeoFluids Consortium Annual Meeting (4.21), Austin, TX.

Reece, J. S., Flemings, P. B., & Germaine, J. T. (2013). Data Report: Permeability, compressibility, and microstructure of resedimented mudstone from IODP Expedition 322, Site C0011. *In* S. Saito, M. B. Underwood, Y. Kubo, & the Expedition 322 Scientists, *Proc. IODP*, 322: Tokyo (Integrated Ocean Drilling Program Management International, Inc.).

Reilly, M. J. (2008). *Deep Pore Pressures and Seafloor Venting in the Auger Basin, Gulf of Mexico*. Master's thesis, Pennsylvania State University. Retrieved from https://etda.libraries.psu.edu/catalog/9148

Reilly, M. J., & Flemings, P. B. (2010). Deep pore pressures and seafloor venting in the Auger Basin, Gulf of Mexico. *Basin Research*, *22*(4), 380–397.

Rice, J. R., & Cleary, M. P. (1976). Some basic stress diffusion solutions for fluid-saturated elastic porous media with compressible constituents. *Reviews of Geophysics*, *14*(2), 227–241.

Rockfield. (2017). ELFEN Forward Modeling User Manual: Rockfield Software Limited.

Rogers, G., & Dragert, H. (2003). Episodic tremor and slip on the Cascadia subduction zone: the chatter of silent slip. *Science*, *300*(5627), 1942–1943.

Rohleder, S. A., Sanders, W. W., Williamson, R. N., Faul, G. L., & Dooley, L. B. (2003). *Challenges of drilling an ultra-deep well in deepwater – Spa prospect*. Paper presented at the SPE/IADC Drilling Conference, Amsterdam, Netherlands.

Roscoe, K. H., & Burland, J. B. (1968). On the generalized stress-strain behaviour of "wet" clay. In J. Heyman & F. A. Leckie (Eds.), *Engineering Plasticity* (pp. 535–609). Cambridge, England: Cambridge University Press.

Roscoe, K. H., Schofield, A. N., & Wroth, C. P. (1958). On the Yielding of Soils. *Geotechnique*, *8*(1), 22–53.

Rubey, W. W., & Hubbert, M. K. (1959). Role of Fluid Pressure in Mechanics of Overthrust Faulting Part II. Overthrust Belt in Geosynclinal Area of Western Wyoming in Light of Fluid-Pressure Hypothesis. *Geological Society of America Bulletin*, *70*(2), 167–205.

Saffer, D. M. (2003). Pore pressure development and progressive dewatering in underthrust sediments at the Costa Rican subduction margin: Comparison with northern Barbados and Nankai. *Journal of Geophysical Research: Solid Earth*, *108*(B5), 2261.

Saffer, D. M. (2007). Pore Pressure within Underthrust Sediment in Subduction Zones. In T. H. Dixon & J. C. Moore (Eds.), *The Seismogenic Zone of Subduction Thrust Faults* (pp. 171–209): Columbia University Press.

Saffer, D. M., & Tobin, H. J. (2011). Hydrogeology and Mechanics of Subduction Zone Forearcs: Fluid Flow and Pore Pressure. *Annual Review of Earth and Planetary Sciences*, *39*(1), 157–186.

Sanford, J., Woomer, J., Miller, J., & Russell, C. (2006). *The K2 Project: A Drilling Engineer's Perspective*. Paper presented at the Offshore Technology Conference, Houston, Texas.

Santagata, M. C., & Kang, Y. I. (2007). Effects of geologic time on the initial stiffness of clays. *Engineering Geology*, *89*(1–2), 98–111.

Sassen, R., Milkov, A. V., Ozgul, E., Roberts, H. H., Hunt, J. L., Beeunas, M. A., et al. (2003). Gas venting and subsurface charge in the Green Canyon area, Gulf of Mexico continental slope: evidence of a deep bacterial methane source? *Organic Geochemistry*, *34*(10), 1455–1464.

Sawyer, A. H., Flemings, P., Elsworth, D., & Kinoshita, M. (2008). Response of submarine hydrologic monitoring instruments to formation pressure changes: Theory and application to Nankai advanced CORKs. *Journal of Geophysical Research*, *113*(B1).

Sayers, C. M., Johnson, G. M., & Denyer, G. (2002). Predrill pore-pressure prediction using seismic data. *Geophysics*, *67*(4), 1286–1292.

Sayers, C. M., Woodward, M. J., & Bartman, R. C. (2002). Seismic pore-pressure prediction using reflection tomography and 4-C seismic data. *The Leading Edge*, *21*(2), 188–192.

Schneider, F., Potdevin, J. L., Wolf, S., & Faille, I. (1996). Mechanical and chemical compaction model for sedimentary basin simulators. *Tectonophysics*, *263*(1–4), 307–317.

Schneider, J. (2011). *Compression and Permeability Behavior of Natural Mudstones*. Doctoral thesis, The University of Texas at Austin.

Schneider, J., Flemings, P. B., Day-Stirrat, R. J., & Germaine, J. T. (2011). Insights into pore-scale controls on mudstone permeability through resedimentation experiments. *Geology*, *39*(11), 1011–1014.

Schowalter, T. T. (1979). Mechanics of secondary hydrocarbon migration and entrapment. *AAPG Bulletin*, *63*(5), 723–760.

Scott, A., Hurst, A., & Vigorito, M. (2013). Outcrop-based reservoir characterization of a kilometer-scale sand-injectite complex. *AAPG Bulletin*, *97*(2), 309–343.

Screaton, E., Saffer, D., Henry, P., & Hunze, S. (2002). Porosity loss within the underthrust sediments of the Nankai accretionary complex: Implications for overpressures. *Geology*, *30*(1), 19–22.

Seldon, B., & Flemings, P. B. (2005). Reservoir pressure and seafloor venting: Predicting trap integrity in a Gulf of Mexico deepwater turbidite minibasin. *AAPG Bulletin*, *89* (2), 193–209.

Sheahan, T. C. (1991). *An Experimental Study of the Time-Dependent Undrained Shear Behavior of Resedimented Clay Using Automated Stress Path Triaxial Equipment*. Doctoral thesis, Massachusetts Institute of Technology.

Shepard, F. P. (1954). Nomenclature based on sand-silt-clay ratios. *Journal of Sedimentary Petrology, Vol. 24*, 151–158.

Shumaker, N., Haymond, D., & Martin, J. (2014). Kinematic linkage between minibasin welds and extreme overpressure in the deepwater Gulf of Mexico. *Interpretation*, *2*(1), SB69-SB77.

Shumaker, N., Lindsay, R., & Ogilvie, J. (2007). Depth-calibrating seismic data in the presence of allochthonous salt. *The Leading Edge*, *26*(11), 1442–1453.

Shumaker, N., & Vernik, L. (2009). The use of "verticalized" stacking velocities to constrain shale properties in west Africa. *The Leading Edge*, *28*(2), 184–188.

Skempton, A. W. (1954). The Pore-Pressure Coefficients A and B. *Geotechnique*, *4*(4), 143–147.

Skempton, A. W. (1969). The consolidation of clays by gravitational compaction. *Quarterly Journal of the Geological Society, 125*(1–4), 373–411.

Smith, A. J., Flemings, P. B., & Fulton, P. M. (2014). Hydrocarbon flux from natural deepwater Gulf of Mexico vents. *Earth and Planetary Science Letters, 395*(0), 241–253.

Smith, D. A. (1966). Theoretical Considerations of Sealing and Non-Sealing Faults. *AAPG Bulletin, 50*(2), 11.

Sone, H., & Zoback, M. D. (2014). Time-dependent deformation of shale gas reservoir rocks and its long-term effect on the in situ state of stress. *International Journal of Rock Mechanics and Mining Sciences, 69*, 120–132.

Stigall, J., & Dugan, B. (2010). Overpressure and earthquake initiated slope failure in the Ursa region, northern Gulf of Mexico. *Journal of Geophysical Research: Solid Earth, 115*, 11.

Stoffa, P. L., Wood, W. T., Shipley, T. H., Moore, G. F., Nishiyama, E., Botelho, M. A. B., et al. (1992). Deepwater high-resolution expanding spread and split spread seismic profiles in the Nankai Trough. *Journal of Geophysical Research: Solid Earth, 97*(B2), 1687–1713.

Stork, C. (1992). Reflection tomography in the postmigrated domain. *Geophysics, 57*(5), 680–692.

Stump, B. B., & Flemings, P. B. (2001). Consolidation State, Permeability, and Stress Ratio as Determined from Uniaxial Strain Experiments on Mudstone Samples from the Eugene Island 330 Area, Offshore Louisiana. In *Pressure regimes in sedimentary basins and their prediction* (Vol. 76, pp. 131–144): AAPG.

Su, K., & Onaisi, A. (2019). *Dealing with the Uncertainty in the Prediction of Fracture Gradient*. Paper presented at the Second EAGE Workshop on Pore Pressure Prediction, Amsterdam, Netherlands.

Suarez-Rivera, R., & Fjær, E. (2013). Evaluating the Poroelastic Effect on Anisotropic, Organic-Rich, Mudstone Systems. *Rock Mechanics and Rock Engineering, 46*(3), 569–580.

Swanston, A. M., Flemings, P. B., Comisky, J. T., & Best, K. D. (2003). Time-lapse imaging at Bullwinkle Field, Green Canyon 65, offshore Gulf of Mexico. *Geophysics, 68*(5), 1470–1484.

Taylor, G. I., & Quinney, H. (1932). The Plastic Distortion of Metals. *Philosophical Transactions of the Royal Society of London. Series A, Containing Papers of a Mathematical or Physical Character, 230*(681–693), 323–362.

Terzaghi, K. (1923). Die Berechnung der Durchlassigkeitzeriffer des Tones aus dem Verlauf der Hydrodymanischen Spannungserscheinungen [The computation of permeability of clays from the progress of hydrodynamic strain]. *Akademie der Wissenschaften in Wien, Sitzungsberichte, Mathematisch-Naturwissenschaftliche Klasse, Part IIa, 132*, 125–138.

Terzaghi, K. (1943). *Theoretical Soil Mechanics*. London: Chapman and Hall.

Terzaghi, K. (1950). Mechanism of landslides. In *Application of Geology to Engineering Practice: Berkey Volume* (pp. 83–123). New York: Geological Society of America.

Thomeer, J. H. M. (1960). Introduction of a Pore Geometrical Factor Defined by the Capillary Pressure Curve. *Journal of Petroleum Technology, 12*(03), 73–77.

Thomsen, L. (1986). Weak elastic anisotropy. *Geophysics, 51*(10), 1954–1966.

Traugott, M. O., & Heppard, P. D. (1994). *Prediction of pore pressure before and after drilling – taking the risk out of drilling overpressured prospects*. Paper presented at the American Association of Petroleum Geologists Hedberg Research Conference, Denver, Colorado.

Tréhu, A. M., Flemings, P. B., Bangs, N. L., Chevallier, J., Gràcia, E., Johnson, J. E., et al. (2004). Feeding methane vents and gas hydrate deposits at south Hydrate Ridge. *Geophysical Research Letters*, *31*(23), L23310.

Tsuji, T., Tokuyama, H., Costa Pisani, P., & Moore, G. (2008). Effective stress and pore pressure in the Nankai accretionary prism off the Muroto Peninsula, southwestern Japan. *Journal of Geophysical Research: Solid Earth*, *113*(B11).

Velde, B. (1996). Compaction trends of clay-rich deep sea sediments. *Marine Geology*, *133* (3–4), 193–201.

Vétel, W., & Cartwright, J. (2010). Emplacement mechanics of sandstone intrusions: insights from the Panoche Giant Injection Complex, California. *Basin Research*, *22* (5), 783–807.

Vigorito, M., & Hurst, A. (2010). Regional sand injectite architecture as a record of pore-pressure evolution and sand redistribution in the shallow crust: insights from the Panoche Giant Injection Complex, California. *Journal of the Geological Society*, *167*(5), 889–904.

Vigorito, M., Hurst, A., Cartwright, J., & Scott, A. (2008). Regional-scale subsurface sand remobilization: geometry and architecture. *Journal of the Geological Society*, *165*(3), 609–612.

Walker, C., Belvedere, P., Petersen, J., Warrior, S., Cunningham, A., Clemenceau, G., et al. (2013). Straining at the Leash: Understanding the Full Potential of the Deep-Water, Subsalt Mad Dog Field, from Appraisal through Early Production. In *New Understanding of the Petroleum Systems of Continental Margins of the World: 32nd Annual* (Vol. 32, pp. 25–64).

Wang, H. (2000). *Theory of Linear Poroelasticity: With Applications to Geomechanics and Hydrogeology*. Princeton, NJ: Princeton University Press.

Wang, Z. (2002). Seismic anisotropy in sedimentary rocks, part 2: Laboratory data. *Geophysics*, *67*(5), 1423–1440.

Warpinski, N. R., Branagan, P., & Wilmer, R. (1985). In-Situ Stress Measurements at U.S. DOE's Multiwell Experiment Site, Mesaverde Group, Rifle, Colorado. *Journal of Petroleum Technology*, *37*(03), 527–536.

Watabe, Y., & Leroueil, S. (2015). Modeling and Implementation of the Isotache Concept for Long-Term Consolidation Behavior. *International Journal of Geomechanics*, *15* (5), A4014006.

Watts, N. L. (1987). Theoretical aspects of cap-rock and fault seals for single- and two-phase hydrocarbon columns. *Marine and Petroleum Geology*, *4*(4), 274–307.

Weatherl, M. H. (2010). *Gulf of Mexico Deepwater Field Development Challenges at Green Canyon 468 Pony*. Paper presented at the SPE Deepwater Drilling and Completions Conference, Galveston, Texas.

Wenk, H.-R., Lonardelli, I., Franz, H., Nihei, K., & Nakagawa, S. (2007). Preferred orientation and elastic anisotropy of illite-rich shale. *Geophysics*, *72*(2), E69–E75.

White, A. J., Traugott, M. O., & Swarbrick, R. E. (2002). The use of leak-off tests as means of predicting minimum in-situ stress. *Petroleum Geoscience*, *8*(2), 189–193.

Wilhelm, R., Franceware, L. B., & Guzman, C. E. (1998). Seismic pressure-prediction method solves problem common in deepwater Gulf of Mexico. *Oil and Gas Journal* *96*(37), 67–75.

Williams, K. E., Redhead, R., & Standifird, W. (2008). Geopressure Analysis in the Subsalt Knotty Head Field, Deepwater Gulf of Mexico. *Gulf Coast Association of Geological Societies Transactions*, *58*, 905–912.

Winker, C. D., & Stancliffe, R. J. (2007a). *Geology of Shallow-Water Flow at Ursa: 1. Setting and Causes*. Paper presented at the Offshore Technology Conference, Houston, Texas.

Winker, C. D., & Stancliffe, R. J. (2007b). *Geology of Shallow-Water Flow at Ursa: 2. Drilling Principles and Practice*. Paper presented at the Offshore Technology Conference, Houston, Texas.

Wood, D. M. (1990). *Soil Behaviour and Critical State Soil Mechanics*: Cambridge University Press.

Woodward, M., Farmer, P., Nichols, D., & Charles, S. (1998). *Automated 3D Tomographic Velocity Analysis of Residual Moveout In Prestack Depth Migrated Common Image Point Gathers*. Paper presented at the 1998 SEG Annual Meeting, New Orleans, Louisiana.

Yang, Y., & Aplin, A. C. (1998). Influence of lithology and compaction on the pore size distribution and modelled permeability of some mudstones from the Norwegian margin. *Marine and Petroleum Geology*, *15*(2), 163–175.

Yang, Y., & Aplin, A. C. (2004). Definition and practical application of mudstone porosity–effective stress relationships. *Petroleum Geoscience*, *10*(2), 153–162.

Yardley, G. S., & Swarbrick, R. E. (2000). Lateral transfer; a source of additional overpressure? *Marine and Petroleum Geology*, *17*(4), 523–537.

Yarger, H., DeKay, L., & Hensel, E. G. (2001). Salt canopy modeling with gravity in deepwater Gulf of Mexico. *SEG Expanded Abstracts*, *20*, 4.

Yilmaz, O. (1987). *Seismic Data Processing*. Tulsa, OK: Society of Exploration Geophysicists.

Yuvancic Strickland, B., Kuhl, E. J., Lee, T. W., Seldon, B. J., Flemings, P. B., & Ertekin, T. (2003). Integration Of Geologic Model And Reservoir Simulation, Popeye Field, Green Canyon 116. *Gulf Coast Association of Geological Societies Transactions*, *53*, 918–932.

Zablocki, M. (2019). *Lithology Based Prediction of Pressure, Fracture Gradient and Strength*. Paper presented at the UT GeoFluids Consortium Annual Meeting (10.03), Austin, TX.

Zhang, J. (2011). Pore pressure prediction from well logs: Methods, modifications, and new approaches. *Earth-Science Reviews*, *108*(1–2), 50–63.

Zoback, M. D. (2007). *Reservoir Geomechanics* (1st ed.). New York, NY: Cambridge University Press.

Zoback, M. D., & Kohli, A. H. (2019). *Unconventional Reservoir Geomechanics* (1st ed.). New York, NY: Cambridge University Press.

Index